All the U.S. Air Force Airplanes, 1907-1983

ALL THE
U.S. AIR FORCE
AIRPLANES,
1907-1983

Andrew W. Waters

HIPPOCRENE BOOKS
New York

HIPPOCRENE BOOKS, INC.
171 Madison Ave.
New York, NY 10016

ISBN 0-88254-582-5

Library of Congress Catalog Card Number 81-86232

Printed in the United States of America

TABLE OF CONTENTS

INTRODUCTION   1

HISTORICAL INDEX SECTION
Air Force Aircraft Designations   3
Major Designers of Air Force Aircraft—1909 to 1979   7
Historical Index of Airplanes   10

HISTORICAL AIRCRAFT SECTION
Attack Planes—WWI (Foreign-built)   59
Attack Planes—WWI to WWII   60
Attack Planes—WWII to 1948   67
Attack Planes—1962 to the present   77

Bombers, Prop-driven—WWI to WWII   83
Bombers, Prop-driven—WWII   90
Bombers, Prop-driven—after WWII   102
Bombers, Jet—1946 to 1962   108
Bombers, Jet—1962 to the present   120

Cargo Transports—1919 to 1962   123
    Transports—1919 to 1925   123
    Cargo Transports—1925 to 1962   124
    Cargo Transports—1962 to the present   164

Electronics Aircraft for the USAF—1962 to the present   170

Fighters—Birth of the Jets   173
Fighters, Jet—WWII to 1962   176
Fighters, Jet—1962 to the present   193

Gliders—1941 to 1962   200
Gyroplanes (Helicopters)   203

Helicopters—1935 to the present   203
Gyroplanes—1935 to 1941   203
Rotating Wing—1941 to 1948   203
Helicopters—1948 to 1962   204
Helicopters—1962 to the present   208

Observation and Liaison Aircraft   214
   Corps Observation—1919 to 1924   214
   Artillery Observation—1919 to 1924   214
   Observation—1924 to 1942   215
   Liaison—1942 to 1962   221
   Observation—1962 to the present   226
Observation Amphibians—1919 to 1962   229
   Corps Observation Amphibians—1919 to 1925   229
   Observation Amphibians—1925 to 1948   229
   Amphibians—1948 to 1962   232

Photography Aircraft—1930 to 1948   234

Pursuits—WWI   237
Pursuits—WWI to WWII   243
Pursuits—WWII   254

Racers—1920 to 1925   271

Strategic Reconnaissance—1962 to the present   274

Tactical Reconnaissance—1978 to the present   275
Trainer Aircraft   276
   The Early Trainers   276
   Trainers—WWI   280
   Trainers—WWI to WWII   283
   Trainers of the Early 1920's   283
   Primary Trainers   284
   Advanced Trainers   285
   Basic Trainers   286
Trainers—WWII   288
Trainers—WWII to the present   296

Utility Planes—1952 to the present   302

"X"—Special Research—1944 to the present   306

Space Vehicles—1974 to the present   312

APPENDICES
Bibliography   316
A Chronology of High Level Command Organizations   318
WWII Command Organizations   319
Aircraft Supplement—U.S. Navy, U.S. Marine, Allied and Enemy
     Aircraft   320
Birth of the Air Force—A Chronology   350
Air Force Bases   363
Historical Brief   371

INDEX   399

Appreciation is extended to all the people at the Air Force Office of Public Affairs-Magazines and Books for their cooperation in making the photographs available. Without their courteous assistance this work would not be complete.

INTRODUCTION

This is a book of United States Air Force Aircraft.

However, it includes all the Army planes from 1907 to 1948, during which period the Air Force was a part of the U. S. Army. But Army aircraft from 1948 to the present are not discussed, except in those cases in which the Air Force and the Army shared the same plane. Likewise, aircraft of the Navy, Marines and Coast Guard are mentioned only in order to indicate their variants of the same airplanes used by the Air Force. See the Appendix entitled ''Chronology of High Level Command Organizations'' for all the USAF progenitors, including the Aeronautical Division, Aviation Section, Airplane Division, Air Service of the AEF, Division of Military Aeronautics, Bureau of Aircraft Production, Army Air Service, Army Air Corps and U. S. Army Air Forces (USAAF), culminating finally in the autonomous United States Air Force (USAF).

The purpose of this book is to present closely related aircraft arranged in a logical order by type and era. Where possible, the airplanes are listed in sequence by number within an alphabetized classification. Separate listings of Air Force aircraft designers and Air Force aircraft designation types are included at the beginning of the Historical Index.

The Historical Index starts with a listing of the Early Planes, from 1909 until just after World War One. From that point the Index lists all aircraft alphabetically by letter-designators of the type: ''A'' represents Attack, ''B'' represents Bomber, ''C'' represents Cargo Transport, etc. Unfortunately, the authorities charged with thinking up letter-designators did not have historians and indexes in mind. For example, Observation planes became Liaison planes, then Observation planes again, and some of them were redesignated into the Utility category. It is virtually impossible to corral them into an alphabetical order. In the Historical Index the Observation and Liaison planes are listed under the Observation section, in sequence according to their period of service, and the Utility planes fall into their position alphabetically.

The reader is advised to study the Table of Contents. Next, the reader is referred to the information contained in the Historical Index. Each aircraft is described with a three-line entry. The first line applies to the column headings at the top of the first page, indicating the designation letter-number, the name of the aircraft (if it had one), variant aircraft (in parentheses) using the same basic airframe, the designer, the wingspan, the length of the aircraft, and the year in which the production model was first delivered for service. The second line makes a general statement about its fame or use. The third line shows some of the characteristics, equipment and performance capabilities. On a few occasions a four-line entry is necessary for aircraft that came in large families, with many variants using the same wings, tail and fuselage but having minor modifications.

The Historical Index lists 524 airplanes with three-line entries. Another 41 planes are indicated in the Bomber and Fighter supplements. An additional 97 variants are cross-referenced but are not individually listed, making a total of 662 aircraft mentioned in the Historical Index.

The Historical Aircraft section employs the same order as the Historical Index, except that the Early Planes are discussed with the Pursuits and Trainers: in the early days, every aeroplane was a Trainer. Each entry presents a brief description of the airplane, its origin, date of delivery, variants, equipment, performance capabilities, the number produced, and—when information is available—its claim to fame and some of the men who flew in it.

The Appendix section contains a Bibliography, a Chronology of High Level Command Organizations, a list of World War Two Command Organizations, and the following additional features:

An Aircraft Supplement presents a selection of U.S. Navy, U.S. Marine Corps, Allied and Enemy aircraft that played significant roles in wartime activities. The discussion of these aircraft is parallel in format to the Historical Aircraft section. These important aircraft are forty in number.

A chronology entitled "Birth of the Air Force" lists significant events leading up to the USAF becoming a separate arm of the military service.

An Air Force Base (AFB) Alphabetical Biography contains a list of many Air Force Bases and presents brief accounts of the persons after whom they were named.

A Historical Brief, following the format of the Historical Index, presents a two-line history of each major and minor aircraft type not presented elsewhere in this book. Some 383 additional airplanes are included in the Historical Brief. The Historical Brief and the Historical Index mesh with each other to form a listing of all the USAF airplanes.

Finally, the Index to the volume as a whole lists the designation number of each aircraft mentioned in this book; the names of all aircraft that had either an official name or an adopted name; and the persons associated with the aircraft. Counting all variants, a total of 1,222 aircraft designations are listed in the Index; nearly 1,000 of these belonged to the USAF or its predecessors.

HISTORICAL INDEX

AIR FORCE AIRCRAFT DESIGNATIONS

TYPE	DESCRIPTION	YEAR
A	Ambulance	1919-1925
A	Attack-Fighter	1924-1948
A	Attack-Bomber	1934-1948
A	Attack (Numbering sequence revised)	1962-today
A	Amphibian	1948-1962
AC	Gunship-Cargo Transport	1962-today
AG	Assault Glider	1942-1948
AO	Artillery Observation	1919-1924
AT	Advanced Trainer	1925-1948
AU	Attack-Utility	1968-today
B	Bombardment	1928-1962
B	Bombardment (Numbering sequence revised)	1962-today
BC	Basic Combat Trainer	1935-1940
BG	Bombardment Glider	1942-1948
BLR	Bombardment-Long Range	1935-1937
BT	Basic Trainer	1930-1948
C	Cargo Transport	1925-1962
C	Cargo Transport (Numbering sequence revised)	1962-today
CG	Cargo Glider	1941-1948
CO	Corps Observation	1919-1924
COA	Corps Observation Amphibian	1919-1925
CV	Cargo Transport VSTOL	1962-today
DB	Day Bombardment	1919-1925
DB	Drone Launch Bomber	1948-1962
DC	Drone Launch Cargo Transport	1948-1962
DF	Aerial Target Director-Fighter	1948-1962
DH	DeHavilland Light Bomber	1918-1930
DOS	Douglas Observation Seaplane	1921-1924
DWC	Douglas World Cruiser	1921-1924

TYPE	DESCRIPTION	YEAR
E	Electronics	1962-today
EC	Electronics Cargo Transport	1948-today
F	Photography	1930-1948
F	Fighter	1948-1962
F	Fighter (Numbering sequence rolled back)	1962-today
FB	Fighter Bomber	1962-today
FM	Fighter Multiplace	1935-1942
FP	Photo Pursuit	1945-1948
G	Gyroplane	1935-1941
G	Glider	1948-1962
GA	Ground Attack	1919-1925
GB	Drone Director Bomber	1948-1962
GC	Drone Director Cargo Transport	1948-1962
GMB	Glenn Martin Bomber	1917-1919
GMP	Glenn Martin Pursuit	1917-1919
H	Helicopter	1948-1962
H	Helicopter (Tri-service designations changed)	1962-today
HB	Heavy Bombardment	1925-1928
HB	Rescue Bomber	1948-1962
HH	Search and Rescue Helicopter	1962-today
HU	Search and Rescue Utility	1962-today
JB	Test Bed Bomber	1948-1962
JC	Test Bed Cargo Transport	1948-1962
JN	Pilot Trainer	1915-1930
KB	Tanker-Bombardment	1948-today
KC	Tanker-Cargo Transport	1948-today
L	Liaison	1942-1962
LB	Light Bombardment	1925-1928
LC	Liaison Cargo Transport	1948-1962
LUSAC	Fighter-Pursuit	1918-1925
MB	Martin Bomber	1918-1925
MB	Morse-Boeing Pursuit	1918-1925
MODEL	Early Planes	1909-1919
NBL	Night Bombardment-Long Range	1919-1925
NBS	Night Bombardment-Short Range	1919-1925
NC	Special Purpose Cargo Transport	1948-1962
NF	Special Purpose Fighter	1948-1962

TYPE	DESCRIPTION	YEAR
O	Observation	1924-1942
O	Observation (Numbering sequence revised)	1962-today
OA	Observation Amphibian	1925-1948
ORENCO	Fighter-Pursuit	1919-1925
OV	Observation-Vertical Take-Off and Landing—VTOL	1962-today
OV	Orbital Vehicle	1976-today
P	Pursuit	1925-1948
PA	Pursuit-Aircooled	1919-1925
PB	Pursuit-Biplace	1935-1942
PC	Personnel Cargo Transport	1948-1962
PG	Pursuit-Ground Attack	1919-1925
PG	Powered Glider	1943-1948
PN	Pursuit-Night	1919-1925
PQ	Pursuit-Aerial Target	1942-1948
PT	Primary Trainer	1924-1948
PW	Pursuit-Watercooled	1919-1925
Q	Aerial Target	1942-1962
QB	Aerial Target Bomber	1948-1962
QF	Aerial Target Fighter	1948-1962
R	Racer	1920-1925
R	Rotating Wing	1941-1948
R	Restricted	1943-1948
R	Reconnaissance	1948-today
RB	Recon Bomber	1948-today
RC	Recon Cargo Transport	1948-today
RF	Recon Fighter	1948-today
RP	Restricted Pursuit	1943-1948
SA	Search Amphibian	1948-today
SCA	Shuttle Carrier Aircraft	1976-today
SH	Search Helicopter	1948-today
SR	Strategic Reconnaissance	1962-today
T	Transport	1919-1925
T	Trainer	1948-today
TA	Trainer-Aircooled	1919-1925
TA	Trainer-Attack	1940-1945
TB	Trainer-Bomber	1945-today
TC	Trainer-Cargo Transport	1945-today
TF	Trainer-Fighter	1941-1948
TG	Trainer-Glider	1941-1948
TH	Trainer-Helicopter	1962-today

TYPE	DESCRIPTION	YEAR
TP	Two-Seat Pursuit	1919-1925
TP	Trainer-Pursuit	1945-1948
TR	Tactical Reconnaissance	1978-today
TW	Trainer-Watercooled	1919-1925
U	Utility-General	1952-today
UC	Utility-Cargo Transport	1943-today
UH	Utility-Helicopter	1952-today
V	VTOL-STOL-VSTOL (Vertical/Short Takeoff/Land)	1954-today
VC	VIP Cargo Transport	1948-today
VCP	Verville-Clark Pursuit	1919-1925
W	Weather Reconnaissance	1948-today
WB	Weather Recon Bomber	1948-today
WC	Weather Recon Cargo Transport	1948-today
X	Experimental Prototype	1925-today
X	Special Research	1948-today
XA	Experimental Prototype-Attack	1925-today
XB	Experimental Prototype-Bombardment	1925-today
XC	Experimental Prototype-Cargo Transport	1925-today
XF	Experimental Prototype-Fighter	1948-today
XP	Experimental Prototype-Pursuit	1925-1948
XS	Supersonic Experiment	1945-1948
Y	Limited Production-Prototype	1928-today
YA	Limited Production-Attack	1928-today
YB	Limited Production-Bombardment	1928-today
YC	Limited Production-Cargo Transport	1928-today
YF	Limited Production-Fighter	1948-today
YP	Limited Production-Pursuit	1928-1948

MAJOR DESIGNERS OF AIR FORCE AIRCRAFT— 1909 TO 1979

Beechcraft—established 1932. Still existing in 1979.

Bell—established 1935. Still exists.

Boeing—established 1916. Still exists.

Cessna—established 1927. Still exists.

Consolidated—established 1923. Became Consolidated-Vultee in 1943.

Consolidated-Vultee—established 1943. Became Convair in 1954.

Convair—established 1954 as a division of General Dynamics. Became General Dynamics-Convair in 1961.

Curtiss—established 1910. Became Curtiss Airplane Division of Curtiss-Wright in 1929. Terminated aircraft production in 1966.

Curtiss-Wright—established 1929. Terminated aircraft production in 1966.

DeHavilland (Canada)—established 1928. Taken over by Canadian government in 1974. Still exists.

Douglas—established 1920. Merged into McDonnell-Douglas 1967.

Fairchild—established 1925. Became Fairchild-Hiller in 1964.

Fairchild-Hiller—established 1964. Added Fairchild-Republic in 1965. Became Fairchild Industries in 1972.

Fairchild Industries—established 1972. Still exists.

Fairchild-Republic—organized 1965 as a division of Fairchild-Hiller. Still exists.

General Dynamics-Convair—established 1961. Became General Dynamics in 1965 with Convair as a subdivision.

General Dynamics—established 1965. Still exists.

Helio—established 1948. Became Helio-GAC in 1969.

Ling-Temco-Vought—established 1961. Reorganized as LTV in 1965. Changed name to Vought in 1976.

Lockheed—established 1916. Became Lockheed-Vega 1941. Absorbed Vega in 1943 and reverted to Lockheed. Still exists.

Martin—established 1909. Became Wright-Martin 1916. Became Martin in 1917. Terminated aircraft production in 1960.

McDonnell—established 1939. Merged into McDonnell-Douglas 1967.

McDonnell-Douglas—established 1967. Still exists.

North American—established 1928. Merged into North American-Rockwell in 1967. Absorbed by Rockwell International 1972.

Northrop—entered military market 1933. Absorbed by Douglas in 1937. Reestablished as Northrop in 1939. Still exists.

Piper—established 1937. Still exists.

Republic—established 1939. Became Fairchild-Republic 1965.

Rockwell International—established 1972. Still exists.

Ryan—established 1922. Became Teledyne-Ryan 1969. Ceased aircraft production 1975.

Seversky—established 1931. Absorbed by Republic in 1939.

Sikorsky—established 1923. Still exists.

Stearman—established 1927. Became Stearman-Boeing in 1938. Absorbed by Boeing in 1941.

Taylorcraft—established 1936. Still exists.

Teledyne-Ryan—established 1969. Ceased aircraft production in 1975.

Thomas-Morse—entered military market 1916. Acquired by Consolidated in 1941.

Vega—established 1937. Became Lockheed-Vega 1941. Absorbed by Lockheed in 1943.

Vultee—established 1939. Acquired Stinson in 1940. Became Consolidated-Vultee in 1943.

Waco—established 1921. Ceased military aircraft production in 1945.

Wright—established 1903. Became Wright-Martin in 1916. Became Dayton-Wright in 1917. Absorbed by Consolidated in 1923. Became Curtiss-Wright in 1929.

HISTORICAL INDEX OF AIRPLANES

NUMBER	NAME-VARIANTS	DESIGNER	SPAN	LENGTH	YEAR

THE EARLY PLANES: 1909-1919

Model-B	Wright Flyer	Wright Brothers	38'	30'	'09

Army's first airplane—Biplane trainer
Dual positions, 60-hp, twin pusher-props, front wheels, 54-mph

Model-D		Curtiss	41'	29'	'11

Army's second airplane design—Tricycle landing gear
Biplane, dual controls, open to the wind, single pusher-prop, 60-mph

Model-H		Burgess	40'	32'	'12

First tractor design—First enclosed fuselage
Biplane, twin fins, four wheels with skids, single seat, 60-mph

Model-G		Curtiss	38'4"	24'	'13

First Curtiss Tractor plane with enclosed fuselage
Biplane, side by side seats, four spoked wheels, 75-mph

Model JN-1		Curtiss	40'2"	26'4"	'14

First of the famous Curtiss "JN" family of trainers
Biplane, 80-hp, tandem open seats, two wheels, tailskid, 84-mph

JN-4	Jenny	Curtiss	43'7"	27'4"	'16

Famous American WWI pilot trainer
Biplane, 90-hp, tandem open seats, dual controls, tailskid, 75-mph

Model-SJ		Standard	43'10"	26'7"	'17

WWI American primary trainer—biplane
Three front wheels, 150-hp, tailskid, tandem open seats, 85-mph

Model S-4		Thomas-Morse	26'6"	19'10"	'17

WWI American advanced pilot solo trainer
Biplane, 80-hp, single seat, one machine gun, 100-mph

NUMBER	NAME-VARIANTS	DESIGNER	SPAN	LENGTH	YEAR

Avro-504K   Avro   36'   29'5"   '17
WWI AEF British primary trainer—biplane
Two open seats, 130-hp, two wheels and a ski up front, 95-mph

Sopwith 1A-2   Sopwith Works   33'6"   25'4"   '17
WWI AEF British two-seat combat trainer
Biplane, 130-hp, open seats, one fixed gun, one flex gun, 105-mph

Breguet-14   Corps d'Armée   Breguet   47'3"   29'2"   '17
WWI AEF French observation attack plane-trainer
Biplane, open seats, 300-hp, two fixed guns, one flex gun, 130-mph

Salmson 2A-2   Salmson   49'   30'   '17
WWI AEF French heavy attack plane-trainer
Biplane, 270-hp, open seats, one fixed gun, two flex guns, 120-mph

Spad-VII   S.P.A.D.   26'8"   20'6"   '17
WWI AEF French fighter-advanced trainer
Biplane, one seat, 150-hp, one fixed gun, 130-mph

Nieuport-11   Bébé   Nieuport   24'6"   19'   '17
First fighter flown in combat by Lafayette Escadrille-AEF Trainer
Biplane, 80-hp, one seat, Lewis gun on top wing, 100-mph

Nieuport-12   Nieuport   26'   21'   '17
WWI AEF French armed combat trainer—Biplane
Side by side or tandem seats, 80-hp, two guns, 95-mph

Nieuport-17   Nieuport   27'   19'6"   '17
First fighter flown over enemy lines by an Army pilot—Trainer
Biplane, 110-hp, one seat, one Vickers gun, one Lewis gun, 105-mph

Nieuport-28   Nieuport   26'3"   20'4"   '18
America's first WWI French combat fighter
Biplane, 170-hp, one seat, two Vickers guns, 122-mph

Spad-XIII   S.P.A.D.   26'10"   20'8"   '18
Eddie Rickenbacker's WWI French dogfighter
Biplane, 220-hp, one seat, two Vickers cowl guns, 140-mph

Sopwith Camel   Sopwith Works   28'   18'9"   '18
WWI AEF British fighter-trainer—Holder of record air victories
Biplane, 130-hp, one seat, two Vickers guns, four bombs, 115-mph

NUMBER	NAME-VARIANTS	DESIGNER	SPAN	LENGTH	YEAR

SE-5   Scout   Royal Aircraft   26'8"   20'11"   '18
WWI AEF British attack-fighter-trainer
Biplane, 210-hp, one seat, two guns, four bombs, 128-mph

Lusac-11   Le Père   Packard   41'7"   25'5"   '18
First Army plane equipped with turbo-supercharged engine
Biplane, 400-hp, 2-seat, 4-gun, 34,510-foot ceiling, 133-mph

HD-1   Hanriot-Macchi 28'7"   19'2"   '17
Italian-built, French-designed fighter of WWI
Biplane, 120-hp, one seat, one gun, 115-mph

ATTACK AIRCRAFT: 1918-1925

DH-4   Flaming Coffin   Dayton-Wright 42'6"   29'11"   '18
First Army attack plane—Only American WWI combat aircraft
Biplane, 400-hp, 2-seat, 4-gun, 220-pounds bombs, 125-mph

DB-1   Gallaudet   67'   44'   '21
First all-American design attack plane—Only "DB" Day Bomber
Low-wing monoplane, 700-hp, 4-guns, 600-pounds bombs

GA-1   Boeing   65'6"   33'8"   '21
First "GA" Ground Attack Air Service aircraft—Boxlike triplane
Two 435-hp pusher engines, 4-man, 4-guns, 250-pounds bombs, 105-mph

GA-2   Boeing   54'   37'   '22
Armor-plated Ground Attack Air Service aircraft—Too heavy
Biplane, 700-hp, 3-Vickers guns, 2-Lewis guns, one-37mm cannon

A—ATTACK: 1925-1948

A-3   Falcon (0-1)   Curtiss   38'   27'2"   '27
First Air Corps production attack plane—Biplane
Two open seats, 6-guns, 300-pounds bombs, 140-mph

YA-8   Shrike (YA-10, A-12) Curtiss   44'   32'3"   '32
First all-metal monoplane with flaps at wing's trailing edge
600-hp engine, two-seat, 5-guns, 400-pounds bombs, 160-mph

YA-11   (P-30, PB-2)   Consolidated 43'11"   29'4"   '34
All-metal low-wing monoplane with enclosed cockpit
Two seats, 675-hp, 5-guns, 300-pounds bombs, 228-mph

NUMBER	NAME-VARIANTS	DESIGNER	SPAN	LENGTH	YEAR
A-12	Shrike (YA-8, YA-10)	Curtiss	44′	32′3″	’33

All-metal low-wing monoplane with streamlined wheel pants
Two open seats, 690-hp, 5-guns, 400-pounds bombs, 175-mph

A-17	Nomad (YA-13, XA-16, A-33)	Northrop-Douglas	47′9″	32′	’36

Low-wing monoplane for Air Corps and export
Enclosed cockpit, 2-3 seats, 5-guns, 600-pounds bombs, 220-mph

YA-18	Shrike (XA-14)	Curtiss	59′6″	42′4″	’37

First Air Corps production twin-engine attack plane
Low-wing, 600-hp, 2-seat, 5-gun, 650-pounds bombs, 238-mph

YA-19		Vultee	50′	38′2″	’39

Foreign-aid low-wing monoplane—Air Corps engine test bed
Enclosed cockpit, 1200-hp, 2-3 seats, 6-gun, 1080-pounds bombs, 230-mph

A-20	Havoc (P-70, F-3)	Douglas	61′4″	47′7″	’41

First USAAF plane to enter European combat in WWII—Twin engines
Tricycle gear, 1600-hp, 3-man, 7-gun, 2600-pounds bombs, 340-mph

A-24	Dauntless (SBD-3, F-24)	Douglas	41′6″	33′	’41

Navy dive bomber adapted as Air Corps attack plane
Low-wing, 1200-hp, 2-man, 4-gun, 1200-pounds bombs, 255-mph

A-25	Shrike-Helldiver (SB2C-1)	Curtiss	49′9″	36′8″	’42

Navy dive bomber converted to Air Corps attack plane
Low-wing, 1750-hp, 2-man, 5-gun, 1400-pounds bombs, 275-mph

A-26	Invader (B-26)	Douglas	70′	49′11″	’43

Fastest WWII attack plane—Twin-engine shoulder-wing monoplane
3-man, 2000-hp, 22-gun, 16-rocket, 6000-pounds bombs, 370-mph

A-27	(AT-6)	North American	42′	29′	’41

First attack plane to fly combat in Pacific during WWII
Low-wing, 775-hp, 2-man, 5-gun, 400-pounds bombs, 250-mph

A-28	Hudson (A-29, C-63, C-111, AT-18)	Lockheed-Vega	65′6″	44′4″	’41

Lend-Lease twin-engine monoplane for Britain and Australia
Mid-wing, 1200-hp, 4-man, 5-gun, 1600-pounds bombs, 250-mph

NUMBER	NAME-VARIANTS	DESIGNER	SPAN	LENGTH	YEAR
A-29	Hudson (A-28, C-63, C-111, AT-18)	Lockheed-Vega	65'6"	44'4"	'41

First plane to sink German U-boat with rockets in WWII
1200-hp, 4-seat, 5-gun, rockets, depth charges, 1600-pounds bombs

| A-30 | Baltimore (XA-22) | Martin | 61'4" | 48'6" | '42 |

Lend-Lease twin-engine monoplane used in Mediterranean
Mid-wing, 1600-hp, 4-man, 12-gun, 2000-pounds bombs, 300-mph

| A-31 | Vengeance (A-35) | Vultee | 48' | 39'9" | '41 |

Lend-Lease monoplane repossessed for Air Corps use—Single engine
Low-wing, 1600-hp, 2-man, 5-gun, 2000-pounds bombs, 275-mph

| A-33 | Nomad (A-17) | Douglas | 47'9" | 32'6" | '42 |

Lend-Lease single-engine monoplane repossessed as combat trainer
Low-wing, 1200-hp, 2-man, 5-gun, 1800-pounds bombs, 250-mph

| A-35 | Vengeance (A-31) | Vultee | 48' | 39'9" | '42 |

Lend-Lease single-engine monoplane for England and Brazil
Low-wing, 1700-hp, 2-man, 7-gun, 2000-pounds bombs, 280-mph

| A-36 | Mustang (P-51) | North American | 37' | 32'3" | '42 |

First USAAF combat variant of single-engined P-51 Mustang
Low-wing, 1325-hp, one seat, 6-gun, 1000-pounds bombs, 310-mph

A—ATTACK: 1962-today

| A-1 | Skyraider (AD-5) | Douglas | 50' | 40' | '63 |

Navy single-engine low-wing monoplane for USAF in Vietnam
2700-hp, 4-cannon, rockets, napalm, 7000-pounds stores, 300-mph

| A-7 | Corsair II | Ling-Temco-Vought | 38'9" | 46'2" | '68 |

Navy single-jet high-wing Strike plane for USAF in Vietnam
20mm cannon, napalm, gun pods, missiles, 13,000-pounds weapons, 680-mph

| A-10 | Thunderbolt II | Fairchild | 57'6" | 53'4" | '75 |

Twin-jet low-wing twin-fin one-seat Close Support plane—520-mph
7-barrel cannon, Laser-guided bombs, missiles, 16,000 pounds weapons

| A-37 | Dragonfly (T-37) | Cessna | 35'11" | 28'3" | '67 |

Twin-jet low-wing monoplane used on COIN operations in Vietnam
2-man, 7.62mm minigun, rockets, napalm, 4855-pounds weapons, 505-mph

NUMBER	NAME-VARIANTS	DESIGNER	SPAN	LENGTH	YEAR

BOMBARDMENT AIRCRAFT: 1918-1925

MB-1	Martin Bomber (GMB, T-1)	Martin	71'5"	46'6"	'18

First all-American designed and built combat biplane
Twin 400-hp engines, 3-man, 5-gun, 1040-pounds bombs, 120-mph

MB-2	Martin Bomber (NBS-1)	Martin-Curtiss	74'2"	42'8"	'20

Biplane bomber that sank battleship to prove Billy Mitchell right
Folding wings, twin engines, 4-seat, 5-gun, 3000-pounds bombs

NBS-1	(MB-2)	Martin-Curtiss	74'2"	42'8"	'20

Only Night Bomber Short-range—Folding wing biplane
Twin engines, 4-seat, 5-gun, 3000-pounds bombs

NBL-1	Barling Bomber	Witteman-Lewis	120'	65'	'23

Experimental Night Bomber Long-range
Triplane, six engines, eight wheels, multi-place, 93-mph

HB—HEAVY BOMBARDMENT: 1925-1928

XHB-1	Cyclops	Huff-Daland	84'7"	59'7"	'26

Only Air Corps single-engine Heavy Bomber experiment
Biplane, 4-man crew, 6-gun, 4000-pounds bombs

LB—LIGHT BOMBARDMENT: 1925-1928

LB-1		Huff-Daland	66'6"	46'2"	'26

Only single-engine "LB" Light Bomber—Biplane
800-hp, 4-man, 5-gun, 1500-pounds bombs, 8-hour endurance

LB-5	Pirate (XLB-3)	Huff-Daland-Keystone	67'	44'8"	'27

First production twin-engine biplane Light Bomber
Triple-fin tail, 5-man crew, 5-gun, 2000-pounds bombs

LB-6	(LB-7 thru LB-14)	Keystone	75'	43'5"	'28

Final series of Keystone twin-engine Light Bombers
Single or twin fins, 530-hp, 5-man, 5-gun, 2000-pounds bombs, 115-mph

B—BOMBARDMENT: 1928-1962

B-2	Condor (C-30)	Curtiss	90'	47'6"	'28

First bomber with gunners in outboard wing nacelles—Biplane
Twin 600-hp engines, 5-man, 6-gun, 4000-pounds bombs, 130-mph

NUMBER	NAME-VARIANTS	DESIGNER	SPAN	LENGTH	YEAR
B-3	(LB-10, B-4, B-5, B-6)	Keystone	74'8"	48'10"	'29

Direct conversion of Keystone "LB" Light Bomber—Single-fin biplane
Twin 525-hp engines, 4-man, 3-gun, 2500-pounds bombs, 120-mph

YB-7	(YO-35)	Douglas	65'3"	46'7"	'32

First twin-engine Monoplane bomber in Air Corps history
Retractable gear, 600-hp, 4-man, 2-gun, 1200-pounds bombs, 180-mph

YB-9		Boeing	76'10"	52'	'32

First all-metal Air Corps bomber—Mid-wing Monoplane
Twin 600-hp engines, 4-man, 2-gun, 2200-pounds bombs, 190-mph

B-10	Martin Bomber (B-12, YB-13, XB-14)	Martin	70'6"	44'9"	'34

First production twin-engine all-metal mid-wing monoplane bomber
Norden bombsight, 775-hp, 3-man, 3-gun, 2260-pounds bombs, 236-mph

B-12	Martin Bomber (B-10)	Martin	70'6"	45'3"	'34

First Army coastal defense float-fitted long range bomber
Retractable gear, 4-man, 3-gun, 2260-pounds bombs, 215-mph

XBLR-1	(XB-15)	Boeing	149'	87'7"	'35

Bomber Long Range prototype of XB-15 hemispheric bomber
Four engines, mid-wing, retractable gear, 35-ton gross weight

XB-15	(XBLR-1, XC-105)	Boeing	149'	87'7"	'37

Largest bomber prototype of the time—Redesignated from XBLR-1
Four engines, 10-man crew, 6-gun, 12,000-pounds bombs, 197-mph

B-17	Flying Fortress (XB-38, YB-40, F-9, YC-108)	Boeing	103'9"	73'2"	'40

Famous WWII four-engine mid-wing monoplane heavy bomber
1200-hp, 10-man, 13-gun, 10,800-pounds bombs, 318-mph

B-18	Bolo (DC-2½, C-39, C-58, B-23)	Douglas	89'6"	57'10"	'37

WWII patrol bomber used in anti-submarine warfare
Twin 1000-hp engines, 6-man, 3-gun, depth charges, 6500-pounds bombs, 215-mph

XBLR-2	(XB-19)	Douglas	212'	132'2"	'38

Bomber, Long Range Prototype of XB-19 global bomber
Four-engine, all-metal, tricycle gear, low-wing monoplane

NUMBER	NAME-VARIANTS	DESIGNER	SPAN	LENGTH	YEAR
XB-19	(XBLR-2)	Douglas	212'	132'2"	'41

First American bomber with wingspan greater than 200 feet
Four 2000-hp engines, 10-man, 11-gun, 3-cannon, 37,000-pounds bombs, 210-mph

B-23	Dragon (B-18, C-67)	Douglas	92'	58'4"	'39

First bomber-trainer with tail gun turret
Twin-engine, low-wing, 6-man, 4-gun, 8000-pounds bombs, 280-mph

B-24	Liberator (C-87, F-7, C-109, AT-22)	Consolidated	110'	67'2"	'41

Famous WWII four-engine bomber—Most-built bomber in history
1200-hp, 10-man, 10-gun, 12,800-pounds bombs, 303-mph

B-25	Mitchell (F-10, AT-24)	North American	67'7"	52'11"	'41

WWII bomber flown by Jimmy Doolittle on the Tokyo raid
1700-hp, 6-man, 18-gun, 75mm cannon, 4000-pounds bombs, 280-mph

B-26	Marauder (AT-23)	Martin	71'	58'3"	'41

First of the two B-26 bombers—Only USAAF torpedo carrier
Twin 2000-hp engines, 7-man, 12-gun, 4000-pounds bombs, 280-mph

B-26	Invader (A-26)	Douglas	70'	51'3"	'48

Korean conflict and Vietnam war tactical light bomber
2000-hp, 3-man, 22-gun, rockets, 4000-pounds bombs, 355-mph

B-29	Superfortress (F-13, C-97, B-50)	Boeing	141'3"	99'	'44

Famous WWII heavy bomber—Dropped first atomic bomb
2220-hp engines, 11-man, 12-gun, 20,000-pounds bombs, 360-mph

KB-29	Superfortress (B-29)	Boeing	141'3"	120'1"	'48

First Air Force inflight refueling tanker—Stretched B-29
Four engines, mid-wing, 6-man crew, no armament, 360-mph

B-32	Dominator (TB-32)	Consolidated-Vultee	135'	82'1"	'44

WWII heavy bomber used sparingly in western Pacific
Four 2200-hp engines, 9-man, 10-gun, 8000-pounds bombs, 355-mph

B-34	Ventura (B-37, C-60)	Lockheed-Vega	65'6"	51'5"	'42

WWII overwater patrol bomber and navigator trainer
Twin 2000-hp engines, 4-man, 10-gun, 3000-pounds bombs, 320-mph

NUMBER	NAME-VARIANTS	DESIGNER	SPAN	LENGTH	YEAR
YB-35	Flying Wing (YB-49)	Northrop	172'	53'	'46

First USAAF long range bomber with no fuselage
Four 3000-hp engines, pusher-props, 6-man, 20-gun, 51,000-pounds bombs

| B-36 | Peacemaker (YB-60, XC-99) | Consolidated-Vultee | 230' | 162'1" | '47 |

Largest bomber in military history—Only 10-engine aircraft
Six 3000-hp radials, 4-jets, 15-man, 16-cannon, 84,000-pounds bombs

| B-37 | Ventura (B-34, O-56, C-60) | Lockheed | 65'6" | 51'5" | '43 |

WWII anti-submarine patrol bomber—twin finned tail
Two 1700-hp engines, 4-man, 8-gun, 2500-pounds bombs, 300-mph

| XB-43 | (XA-42, XB-42) | Douglas | 71'2" | 51'5" | '46 |

First USAAF jet bomber prototype in history
Twin turbojets, 3-man, no guns, 8000-pounds bombs, 500-mph

| B-45 | Tornado | North American | 89' | 75'4" | '48 |

First USAF production jet bomber—First combat jet bomber
Four-jet, high-wing, 4-man, 2-tail guns, 22,000-pounds bombs, 580-mph

| B-47 | Stratojet (YB-56) | Boeing | 116' | 107' | '51 |

World's first swept-wing heavy jet bomber—Landing chute
Six-jet, high-wing, 3-man, 4-cannon, 20,000-pounds bombs, 620-mph

| KB-47 | Stratojet (B-47) | Boeing | 116' | 112' | '53 |

World's first jet-to-jet air refueling tanker
Six-jet, high-wing, bicycle landing gear, outriggers for balance

| YB-49 | Flying Wing (YB-35) | Northrop | 172' | 53' | '47 |

First USAF jet bomber with no fuselage
Eight-jet, 6-man, no guns, 16,000-pounds bombs, 510-mph

| B-50 | Superfortress (B-29D) | Boeing | 141'3" | 99' | '47 |

World's first nonstop flight around the world by air refueling
Four 3500-hp engines, 10-man, 13-gun, 28,000-pounds bombs, 380-mph

| KB-50 | Superfortress (B-50D) | Boeing | 141'3" | 105'1" | '52 |

Inflight refueling tanker for tactical fighters
Four radials, two jets, 6-man, no armament, 444-mph

| B-52 | Stratofortress | Boeing | 185' | 152'9" | '55 |

Most durable bomber in history—Dropped H-bomb at Bikini
Eight-jet, 6-man, 4-gun, 28,250-pounds bombs, 660-mph

NUMBER	NAME-VARIANTS	DESIGNER	SPAN	LENGTH	YEAR

B-57   Intruder (Canberra)   Martin   64'   65'6"   '53
First non-American designed USAF bomber—British exchange
One or 2-man, 8-gun, 16-rocket, 6000-pounds bombs, 580-mph

RB-57D   Intruder   Martin   106'   67'10"   '54
Strategic reconnaissance version of B-57 with stretched wing
Twin-jet, one or 2-man, tricycle gear, no armament, 630-mph

RB-57F   Intruder   General Dynamics   122'   63'10"   '64
Strategic reconnaissance version of B-57 with extended wing
Two turbofans, two turbojets, 93,000-foot altitude, 10-hr endurance

B-58   Hustler   Convair   57'   97'   '56
First supersonic bomber—First weapons system aircraft
Four-jet, delta wing, 3-man, one cannon, nuclear bomb, 1380-mph

YB-60   (B-36)   Consolidated-Vultee   206'5"   175'2"   '52
Swept-wing all-jet variant of B-36 worldwide bomber
Eight-jet, 11-man, ten-gun, 72,000-pounds bombs, 505-mph

B-61   Matador (TM-61)   Martin   28'7"   39'6"   '51
Pilotless tactical missile assigned a ''B'' designation
Single jet, nuclear war head, 620-mile range, 600-mph

B-66   Destroyer   Douglas   72'6"   75'   '56
Participated in operation to drop H-bomb at Bikini
Twin-jet, high-wing, 3-man, 2-cannon, 15,000-pounds bombs, 630-mph

RB-66   Destroyer   Douglas   72'6"   75'   '54
Vietnam war all-weather night photo-recon bomber
Twin-jet, high swept wing, 7-man, radomes, chaff dispenser

XB-70   Valkyrie   North American   105'   185'   '64
Only triplesonic bomber—Highest numbered manned bomber
Six-jet, delta wing, foreplane, twin-fin, tailfirst, 2300-mph

B—BOMBARDMENT: 1962-today

B-1   Rockwell   136'8"   150'3"   '74
First variable-wing heavy bomber in Air Force history
4-turbofans, 4-man, 12-warheads, 115,000-pounds weapons, 1400-mph

NUMBER	NAME-VARIANTS	DESIGNER	SPAN	LENGTH	YEAR
B-26K	Counter-Invader (B-26)	On Mark	71'6"	51'10"	'64

Vietnam war COunter-INsurgency (COIN) attack-bomber
Twin-engine, 15-gun, 14-rocket, 11,000-pounds ordnance, 355-mph

NUMBER	NAME-VARIANTS	DESIGNER	SPAN	LENGTH	YEAR
FB-111	(TFX, F-111A)	General Dynamics	70'	73'6"	'69

First variable-wing jet fighter-bomber for SAC
2-turbofans, 2-man, 6-nuclear bombs, 31,500-pounds weapons, 1450-mph

SUPPLEMENT TO BOMBERS

EXPERIMENTAL JET BOMBERS THAT DID NOT REACH PRODUCTION

NUMBER	DESIGNER	
XB-42	Douglas	Mixmaster—2 turbojets, 2 turboprops—1944
XB-46	Consolidated	Four jet engines—1947
XB-48	Martin	Six jet engines—1947
XB-51	Martin	First 3-jet, variable-swept wing Bomber—1949
XB-53	Convair	Three turbojets—contract cancelled
XB-55	Boeing	4 turbojets, 4 turboprops—cancelled
XB-56	Boeing	B-47 with engine change—cancelled

PILOTLESS MISSILES THAT WERE LATER CATEGORIZED AS MISSILES

B-62	Northrop	Snark missile—became SM-62
B-63	Bell	Rascal missile—GAM-63 (X-9)
B-64	North American	Navaho missile—XM-64 (X-10)
B-65	Convair	Atlas missile—SM-65 (X-11)
B-67	Radioplane	Crossbow missile—GAM-67
SB-68	Martin	Titan missile—SM-68
RB-69	Lockheed	Neptune missile—P2V-5
B-72	McDonnell	Quail missile—GAM-72
B-75	Douglas	Thor missile—SM-75
B-76	Martin	Mace missile—TM-76
B-77	North American	Hound Dog missile—GAM-77
B-78	Chrysler	Jupiter missile—SM-78
B-80	Boeing	Minuteman missile—SM-80
B-83	Martin	Bullpup missile—GAM-83
B-87	Douglas	Skybolt missile—GAM-87. Highest number assigned in the ''B'' category.

NUMBER	NAME-VARIANTS	DESIGNER	SPAN	LENGTH	YEAR

CARGO TRANSPORTS
T—TRANSPORT: 1919-1925

| T-1 | (GMB, GMP, MB-1) | Martin | 71′5″ | 44′10″ | '19 |

Army's first officially designated transport plane
Two engine, twin fin, 10-passenger, cabin biplane, 105-mph

| T-2 | (A-2) | Fokker | 79′8″ | 49′1″ | '23 |

America's first nonstop coast to coast flight—First monoplane
1-engine, high-wing monoplane—Largest airplane of that day, 110-mph

A—AMBULANCE: 1919-1925

| A-2 | (T-2) | Fokker | 79′8″ | 49′1″ | '23 |

Army's first airborne ambulance—Conversion from T-2 transport
One engine, high-wing cabinplane, 2-wheels, rear skid, 110-mph

C—CARGO TRANSPORT: 1925-1962

| C-1 | (O-2) | Douglas | 39′8″ | 29′6″ | '25 |

Army's first cargo and personnel transport
Single engine, cabin biplane, 2-man, 8-passenger, 115-mph

| C-2 | Fokker Trimotor (XC-5, C-7) | Fokker | 74′2″ | 48′4″ | '26 |

Lt Comdr Byrd and Floyd Bennett North Pole crossing plane
Three engine, high-wing cabinplane, 10-passenger, 110-mph

| C-4 | Ford Trimotor (C-3, C-9) | Ford | 77′11″ | 51′ | '29 |

Adm Byrd's plane on first crossing of South Pole
3-engine, high-wing cabinplane, leg pants, 10-passenger, 145-mph

| C-6 | | Sikorsky | 71′6″ | 40′3″ | '29 |

Army's first amphibian cargo transport
Two engine, twin-fin, high-wing monoplane, 12-seat, 185-mph

| C-8 | (F-1, UC-96) | Fairchild | 50′ | 33′ | '30 |

Double-duty plane used for transport and photography
Single engine, high-wing cabinplane, 3-man, 7-passenger, 140-mph

| YC-14 | (XC-15) | Fokker | 59′ | 43′3″ | '31 |

Cargo transport and airborne ambulance with parasol wing
Single engine, high-wing cabinplane, 6-passenger, 135-mph

NUMBER	NAME-VARIANTS	DESIGNER	SPAN	LENGTH	YEAR
C-21	Dolphin (OA-3, C-26, C-29)	Douglas	60'	43'10"	'32

First Douglas amphibian cargo transport
Two engine high-wing monoplane, 7-place, 140-mph

| C-26 | Dolphin (OA-4, C-21, C-29) | Douglas | 60' | 43'10" | '32 |

Treasury Department Mexican border patrol plane
Two engine, high-wing monoplane, 8-place, 550-mile range, 140-mph

| C-27 | Airbus | Bellanca | 65' | 42'9" | '32 |

Civilian airliner adapted to military standards
Single engine, high-wing monoplane, fancy pants, 12-passenger

| C-29 | Dolphin (C-21, C-26) | Douglas | 60' | 43'10" | '33 |

Direct conversion from C-26 with engines changed
Twin engine, high-wing monoplane, 8-place, 150-mph

| C-30 | Condor (B-2) | Curtiss | 90' | 47'6" | '33 |

Admiral Byrd's plane on second Antarctic expedition
Twin engine, biplane, twin-fin, 15-passenger, 4000-pounds cargo, 130-mph

| C-32 | (DC-2) | Douglas | 85' | 61'6" | '36 |

First of the famous Douglas "DC" family—4,000-pound cargo load
Twin engine, low-wing, personnel transport, 14-seat, 180-mph

| C-33 | (DC-2) | Douglas | 85' | 61'6" | '36 |

Army's first 200-mph-plus cargo transport
Twin engine, low-wing, 14-seat, 5000-pounds cargo, 205-mph

| YC-34 | (C-32, DC-2) | Douglas | 85' | 61'6" | '36 |

A C-32 with different internal arrangements
Twin engine, low-wing, personnel transport, 16-seat, 180-mph

| XC-35 | Electra (UC-36, UC-37) | Lockheed | 55' | 39'8" | '36 |

Won Collier Trophy for aircraft achievement in 1937
World's first cargo transport with pressurized cabin

| UC-36 | Electra (UC-37) | Lockheed | 55' | 38'7" | '37 |

Amelia Earhart's around the world airplane
Twin engine, twin-fin, low-wing, 2-crew, 10-passenger, 205-mph

NUMBER	NAME-VARIANTS	DESIGNER	SPAN	LENGTH	YEAR
UC-37	Electra (UC-36)	Lockheed	55'	38'7"	'37

Military adaptation of Lockheed Model-10A civilian airliner
Twin engine, twin-fin, low-wing, 2-crew, 8-passenger, 220-mph

C-38	(C-33, DC-2½)	Douglas	85'	61'6"	'37

Cargo transport with wings of a DC-2 and fuselage of a DC-3
Two engine, low-wing, cargo door, 16-seat, 6000-pound payload, 210-mph

C-39	(C-38, XC-41, XC-42, DC-2½)	Douglas	85'	61'6"	'39

Same as C-38 with engines changed
Two engine, low-wing, cargo door, 16-seat, 6,000-pound payload, 210-mph

UC-40	Electra	Lockheed	49'6"	36'4"	'38

Lockheed civilian Model-12A Electra adapted for the Air Corps
Twin engine, twin-fin, low-wing, 2-crew, 7-passenger, 220-mph

UC-43	Traveler	Beechcraft	32'	25'9"	'41

WWII staggerwing light cargo transport
Single engine, cabin biplane, streamlined spats, 5-seat, 210-mph

UC-45	Expediter (AT-7, AT-11, F-2)	Beechcraft	47'8"	34'3"	'40

Popular WWII multi-purpose light cargo, utility transport
Two engine, low-wing cabinplane, 2-crew, 6-passenger, 220-mph

C-46	Commando (XC-55)	Curtiss	108'1"	76'4"	'42

Builder of the famous ''air bridge'' over the Himalayan ''Hump''
Four-crew, 50-troop, 33-litter, 16,000-pounds freight, 260-mph

C-47	Skytrain-Gooney Bird (DC-3)	Douglas	95'6"	63'9"	'42

Famous WWII cargo transport used on all fighting fronts
3-crew, 27-troop, 24-litter, 7,500-pounds cargo, 230-mph

C-48	(DC-3)	Douglas	95'6"	63'9"	'41

Civilian personnel transport commandeered for WWII
Twin engine, low-wing, 21-seat, 14-berth, 220-mph

C-49	(DC-3)	Douglas	95'6"	63'9"	'41

Troop carrier-sleeper commandeered for WWII
Twin engine, low-wing, 24-bench seats, 14-berth, 220-mph

C-50	(C-49, DC-3)	Douglas	95'6"	63'9"	'41

Like the C-49 with engine change
Twin engine, low-wing, 24-bench seats, 14-berth, 220-mph

NUMBER	NAME-VARIANTS	DESIGNER	SPAN	LENGTH	YEAR
C-51	(C-49, DC-3)	Douglas	95'6"	63'9"	'41

Like the C-49 with engine change
Twin engine, low-wing, 24-bench seats, 14-berth, 220-mph

C-52	(C-49, DC-3)	Douglas	95'6"	63'9"	'41

Like the C-49 with engine change
Twin engine, low-wing, 24-bench seats, 14-berth, 220-mph

C-53	Skytrooper (C-47)	Douglas	95'6"	63'9"	'41

WWII troop carrier with extra tanks for long range
2-engine, 42-passenger, 26-troop, 4000-pounds cargo, 230-mph

C-54	Skymaster (YC-116, DC-4)	Douglas	117'6"	93'11"	'42

Largest WWII mass produced cargo transport—Tricycle gear
4-engine, low-wing, 50-seat, 26-stretcher, 14,000-pound load, 285-mph

C-56	Lodestar (C-60)	Lockheed-Vega	65'6"	49'10"	'41

Lockheed Model-18 executive monoplane impressed for WWII
Two-engine, mid-wing, twin-fin, 7-tables and seats, 255-mph

C-57	Lodestar (C-59)	Lockheed-Vega	65'6"	49'10"	'41

Lockheed Model-18 personnel transport
Two engine, mid-wing, twin-fin, 14-17 passenger, 270-mph

C-58	Bolo (B-18, DC-2)	Douglas	89'6"	57'10"	'42

Battlefield adaptation of B-18 bomber as a cargo transport
Twin engine, low-wing, 18-seat, 6000-pounds cargo, 215-mph

C-59	Lodestar (C-57)	Lockheed-Vega	65'6"	49'10"	'42

Like C-57 for export to Britain
Two engine, mid-wing, twin-fin, 14-17 passenger, 270-mph

C-60	Lodestar (C-56, B-34)	Lockheed-Vega	65'6"	49'10"	'42

Fastest and most-produced of the Lockheed Lodestar family
Twin engine, mid-wing, twin-fin, 7-VIP seats, 18-troop, 270-mph

UC-61	Forwarder (UC-86)	Fairchild	36'4"	23'9"	'41

WWII light cargo transport used by America and Britain
Single engine, high-wing cabinplane, 4-seat, 130-mph

C-63	Hudson (A-29, AT-18, C-111)	Lockheed-Vega	65'6"	44'4"	'41

WWII cargo transport for America and England
Two-engine, twin-fin, 3-crew, 17-passenger, 4000-pound load, 250-mph

NUMBER	NAME-VARIANTS	DESIGNER	SPAN	LENGTH	YEAR

C-64 Norseman Noorduyn 51'6" 32'4" '42
Canadian light transport acquired on Reverse Lend-Lease
Single engine, high-wing, land plane with optional floats, 160-mph

C-66 Lodestar (C-60) Lockheed 65'6" 49'10" '42
Last of the Lockheed family of Model-18 Lodestars
Two engine, mid-wing, twin-fin, 11-VIP seats, 270-mph

C-67 Dragon (B-23) Douglas 92' 58'4" '42
Cargo transport and glider tug converted from B-23 bomber
Twin engine, low-wing, 21-seat, 8000-pounds cargo, 280-mph

C-68 (DC-3A) Douglas 95' 64'5" '42
Commercial airliner commandeered for WWII
Twin engine, low-wing, 21-passenger, 10,000-pounds cargo, 230-mph

C-69 Constellation Lockheed 123' 94'11" '43
Largest and fastest WWII cargo transport—Limited production
4-engine, low-wing, 3-fin, 64-seat, 82,000-pound gross weight, 330-mph

UC-70 Nightingale Howard 38' 26' '42
Light cargo transport impressed for duty in WWII
Single engine, high-wing, fancy pants, 5-seat, 175-mph

UC-71 Executive Spartan 39' 26'10" '42
Light cabinplane used for staff transport and station taxi
Single engine, low-wing cabinplane, 5-seat, 210-mph

UC-72   Waco 34'9" 27'10" '42
Light cabinplane staff transport and station hack
Single engine, cabin biplane, fancy wheels, 5-seat, 185-mph

C-73   Boeing 74' 51'7" '42
First to have rubber de-icer boots and variable-pitch props
Two engine, low-wing monoplane, 10-passenger, 200-mph

C-74 Globemaster (C-124) Douglas 173'3" 124'2" '45
Huge cargo transport—Too late for WWII—Limited production
4-engine, low-wing, 125-passenger, 60,000-pounds cargo, 310-mph

C-75 Stratoliner Boeing 107' 74'4" '42
First production model cargo transport with pressurized cabin
4-engine, low-wing, 33-seat, 25-bunk, 6600-pounds cargo, 250-mph

NUMBER	NAME-VARIANTS	DESIGNER	SPAN	LENGTH	YEAR
UC-78	Bobcat (AT-8, AT-17)	Cessna	41'11"	42'9"	'41

WWII light personnel transport
Twin engine, low-wing cabinplane, 5-seat, 195-mph

UC-81	Reliant (AT-19)	Stinson-Vultee	41'11"	27'11"	'42

WWII private transport commandeered for the war
Single engine, gull-wing, fancy spats, 5-seat, 175-mph

C-82	Packet (C-119)	Fairchild	106'6"	77'1"	'45

Troop carrier and cargo freighter too late for WWII
Twin-boom, twin-fin, high-wing, 42-seat, 11,500-pounds cargo, 250-mph

UC-83	Coupe (L-4)	Piper	36'2"	22'6"	'43

Light communications aircraft drafted by USAAF
Single engine, high-wing cabinplane, 3-seat, 100-mph

C-84	(DC-3B)	Douglas	95'	64'5"	'42

Commercial airliner commandeered by USAAF for WWII
Twin engine, low-wing, 28-passenger, 10,000-pounds cargo, 230-mph

UC-86	Forwarder (UC-61)	Fairchild	36'4"	23'9"	'42

WWII light cargo transport impressed for service
Single engine, high-wing cabinplane, 4-seat, 130-mph

C-87	Liberator (B-24, AT-22, C-109)	Consolidated	110'	66'4"	'42

Primary long-range heavy cargo transport of WWII
5-crew, 25-seat, 10-berth, 12,000-pound load, 2900-mile range, 305-mph

UC-95	Grasshopper (L-2)	Taylorcraft	35'5"	22'9"	'42

Light communications aircraft commandeered for WWII
Single engine, high-wing cabinplane, 2-seat, 90-mph

UC-96	(C-8, F-1)	Fairchild	50'	33'	'42

Double-duty light passenger or photography plane
Single engine, high-folding-wing, 3-crew, 7-seat, 140-mph

C-97	Stratofreighter (B-29, B-50)	Boeing	141'3"	110'4"	'49

Blending of B-29 and B-50 into a cargo transport
134-troop, 83-litter, 53,000-pound load, 4300-mile range, 375-mph

NUMBER	NAME-VARIANTS	DESIGNER	SPAN	LENGTH	YEAR
KC-97	Stratofreighter (B-29, B-50)	Boeing	141′3″	115′	'51

First USAF cargo tanker—Last Boeing piston-engine design
5-crew, 1400-gallon transfer load, 4300-mile range, 375-mph

| XC-99 | (B-36) | Consolidated-Vultee | 230′ | 185′ | '49 |

World's largest prop-driven cargo transport experiment
6-engine, 400-troop, 300-stretcher, 101,000-pound load, 300-mph

| UC-100 | Gamma (A-17) | Northrop | 48′ | 32′ | '42 |

Adapted from Howard Hawks' famous "Skychief" speed plane
Single engine, 2-seat, low-wing, 2500-mile range, 230-mph

| UC-101 | Vega (YC-12, YC-17) | Lockheed | 41′ | 27′8″ | '42 |

Adaptation from Wiley Post's famous around-the-world plane
Single engine, high-wing cabinplane, fancy pants, 7-seat

| XC-105 | (XB-15) | Boeing | 149′ | 87′11″ | '43 |

Heavy bomber modified as transport with cargo door and hoist
4-engine, mid-wing, 64-passenger, 21,000-pounds cargo, 195-mph

| YC-108 | Flying Fortress (B-17) | Boeing | 103′9″ | 73′10″ | '41 |

General MacArthur's armed VIP personnel transport
4-engine, mid-wing, 38-seat, 4-gun, 2500-mile range, 315-mph

| C-109 | Liberator (B-24, C-87) | Consolidated | 110′ | 67′2″ | '43 |

Fuel ferry that flew gas over the "Hump" to B-29's in China
4-engine, 5-man, 3000-gallon capacity, 2900-mile range, 305-mph

| C-111 | Super Electra (C-63) | Lockheed | 65′6″ | 44′4″ | '44 |

Cargo transport commandeered by USAAF for WWII
2-engine, mid-wing, twin-fin, 3-crew, 17-seat, 2500-pound load, 250-mph

| YC-116 | Skymaster-II (C-54, DC-4) | Douglas | 117′6″ | 93′11″ | '46 |

Heavier C-54 variant with engine change and thermal de-icers
4-engine, 50-seat, 26-stretcher, 14,000-pounds cargo, 260-mph

| C-117 | Skytrain-II (C-47) | Douglas | 90′ | 67′9″ | '45 |

Last of the Douglas DC-3 family of military cargo transports
2-engine, low-wing, 3-crew, 18-seat, 1600-mile range, 230-mph

NUMBER	NAME-VARIANTS	DESIGNER	SPAN	LENGTH	YEAR
C-118	Liftmaster (XC-112, DC-6)	Douglas	117'6"	105'7"	'46

MATS trans-oceanic cargo and personnel transport
76-seat, 60-litter, 27,000-pound load, 4910-mile range, 370-mph

C-119	Flying Boxcar (C-82, XC-120)	Fairchild	109'4"	85'	'47

Korean war heavy freight and personnel transport
5-crew, 62-seat, 20,000-pound load, 20-paracans, 280-mph

C-121	Super Constellation	Lockheed	123'5"	116'2"	'49

Vietnam war computerized electronics cargo transport
4-engine, 3-fin, 106-passenger, 47-litter, 40,000-pound load, 380-mph

YC-122	Cargo Glider (CG-18, X-18)	Chase	95'8"	61'8"	'48

Powered light assault-invasion cargo transport
All-metal, two-engine, high-wing monoplane, 60-troop, 235-mph

XC-123	Avitruc (G-20)	Chase	105'	71'	'49

Powered assault-invasion cargo transport
All-metal, two-engine, high-wing monoplane, 60-troop, 245-mph

XC-123A		Chase	110'	76'	'51

First American all-jet cargo transport experiment
Four J47 turbojets, high-wing, tricycle gear, 67-troop

C-123B	Provider (YC-134)	Chase-Fairchild-Stroukoff	110'	76'3"	'55

Vietnam war jet-assisted intra-theater cargo transport
2-engine (JATO), 60-troop, 50-litter, 30,000-pound payload, 245-mph

C-124	Globemaster-II (C-74)	Douglas	174'2"	130'5"	'50

Korean war heavy cargo transport-mercyship
8-crew, 222-troop, 136-litter, 52-patients, 15-medics, 304-mph

C-125	Raider	Northrop	86'6"	67'	'50

Powered assault transport—Arctic rescue plane—Mechanics trainer
3-engine, high-wing, STOL, 32-troop, 11,000-pounds cargo, 207-mph

LC-126		Cessna	36'2"	27'4"	'49

Light cargo transport and instrument trainer—Arctic rescue plane
Single engine, high-wing cabinplane, 2-crew, 3-passenger, 175-mph

NUMBER	NAME-VARIANTS	DESIGNER	SPAN	LENGTH	YEAR
C-130	Hercules	Lockheed	132'7"	97'9"	'56

Vietnam war STOL-JATO cargo transport
4-5 crew, 92-seat, 64-troop, 74-litter, 40,000-pounds cargo, 385-mph

C-131	Samaritan (T-29)	Convair	105'4"	79'2"	'54

First airborne hospital with pressurized cabin
2-engine, low-wing, 48-seat, 36-troop, 27-litter, 4-crew, 295 mph

C-133	Cargomaster	Douglas	179'8"	157'7"	'57

Largest prop-driven production cargo transport in USAF history
4-engine, 10-man, 200-seat, 100,000-pound load, 4030-mi range, 360-mph

YC-134	Pantobase (YC-123E)	Stroukoff	110'	82'1"	'58

Amphibian STOL variant of Fairchild C-123 Provider assault plane
2-engine, high-wing, 1600-mile range with 24,000-pound load, 240-mph

C-135	Stratolifter (707)	Boeing	130'10"	134'6"	'61

First all-jet production strategic cargo transport
4-jet, 5-crew, 126-troop, 44-litter, 89,000-pounds cargo, 585-mph

KC-135	Stratotanker (707)	Boeing	130'10"	136'3"	'57

First all-jet inflight refueling tanker for USAF
Four underwing turbojets, 120,000-pounds transfer fuel, 585-mph

VC-137	(707)	Boeing	130'10"	144'6"	'59

Presidential "Air Force-One"—VIP carrier—Special Air Missions
4-turbofans, 12-crew, 14-plush seats, conference tables, 625-mph

C-140	Jetstar (XT-40)	Lockheed	54'6"	60'6"	'61

Light utility cargo transport for AACS and SAM
4-jet, 2-crew, 10-seat, 3000-pound load, 2000-mile range, 570-mph

C-141A	Starlifter	Lockheed	159'11"	145'	'65

Powerful strategic cargo transport—Operation "Reforger"
4-crew, 154-seat, 123-troop, 80-litter, 70,850-pound load, 570-mph

YC-141B	Starlifter	Lockheed	159'11"	168'4"	'77

Stretched C-141A with two fuselage extension plugs
Inflight refueling capability, 89,150-pound payload, 570-mph

XC-142		LTV-Hiller-Ryan	67'6"	58'1"	'64

Tilting-wing cargo transport for VTOL and STOL experiments
4-engine, 2-crew, 4-medic, 32-troop, 24-litter, 8000-pound load, 430-mph

NUMBER	NAME-VARIANTS	DESIGNER	SPAN	LENGTH	YEAR

C—CARGO TRANSPORT: 1962-today

| C-5 | Galaxy (XC-4) | Lockheed | 222'9" | 247'10" | '69 |

World's largest jet cargo transport—Operation "Reforger"
4-jet, 345-troop, 265,000-pound payload, 6500-mile range, 570-mph

| VC-6 | King Air (U-21) | Beechcraft | 45'11" | 35'6" | '65 |

VIP and SAM (Special Air Missions) light cargo transport
2-engine, low-wing, cabinplane, 8-seat, 1575-mile range, 280-mph

| C-7 | Caribou (CV-2) | DeHavilland | 95'8" | 72'7" | '67 |

Army CV-2 STOL tactical transport transferred to USAF
2-engine, high-wing, 32-troop, 22-litter, 6000-pounds cargo, 215-mph

| C-8 | Buffalo (CV-7) | Bell-DeHavilland | 96' | 77'4" | '67 |

Air Cushion Landing System (ACLS-STOL) experiment—Army CV-7
2-engine, high-wing, T-tail, 3-crew, 41-seat, 24-litter, 290-mph

| C-9 | Nightingale (DC-9) | McDonnell-Douglas | 93'5" | 119'4" | '68 |

Vietnam war airborne ambulance, hospitalship, cargo transport
2-jet, 3-crew, 45-patient, 40-litter, 5-medic, 570-mph

| KC-10 | Extender (DC-10) | McDonnell-Douglas | 155'4" | 181'7" | '80's |

Advanced Tanker Cargo Aircraft (ATCA) to replace KC-135
3-jet, 380-passenger, 170,000-pound load, 7000-mile range, 610-mph

| C 12 | Super King Air | Beechcraft | 54'6" | 43'10" | '75 |

Light executive cabinplane for Military Assistance Groups (MAG)
Twin-engine, low-wing, T-tail, 2-crew, 8-passenger, 330-mph

| YC-14 | AMST-STOL | Boeing | 129' | 131'8" | '76 |

Advanced Medium STOL Transport (AMST) experimental plane
Two-jet, 150-seat, 81,000-pounds cargo, 3190-mile range, 505-mph

| YC-15 | AMST-STOL | McDonnell-Douglas | 110'4" | 124'3" | '75 |

AMST (Advanced Medium STOL Transport) experimental prototype
Four-jet, 150-seat, 62,000-pound load, 2990-mile range, 500-mph

| AC-47 | Gunship-Gooney Bird (C-47) | Douglas | 95' | 64'5" | '65 |

Vietnam close support plane—code named "Shadow"
3-miniguns firing 6,000 rounds per minute, 54,000-rounds, 220-mph

NUMBER	NAME-VARIANTS	DESIGNER	SPAN	LENGTH	YEAR
AC-119	Gunship-Flying Boxcar (C-119)	Fairchild-Hiller	109'4"	85'	'68

Vietnam war close support aircraft—code named "Spooky"
Four 7.62mm miniguns, two 20mm cannon, two jet-boosters, 280-mph

AC-130	Gunship-Hercules (C-130)	Lockheed-LTV	132'7"	97'9"	'70

Heavily-armed Vietnam war close support attack plane—Infrared
Four 20mm multi-barrel cannon, four 7.62mm miniguns, 385-mph

E—ELECTRONICS: 1962—today

E-3	Sentry (EC-137D, 707)	Boeing	145'9"	152'11"	'75

Computerized Airborne Warning And Control System (AWACS)
17-crew, 30-foot rotodome, 300-mile radar, 7475-mile range, 725-mph

E-4	Doomsday (747)	Boeing	195'8"	231'4"	'74

Computerized Advanced Airborne Command Post (AACP)
4-jet, 164,000-pounds thrust, 3-deck, 6000-mile range, 630-mph

F—FIGHTER: 1948-1962

F-24	Dauntless (A-24)	Douglas	41'6"	33'	'48

A-24 attack plane carried over for fighter-bomber experiments
1200-hp engine, 2-man, 4-gun, 1200-pounds bombs, 255-mph

F-38	Lightning (P-38)	Lockheed	52'	37'10"	'48

Carried over for experimental purposes
Two 1425-hp engines, 5-gun, 10-rocket, 3200-pounds bombs, 420-mph

F-40	Warhawk (P-40)	Curtiss	37'4"	31'9"	'48

Carried over for Air National Guard
1200-hp engine, 6-gun, 1500-pounds bombs, 375-mph

F-47	Thunderbolt-Jug (P-47)	Republic	40'9"	36'2"	'48

Used by TAC, SAC, ADC and Air National Guard
2800-hp engine, 8-gun, 10-rocket, 2500-pounds bombs, 430-mph

F-51	Mustang (P-51)	North American	37'	32'3"	'48

Korean war and Vietnam war fighter-bomber close support
6-gun, rockets, missiles, napalm, 4500-pounds bombs, 440-mph

F-59	Airacomet (P-59, XF-83)	Bell	45'6"	38'2"	'48

America's first jet fighter-transition trainer-test bed
2-jet, mid-wing, tricycle gear, 3-gun, one cannon, 410-mph

NUMBER	NAME-VARIANTS	DESIGNER	SPAN	LENGTH	YEAR
RF-61	Black Widow (P-61, F-15)	Northrop	66'	48'11"	'48

Nightfighter and photo-recon plane for ADC
Two 2800-hp engines, 4-gun, 4-cannon, cameras, bomb racks, 430-mph

QF-63	Kingcobra (RP-63)	Bell	38'2"	32'8"	'48

Manned aerial target for gunnery training—Nicknamed "Pinball"
1325-hp engine, armor-plated, verification lights, 410-mph

F-80	Shooting Star (P-80, T-33)	Lockheed	38'11"	34'6"	'48

Korean war fighter—Scored first jet air victory in USAF history
One jet, low-wing, 6-gun, 10-missile, 4000-pounds bombs, 605-mph

F-82	Twin Mustang (P-82)	North American	51'3"	39'1"	'48

Korean war fighter—Scored first air victory in Korean war
Two 2200-hp engines, 14-gun, 25-rocket, 7200-pounds bombs, 475-mph

F-84	Thunderjet (P-84)	Republic	36'4"	38'5"	'48

First tactical fighter-bomber to carry a nuclear weapon
One jet, low-wing, 6-gun, 32-rocket, 4000-pounds bombs, 620-mph

F-84F	Thunderstreak (XF-96)	Republic	33'6"	43'4"	'52

First swept-wing tactical fighter-bomber for the Air Force
Single jet, 6-gun, 6000-pounds bombs, 1000-mile range, 695-mph

RF-84F	Thunderflash	Republic	33'7"	47'8"	'54

Satellite photo-recon plane carried by B-36 mothership
Single jet, 4-gun, 6-camera, 2200-mile range, 680-mph

F-86	Sabre Jet (P-86, YF-93, YF-95)	North American	37'1"	37'6"	'48

Korean war fighter—First to break sound barrier in a dive
One jet, low-wing, 6-gun, 16-rocket, 2000-pounds bombs, 700-mph

F-89	Scorpion (P-89)	Northrop	59'8"	53'10"	'50

Fighter-interceptor—Last with span greater than length
Two jets, 2-crew, 6-cannon, 52-rocket, 3200-pounds bombs, 635-mph

F-94	Starfire (TF-80C, XF-97)	Lockheed	42'5"	44'6"	'49

Nightfighter-interceptor—First with afterburner
Single jet, low-wing, 2-crew, 4-gun, 48-rocket, 585-mph

F-100	Super Sabre (YF-107)	North American	36'7"	45'3"	'53

Vietnam war fighter-bomber—First to cruise at supersonic speed
4-cannon, 7500-pounds weapons, rockets, missiles, napalm, 860-mph

NUMBER	NAME-VARIANTS	DESIGNER	SPAN	LENGTH	YEAR

F-101   Voodoo (XF-88)   McDonnell   39′8″   69′3″   '54
Long-range interceptor-fighter-bomber-reconnaissance plane
Two jets, low-wing, T-tail, 2-man, 4-cannon, 15-missile, 1050-mph

F-102   Delta Dagger   Convair   38′2″   68′5″   '55
First interceptor with complete manned weapons system
One jet, delta-wing, 6-missile, 24-rocket, 1450-mile range, 825-mph

F-104   Starfighter   Lockheed   21′1″   54′9″   '56
"Missile with a man in it"—First Mach-2 fighter-interceptor
One cannon, 4-missile, 38-rocket, 4000-pounds weapons, 1450-mph

F-105   Thunderchief   Republic   35′   64′   '57
Vietnam war fighter-bomber—First Mach-2 fighter-bomber
14,000-pounds ordnance, 4-missile, one 6-barrel cannon, 1400-mph

F-106   Delta Dart   Convair   38′1″   70′7″   '59
Mach-2 interceptor—Semi-Automatic Ground Environment (SAGE)
4-missile, 2-rocket, one Gatling gun, 2400-mile range, 1530-mph

F-111   Tactical Fighter (FB-111) General   63′   72′2″   '65
                                                          Dynamics
Vietnam war fighter-bomber—First variable-geometry wing
2-jet, 2-man, 1-cannon, 2-missile, 27,000-pounds weapons, 1665-mph

F—FIGHTER: 1962-today

F-4   Phantom (F-110)   McDonnell   38′4″   58′3″   '63
Vietnam war fighter—First all-supersonic air victory in history
2-jet, 2-man, 11,000-pounds bombs, 1-cannon, 8-missile, 1585-mph

F-5   Freedom Fighter-   Northrop   25′3″   47′2″   '63
              Tiger II (T-38)
Free-world fighter-interceptor—Military Assistance Program (MAP)
Two jets, 1-2 man crew, 2-cannon, 6200-pounds weapons, 1000-mph

YF-12   Blackbird (SR-71)   Lockheed   55′8″   107′4″   '64
World's fastest fighter—First triple-sonic fighter-interceptor
2-jet, 18 nuclear-tipped missiles, 100,000-foot ceiling, 2100-mph

F-15   Eagle   McDonnell-   42′10″   63′10″   '74
                               Douglas
Air superiority fighter and attack aircraft—HUD system
2-jet, rotary cannon, 8-missile, 16,000-pounds weapons, 920-mph

NUMBER	NAME-VARIANTS	DESIGNER	SPAN	LENGTH	YEAR
F-16	Fighting Falcon	General Dynamics	32'10"	46'6"	'76

Air combat dogfighter with Pave Penny Laser pod—HUD system
One jet, 15,200-pounds weapons, one cannon, 6-missile, 1500-mph

SUPPLEMENT TO FIGHTERS
EXPERIMENTAL JET FIGHTERS THAT DID NOT REACH PRODUCTION

NUMBER	DESIGNER	
XP-81	Convair	Twin-jet pursuit of 1945
XP-83	Bell	Advanced version of P-59 in 1945
XF-85	McDonnell	Parasite fighter for B-36 in August 1948
XP-87	Curtiss	Four-jet fighter of March 1948
XF-88	McDonnell	Twin-jet penetration fighter of 1948
XF-90	Lockheed	Twin-jet penetration fighter of 1949
XF-91	Republic	Interceptor of 1949
XF-92	Consolidated-Vultee	Delta-wing interceptor of 1949
YF-93	North American	Original designation of F-86C
YF-95	North American	Original designation of F-86D
XF-96	Republic	Original designation of F-84F
XF-97	Lockheed	Original designation of F-94C
XF-98	Hughes	Pilotless GAR Falcon air-to-air missile
XF-99	Boeing	Pilotless IM-99 BOMARC guided missile
XF-103	Republic	Delta-wing interceptor—Cancelled
YF-107	North American	Variant of F-100B
XF-108	North American	Delta-wing Mach-3 fighter—Cancelled
XF-109	Bell (VSTOL)	Vertical-Short Takeoff and Landing
XF-110	McDonnell	Original designation of F-4C Phantom

CG—CARGO GLIDER: 1941-1948

CG-4	(CG-15, G-4)	Waco	83'8"	48'4"	'42

Most-produced WWII powerless assault glider
High-wing, 15-troop, 3800-pound load, 120-mph max towing speed

CG-13	(G-13)	Waco	85'7"	54'3"	'43

America's largest WWII assault cargo glider
High-wing monoplane, 30-troop, 10,200-pound payload, 205-mph

CG-15	Hadrian (CG-4A, G-15, PG-3, G-3)	Waco	62'2"	48'4"	'45

Improved CG-4 with decreased span and increased payload
High-wing monoplane, 16-troop, 4300-pound payload, 180-mph

NUMBER	NAME-VARIANTS	DESIGNER	SPAN	LENGTH	YEAR
XCG-20	(XG-20, C-123)	Chase	110'	75'9"	'47

Postwar powered assault cargo transport—Redesignated C-123
2-engine, 2-crew, 67-troop, 16,000-pound payload, 230-mph

MK-I	Horsa	Airspeed	88'	67'	'42

WWII British invasion glider acquired on Reverse Lend-Lease
High-wing monoplane, 28-troop, 7120-pound payload, 150-mph

TG—TRAINER-GLIDER: 1941-1948

TG-1	Cinema	Frankfort	46'3"	23'2"	'41

First Air Corps monoplane Trainer-Glider
High-wing sailplane, 2 seats, 2-way radio, landing skids, 80-mph

TG-2		Schweizer	49'	25'	'41

Limited production sailplane adapted as glider
Two tandem seats, dual controls, landing skids

TG-3		Schweizer	54'	27'7"	'42

Civilian sailplane adapted as glider for USAAF
Two tandem seats, dual controls, landing skids, 100-mph

TG-4	Yankee Doodle	Laister-Kauffman	50'	22'	'42

Civilian mid-wing glider produced for USAAF
Two tandem seats, dual controls, 2-way radio, 2 wheels, tailskid

TG-5	Defender (L-3, TG-33)	Aeronca	35'	23'10"	'42

Unpowered L-3 Liaison plane with long nose and new landing gear
High-wing, 3 wheels, 3 seats with full controls, 130-mph

TG-6	Grasshopper (L-2)	Taylorcraft	35'5"	25'	'42

Unpowered L-2 Liaison plane with new nose, tail and landing gear
High-wing, 3 tandem seats, 3 wheels, nose skid, 2-way radio, 115-mph

TG-8	Cub (L-4)	Piper	35'3"	23'2"	'42

Unpowered L-4 Liaison plane—New nose, tailfin and landing gear
High-wing, 3 seats with full controls, 3 wheels, 2-way radio, 120-mph

TG-32	(LNE-1)	Pratt	40'	26'	'43

Unpowered Navy LNE-1 glider trainer transferred to USAAF
Two side-by-side seats, dual controls, 3 wheels

XTG-33	Defender (TG-3, L-3)	Aeronca	35'	23'6"	'44

Highest number assigned in the "TG" category
Experimental with pilot lying on his belly

NUMBER	NAME-VARIANTS	DESIGNER	SPAN	LENGTH	YEAR

HELICOPTERS

G—GYROPLANE: 1935-1941

| YG-1 | (XR-2) | Kellett | 40' | 28' | '35 |

First Army helicopter—Only "G" designated gyroplane
Single 225-hp engine, two tandem seats, no tail rotor, 125-mph

R—ROTATING WING: 1941-1948

| XR-2 | (YG-1, XR-3, YO-60) | Kellett | 40' | 28' | '41 |

Constant speed rotor—Direct conversion from YG-1C Gyroplane
225-hp engine, tractor-prop and overhead rotor, 2-seat, 125-mph

| R-4 | (H-4) | Vought-Sikorsky | 38' | 48'2" | '42 |

First Sikorsky helicopter—WWII coastal patrol chopper
180-hp engine, 2-crew, 130-mile range, 75-mph

| R-5 | (H-5) | Vought-Sikorsky | 48' | 57'1" | '44 |

First helicopter for Air Rescue Service (ARS) in WWII
450-hp engine, 2-crew, 2-litter, 360-mile range, 105-mph

| R-6 | (H-6) | Sikorsky | 38' | 38'3" | '44 |

Improved R-4 for WWII overwater surveillance
225-hp engine, 2-crew, 150-mile range, 95-mph

| YR-12 | (H-12) | Bell | 47'6" | 41'7" | '47 |

First Bell helicopter—Limited production for trials
600-hp engine, 5-seat, 6800-pound gross weight, 100-mph

| YR-13 | Sioux (H-13) | Bell | 35'1" | 27'4" | '47 |

Limited production helicopter for evaluation
175-hp engine, 3-seat, 220-mile range, 95-mph

H—HELICOPTER: 1948-1962

| H-4 | (R-4) | Vought-Sikorsky | 38' | 48'2" | '48 |

Direct redesignation from the R-4 Rotating Wing
200-hp engine, 2-crew, 170-mile range, 90-mph

| H-5 | (R-5) | Vought-Sikorsky | 48' | 57'1" | '48 |

Korean conflict utility and rescue helicopter
450-hp engine, pontoons, rescue hoist, 2-crew, 2-litter, 105-mph

NUMBER	NAME-VARIANTS	DESIGNER	SPAN	LENGTH	YEAR

H-6   (R-6)   Sikorsky   38'   38'3"   '48
Direct redesignation from the R-6 Rotating Wing helicopter
240-hp engine, 2-crew, 160-mile range, 100-mph

YH-12   (YR-12)   Bell   47'6"   41'7"   '48
Limited production helicopter for service test programs
600-hp engine, 2-crew, 8-passenger, 100-mph

H-13   Sioux (YR-13, UH-13)   Bell   35'1"   27'4"   '48
Korean war utility and rescue helicopter
200-hp engine, 3-crew, 2-litter, 240-mile range, 100-mph

H-19   Chickasaw   Sikorsky   53'   42'2"   '49
               (SH-19, HH-19)
Korean war rescue and utility helicopter
700-hp engine, 2-crew, 8-troop, 6-litter, 105-mph

H-21   Workhorse (HH-21)   Piasecki   44'6"   86'4"   '52
First double-tandem rotor helicopter for USAF
1250-hp engine, 2-crew, 12-litter, 14-troop, 125-mph

H-43   Huskie (HH-43)   Kaman   51'6"   25'   '58
First crash rescue and airborne firefighting helicopter
600-hp engine, 2-crew, 4-fireman, 4-litter, 115-mph

H—HELICOPTER: 1962-today

HH-1   Iroquois (UH-1F, TH-1F) Bell   48'   57'1"   '64
Vietnam war combat and rescue helicopter
2-crew, 14-troop, 6-litter, 4-gun, 3900-pound payload, 125-mph

HH-3   Jolly Green Giant   Sikorsky   62'   57'3"   '63
Vietnam war rescue and mercyship helicopter
1500-hp, 30-troop, 15-litter, 5000-pound payload, 165-mph

UH-13   Sioux (H-13)   Bell   35'1"   27'4"   '62
Vietnam war utility and rescue helicopter
250-hp engine, 3-crew, 2-litter, 250-mile range, 105-mph

HH-19   Chickasaw (H-19)   Sikorsky   53'   42'2"   '62
Vietnam war rescue and utility helicopter
800-hp engine, 2-crew, 8-troop, 8-litter, 115-mph

HH-21   Workhorse (H-21)   Piasecki-Vertol   44'6"   86'4"   '62
Vietnam war tandem rotor combat and rescue helicopter
1425-hp engine, 2-crew, 12-litter, 14-troop, 135-mph

NUMBER	NAME-VARIANTS	DESIGNER	SPAN	LENGTH	YEAR
HH-43	Huskie (H-43)	Kaman	51'6"	25'	'62

Crash rescue and airborne firefighting helicopter
860-hp engine, 2-crew, 4-firemen, 4-litter, 125-mph

HH-53	Sea Stallion	Sikorsky	72'3"	67'2"	'67

Heavy-lift rescue and utility helicopter for ARRS—Twin engine
3080-hp, 2-gun, 38-troop, 24-litter, 4-medic, 8000-pound load, 185-mph

OBSERVATION AND LIAISON AIRCRAFT
CO—CORPS OBSERVATION: 1919-1924

CO-4	(AO-1)	Fokker	41'10"	29'8"	'22

First post-WWI military plane built by Fokker in Holland
180-hp engine, biplane, two tandem seats, tailskid, 130-mph

AO—ARTILLERY OBSERVATION: 1919-1924

AO-1	(CO-4)	Fokker-Atlantic	41'10"	29'8"	'23

Only aircraft to carry the "AO" designation
180-hp engine, biplane, two tandem seats, tailskid, 130-mph

O—OBSERVATION: 1924-1942

O-1	Falcon (O-11, XO-12, O-13, XO-16, XO-18, YO-26, O-39, A-3, XA-4, XBT-4)	Curtiss	38'	28'4"	'25

First in a long line of Curtiss Falcon Observation biplanes
435-hp engine, 2-seat, 4-gun, 600-mile range, tailskid, 135-mph

O-2	(YO-6, O-7, O-8, O-9, XO-14, O-22, O-25, O-29, O-32, O-38, C-1, XA-2, BT-1, BT-2)	Douglas	39'8"	29'6"	'24

First in a large family of Douglas Observation biplanes
400-hp engine, 2-seat, 3-gun, 500-mile range, tail wheel, 130-mph

O-5		Douglas	50'	35'2"	'24

Douglas World Cruiser (DWC)—Around-the-world biplane
420-hp engine, float-fitted, 2-seat, 3-gun, tailskid, 100-mph

YO-6	(O-2)	Thomas-Morse	40'	30'	'25

First experiment with an all-metal Observation biplane
400-hp engine, 2-seat, 3-gun, 500-mile range, tail wheel, 130-mph

NUMBER	NAME-VARIANTS	DESIGNER	SPAN	LENGTH	YEAR

O-17 Courier (PT-3, XPT-8) Consolidated 34'6" 28'1" '28
Air Corps Observation biplane and combat trainer
220-hp engine, 2-seat, one gun, 4-hour endurance, tailskid, 100-mph

O-19 (YO-20, XO-21, YO-23) Thomas-Morse 39'9" 28'4" '28
First successful all-metal biplane—Open cockpits
450-hp engine, 2-seat, 2-gun, 950-rounds ammo, tail wheel, 135-mph

YO-27 (XB-8) Fokker 64' 47' '31
First Monoplane Observation aircraft—Enclosed cockpits
2-engine, 3-seat, high-wing, tail wheel, 160-mph

O-31 (O-43, O-46) Douglas 45'8" 33'10" '31
All-metal butterfly-wing monoplane—Enclosed cockpits
675-hp engine, 2-seat, 2-gun, tail wheel, sliding canopy, 190-mph

O-32 (O-2, BT-2) Douglas 40' 31'2" '31
Biplane observer-trainer with open cockpits
450-hp engine, 2-seat, tail wheel, no armament, 135-mph

YO-35 (XO-36, YB-7) Douglas 65'3" 46'7" '32
Twin engine high-wing monoplane—Retractable undercarriage
Part-metal, 600-hp engines, 2-gun, 3-wheel, 180-mph

O-47 North American 46'4" 33'7" '37
Revolutionary advanced single-engine mid-wing monoplane
975-hp, 3-crew, long canopy, 2-gun, retractable gear, 220-mph

O-49 Vigilant (L-1) Stinson 50'11" 34'3" '40
Reversal of philosophy to light low-powered unarmed aircraft
Single 295-hp engine, 2-seat, high-wing cabinplane, no arms, 120-mph

O-52 Owl Curtiss 43'10" 25'5" '40
All-metal single-engine high-wing monoplane—WWII trainer
600-hp, 3-crew, long canopy, one gun, retractable gear, 210-mph

O-57 Grasshopper (L-2) Taylorcraft 35'5" 22'9" '41
High-wing light unarmed cabinplane for liaison duty
Single 65-hp engine, 2-seat, 1300-pounds gross weight, 90-mph

O-58 Defender (L-3) Aeronca 35' 21' '41
High-wing light unarmed cabinplane for liaison duty
Single 65-hp engine, 2-seat, 190-mile range, 85-mph

NUMBER	NAME-VARIANTS	DESIGNER	SPAN	LENGTH	YEAR

O-59   Cub (L-4)   Piper   35'3"   22"   '41
High-wing light unarmed cabinplane for liaison duty
Single 65-hp engine, 2-seat, 1220-pounds gross weight, 85-mph

YO-60   (XR-2)   Kellett   40'   28'   '42
First Air Corps observation helicopter—Two tandem seats
225-hp engine, tractor-prop, overhead rotor, 125-mph

O-62   Sentinel (L-5)   Stinson-Vultee   34'   24'1"   '42
High-wing cabinplane for ambulance—Photo-recon—Trainer
Single 185-hp engine, 2-seat, 420-mile range, 130-mph

XO-63   Cadet (L-6)   Interstate   35'6"   23'6"   '42
Highest numbered observation plane—Redesignated as XL-6
100-hp engine, 2-seat, cabinplane, 1650-pound gross, 105-mph

L—LIAISON: 1942-1962

L-1   Vigilant (O-49)   Stinson   50'11"   34'3"   '42
WWII ambulance—Glider pickup trainer—Amphibian rescue
295-hp engine, high-wing cabinplane, two tandem seats, 120-mph

L-2   Grasshopper (O-57, Taylorcraft   35'5"   22"9"   '42
              UC-95, TG-6)
WWII artillery spotter—Glider pilot trainer—Cargo Transport
65-hp engine, high-wing cabinplane, two tandem seats, 90-mph

L-3   Defender (O-58, TG-5) Aeronca   35'   21'   '42
WWII light communications Liaison aircraft—Dual controls
65-hp engine, high-wing cabinplane, two tandem seats, 85-mph

L-4   Cub (O-59, UC-83,   Piper   35'3"   22'   '42
              TG-8)
WWII most-produced Liaison spotter-trainer-transport plane
65-hp engine, high-wing cabinplane, two tandem seats, 85-mph

L-5   Sentinel (O-62, L-9) Stinston-Vultee 34'   24'1"   '42
WWII airborne ambulance—Photo-recon plane—Trainer
185-hp engine, high-wing cabinplane, two tandem seats, 105-mph

L-6   Cadet (O-63)   Interstate   35'6"   23'6"   '43
WWI light utility-communications-trainer aircraft
115-hp engine, high-wing cabinplane, two tandem seats, 105-mph

NUMBER	NAME-VARIANTS	DESIGNER	SPAN	LENGTH	YEAR
L-9	Voyager (YO-54, L-5, AT-19)	Stinson-Vultee	34′	23′	'42

Commercial light transport commandeered for WWII
90-hp engine, high-wing cabinplane, three seats, 100-mph

L-13		Consolidated-Vultee	40′6″	31′9″	'47

General utility triphibian cabinplane with high folding-wing
245-hp engine, 6-seat, camera, skis, floats, wheels, 115-mph

L-20	Beaver (U-6)	DeHavilland	48′	30′4″	'52

Korean triphibian known affectionately as the ''General's Jeep''
450-hp engine, 2-crew, 6-troop, 4-litter, 1000-pound payload, 160-mph

L-21	Super Cub (U-7)	Piper	35′3″	22′6″	'51

Korean war light Liaison aircraft for Army—Later became USAF
Single engine, high-wing, 2-seat, 750-mile range, 125-mph

L-26	Commander (U-4)	Aero Design	49′6″	35′11″	'56

Utility-staff transport-Presidential plane-communications
Two 350-hp engines, high-wing, 2-crew, 9-passenger, 250-mph

L-27	Administrator (U-3)	Cessna	36′11″	29′3″	'57

Light administrative-communications-cargo transport aircraft
Two 260-hp engines, tricycle gear, 6-seat, 600-pound payload, 240-mph

L-28	Super Courier (U-10)	Helio-GAC	39′	30′9″	'58

General utility light transport—Last of the ''L'' category
295-hp engine, 2-crew, 5-troop, 1000-pound cargo load, 160-mph

O—OBSERVATION: 1962-today

O-1	Bird Dog (L-19)	Cessna	36′	25′10″	'63

Vietnam war multi-purpose combat aircraft
213-hp engine, 2-crew, 2-gun, 2-bomb, 4-rocket, 115-mph

O-2	Skymaster	Cessna	38′2″	29′9″	'67

Vietnam war psychological warfare plane—COIN operations
Two 210-hp push-pull engines, 7.62mm gun, rockets, 205-mph

OV-10	Bronco	Rockwell	40′	41′7″	'68

Most versatile light combat aircraft of the Vietnam war
750-hp engines, 8-seat, 4-gun, 2-missile, 500-pounds bombs, 280-mph

NUMBER	NAME-VARIANTS	DESIGNER	SPAN	LENGTH	YEAR

OBSERVATION AMPHIBIANS
COA—CORPS OBSERVATION AMPHIBIAN: 1919-1925

COA-1	(OA-1)	Loening	45'	34'7"	'23

First Army amphibian—Only aircraft in the ''COA'' category
400-hp engine, biplane, 2-crew, 3-gun, 3½-hour endurance, 120-mph

OA—OBSERVATION AMPHIBIAN: 1925-1948

OA-1	(COA-1, OA-2)	Loening	45'	34'7"	'25

Direct redesignation from COA-1—Tandem open seats
400-hp engine, biplane, 2-crew, 3-gun, 3½-hour endurance, 120-mph

OA-2	(OA-1)	Loening	45'	34'10"	'29

Like the OA-1 with engine change and a wing gun
480-hp engine, biplane, 2-crew, 3-gun, 585-mile range, 125-mph

OA-3	Dolphin (C-21)	Douglas	60'	43'10"	'33

First monoplane amphibian—Enclosed cockpit—Wooden wings
Two 300-hp engines, 7-seat, no armament, 550-mile range, 140-mph

OA-4	Dolphin (C-26, XOA-7)	Douglas	60'	43'10"	'34

Monoplane amphibian—Experimental tricycle landing gear
Two 350-hp engines, 8-seat, 550-mile range, 150-mph

YOA-8	(XOA-11)	Sikorsky	86'	51'2"	'37

WWII observation amphibian—Cargo transport—Overwater rescue
Two 750-hp engines, 11-seat, 775-mile range, 185-mph

OA-9	Goose (OA-13, A-9)	Grumman	49'	38'4"	'38

WWII observation amphibian—Cargo transport—Overwater rescue
Two 450-hp engines, 6-seat, 700-mile range, 195-mph

OA-10	Catalina (PBY-5, A-10)	Consolidated-Vultee	104'	63'10"	'42

WWII search and rescue—Amphibian patrol bomber
6-crew, 5-gun, 4000-pounds bombs, 2550-mile range, 190-mph

OA-13	Goose (OA-9)	Grumman	49'	38'4"	'42

WWII observation amphibian—Cargo transport—Overwater rescue
Two 450-hp engines, 6-seat, 700-mile range, 195-mph

OA-14	Widgeon	Grumman	40'	31'1"	'42

WWII utility observation amphibian—Overwater rescue
Two 200-hp engines, 5-seat, 650-mile range, 150-mph

NUMBER	NAME-VARIANTS	DESIGNER	SPAN	LENGTH	YEAR

A—AMPHIBIAN: 1948-1962

A-9   Goose (OA-9)   Grumman   49'   38'4"   '48
Redesignated OA-9—Amphibian overseas transport—Overwater rescue
Two 450-hp engines, 6-seat, 700-mile range, 195-mph

A-10   Catalina (PBY-5,   Consolidated-   104'   63'10"   '48
OA-10)   Vultee
Redesignated OA-10—Search and rescue amphibian
Two 1200-hp engines, 6-crew, 2550-mile range, 190-mph

SA-16   Albatross (HU-16)   Grumman   80'   60'8"   '49
Korean war rescue triphibian—Endurance time 23 hours
7-crew, 15-survivor, 12-litter, 2700-mile range, 265-mph

PHOTOGRAPHY AIRCRAFT OF WORLD WAR II
F—PHOTOGRAPHY: 1930-1948

F-1   (C-8, C-96)   Fairchild   50'   33'   '30
First officially designated photography plane—Folding wing
410-hp engine, high-wing monoplane, 3-crew, 140-mph

F-2   Expediter (C-45)   Beechcraft   47'8"   34'3"   '41
WWII utility cargo transport specialized for photo-recon duty
2 multi-lens mapping cameras, trimetrogon system, fuselage ports

F-3   Havoc (A-20)   Douglas   61'4"   47'7"   '40
America's first combat photo-recon attack aircraft
T-3A bomb-bay camera, rear fuselage cameras, 5-gun armament

F-4   Lightning (P-38E)   Lockheed-Vega   52'   37'10"   '42
WWII first high-speed photo-recon surveillance pursuit
Four K-17 cameras, drift sight, auto-pilot, 395-mph

F-5   Lightning (P-38G)   Lockheed-Vega   52'   37'10"   '42
WWII most-used photo-recon pursuit—1,254 aircraft
Varying camera arrangements, inter-cooler, 2-seat experiment

F-6   Mustang (P-51, FP-51,   North American   37'   32'3"   '42
RF-51)
WWII armed photo-recon pursuit—Became RF-51 for Korea
Two K-24 cameras, one K-17 and one K-22 camera, 4-cannon

F-7   Liberator (B-24)   Consolidated   110'   67'2"   '42
Long-range photo-recon bomber—Extra tanks and special cabin
Eleven cabin cameras, 3 nose cameras, 6 bomb-bay cameras

NUMBER	NAME-VARIANTS	DESIGNER	SPAN	LENGTH	YEAR
F-8	Mosquito (PR-XVI)	DeHavilland	54'2"	40'10"	'43

Canadian-built English photo-recon fighter-bomber for WWII
Two 1300-hp engines, high-wing, 2-crew, 2700-mile range, 365-mph

F-9	Flying Fortress (B-17, FB-17, RB-17)	Boeing	103'9"	73'2"	'42

WWII long-range photo-recon bomber
Trimetrogon nose cameras, bomb-bay and rear fuselage cameras

F-10	Mitchell (B-25)	North American	67'7"	52'11"	'41

Disarmed WWII photo-recon medium bomber with extra tanks
Trimetrogon camera system in fuselage and a chin fairing

F-13	Superfortress (B-29, FB-29, RB-29)	Boeing	141'3"	99'	'44

WWII long-range photo-recon bomber with extra bomb-bay tanks
Variations of K-18 or K-22 cameras in nose, fuselage and tail

F-14	Shooting Star (P-80, FP-80, RF-80)	Lockheed	38'11"	34'6"	'44

America's first all-jet photo-recon pursuit
Cameras replaced guns in a lengthened nose

F-15	Reporter (P-61, RF-61)	Northrop	66'	48'11"	'46

Postwar photo-recon pursuit—Last "F" Photography aircraft
Cameras located in a special tandem-seat central nacelle

PURSUIT-FIGHTER: 1919-1925

Orenco-D		Curtiss	33'	21'6"	'19

First all-American designed and built dogfighter
300-hp engine, biplane, open cockpit, tailskid, 2-gun, 135-mph

MB-3		Morse-Boeing	26'	20'	'20

First Army pursuit-fighter produced in quantity
300-hp engine, biplane, 2.3-hour endurance, 2-gun, 140-mph

PN-1		Curtiss	30'10"	23'6"	'21

First Army Pursuit—Only "PN" for Pursuit-Night aircraft
Experimental standard open cockpit biplane with tailskid

PA-1		Loening	28'	19'9"	'22

Only aircraft designated "PA" for Pursuit Air-cooled
Experimental standard open cockpit biplane with tailskid

NUMBER	NAME-VARIANTS	DESIGNER	SPAN	LENGTH	YEAR
PG-1		Aeromarine	40′	24′6″	'23

Only Army plane designated "PG" for Pursuit Ground-attack
Experimental standard biplane, 37-mm cannon, K-2 verify camera

PW-1	(VCP-2)	Engineering Division	32′	22′6″	'21

First "PW" for Pursuit Water-cooled aircraft
Standard biplane, 350-hp engine, 2-gun, 135-mph

PW-7	(Model D.XI)	Fokker	38′4″	23′11″	'23

German fighter purchased from Fokker factory in Holland
Standard biplane, 440-hp engine, tailskid, 2-gun, 150-mph

PW-8	(P-1)	Curtiss	32′	22′6″	'23

First coast-to-coast, dawn-to-dusk flight by Lt. Russ Maughan
Biplane, 440-hp engine, 2-gun, ski-fitted, 170-mph

PW-9	(XAT-3)	Boeing	32′1″	22′10″	'24

First Boeing pursuit—Last of the "PW" category
Biplane, 435-hp engine, 2-gun, open seat, tailskid, 155-mph

TP-1	(XCO-5)	Engineering Division	32′6″	23′6″	'24

Only Army aircraft designated "TP" for Two-seat Pursuit
Standard biplane, 400-hp engine, open seats, tailskid

P—PURSUIT: 1925-1948

P-1	Hawk (PW-8, AT-4, AT-5)	Curtiss	32′	22′6″	'25

First of the famous family of "Curtiss Hawks"
435-hp engine, open seat, spoked wheels, tailskid, 2-gun, 160-mph

P-2	Hawk	Curtiss	31′7″	22′10″	'26

First Hawk to experiment with turbo-supercharger
500-hp engine, 2-gun, 33,000-foot ceiling, 175-mph

P-3	Hawk	Curtiss	31′7″	22′5″	'27

First Hawk with a radial engine
410-hp uncowled engine, solid rim wheels, 2-gun, 155-mph

P-5	Hawk	Curtiss	31′6″	23′8″	'27

Turbo-supercharged Hawk—Poor low-level performance
435-hp engine, open seat, tailskid, 2-gun, 165-mph

NUMBER	NAME-VARIANTS	DESIGNER	SPAN	LENGTH	YEAR
P-6	Hawk	Curtiss	31'6"	23'7"	'28

Won 1st and 2nd place in 1927 National Air Races
600-hp engine, streamlined wheel spats, tail wheel, 2-gun, 198-mph

P-12	(F4B-1)	Boeing	30'	20'5"	'29

Army-Navy inter-service pursuit—Mail carrier
500-hp engine, tail wheel, 2-gun, 232-pounds bombs, 187-mph

YP-16	(PB-1)	Berliner-Joyce	34'	28'10"	'32

First two-seat biplane pursuit since the TP-1 of 1924
600-hp engine, tandem seats, 3-gun, 250-pounds bombs, 175-mph

PB-1	(YP-16)	Berliner-Joyce	34'	28'10"	'35

Only "PB" Pursuit Biplace biplane—Redesignation from P-16
600-hp engine, two open seats, dual controls, tailskid, 175-mph

P-26	Peashooter	Boeing	27'11"	23'7"	'32

First all-metal monoplane pursuit—Oldest pursuit in WWII
500-hp engine, open cockpit, 2-gun, 200-pounds bombs, 234-mph

P-30	(PB-2, A-11)	Consolidated	43'11"	29'4"	'34

All-metal low-wing monoplane—Retractable gear—Closed canopy
675-hp supercharged engine, tail wheels, 2-seat, 3-gun, 270-mph

PB-2	(P-30, A-11)	Consolidated	43'11"	30'	'35

Only "PB" Pursuit Biplace monoplane—Redesignation from P-30
700-hp supercharged engine, 2-seat, 3-gun, 275-mph

P-35	(XP-41, P-43)	Seversky	36'	25'2"	'37

First one-seat pursuit with enclosed cockpit—Retractable gear
1050-hp engine, low-wing, 2-gun, 210-pounds bombs, 290-mph

P-35A	Guardsman (AT-12)	Republic	36'	26'10"	'40

WWII pursuit caught on ground during raids in Philippines
1200-hp engine, 4-gun, 350-pounds bombs, 950-mile range, 320-mph

P-36	Hawk (YP-37, P-40)	Curtiss	37'4"	28'6"	'37

WWII pursuit—Scored first air victory at Pearl Harbor
1200-hp engine, 2 cowl guns, 2 wing guns, 300-mph

Hawk-75	Hawk-Mohawk	Curtiss	37'4"	28'7"	'37

Downed first German plane over France in WWII—By French pilot
875-hp engine, 6-gun, 300-pounds bombs, 180-mph

NUMBER	NAME-VARIANTS	DESIGNER	SPAN	LENGTH	YEAR
YP-37	Hawk (P-36)	Curtiss	37'4"	31'6"	'39

First Air Corps pursuit to exceed 300-mph—Lost out to P-40
1040-hp turbo-supercharged engine, 2-gun, 320-mph

| YFM-1 | Airacuda | Bell | 69'10" | 44'10" | '39 |

Only Air Corps pursuit designated ''FM'' for Fighter Multi-place
Two 1150-hp pusher engines, 5-crew, 2-gun, bomb racks, 270-mph

| P-38 | Lightning (F-4, F-5, F-38) | Lockheed-Vega | 52' | 37'10" | '41 |

Famous WWII Pursuit—Won fame in Pacific campaign
Two 1425-hp engine, 5-gun, 10-rocket, 3200-pounds bombs, 420-mph

| P-39 | Airacobra (P-63) | Bell | 34' | 29'3" | '41 |

WWII pursuit—Noted for ground attack missions
1200-hp engine, 6-gun, one cannon, one 500-pound bomb, 400-mph

| P-40 | Warhawk (P-36, F-40) | Curtiss | 37'4" | 31'9" | '41 |

WWII pursuit—Gained fame with ''Flying Tigers'' in China-Burma
1200-hp engine, 6-gun, 1500-pounds bombs, 375-mph

| P-43 | Lancer (P-35, XP-41) | Republic | 36' | 28'6" | '40 |

WWII fighter for China—Not flown in combat by Americans
1200-hp engine, low-wing, 4-gun, 120-pounds bombs, 360-mph

| MK-V | Spitfire | Supermarine | 36'10" | 29'11" | '40 |

WWII British fighter flown in combat by American Eagle Squadrons
1470-hp engine, 8-gun, 2-cannon, 470-mile range, 370-mph

| P-47 | Thunderbolt-Jug (F-47) | Republic | 40'9" | 36'2" | '42 |

Famous WWII fighter—Won fame over Europe
2800-hp engine, 8-gun, 10-rocket, 2500-pounds bombs, 430-mph

| P-51 | Mustang (F-6, A-36, F-51) | North American | 37' | 32'3" | '43 |

Famous WWII fighter—Won fame over Europe
1380-hp engine, 4-8 guns, 10-rocket, 2000-pounds bombs, 440-mph

| P-59 | Airacomet (F-59) | Bell | 45'6" | 38'2" | '44 |

America's first jet aircraft—Not used in combat
Two turbojets, 3 guns, one cannon, 1000-pounds bombs, 410-mph

NUMBER	NAME-VARIANTS	DESIGNER	SPAN	LENGTH	YEAR
P-61	Black Widow (F-15, RF-61)	Northrop	66'	48'11"	'43

WWII nightfighter—Struck last blow of the war
Two 2800-hp engines, 4-gun, 4-cannon, 6400-pounds bombs, 430-mph

P-63	Kingcobra (P-39, QF-63)	Bell	38'2"	32'8"	'43

Manned aerial target that could be shot at—Not used in combat
1325-hp engine, 4-gun, 6-rocket, 1500-pounds bombs, 410-mph

YP-64	(AT-6)	North American	37'3"	27'	'40

WWII export fighter—Not used in combat by U. S.
875-hp engine, 2-gun, 2-cannon, 550-pounds bombs, 270-mph

P-66	Vanguard	Vultee	36'	28'5"	'41

WWII fighter for China—Not flown in combat by Americans
1200-hp engine, 2 nose guns, 4 wing guns, 410-mph

P-70	Night Havoc (A-20)	Douglas	61'4"	47'7"	'42

WWII nightfighter—Conversion from A-20 Havoc Attack plane
Two 1600-hp engines, 6 nose guns or 4 belly cannon, 330-mph

P-80	Shooting Star (F-14, T-33, F-80)	Lockheed	38'11"	34'6"	'45

America's first operational combat-ready jet fighter
One jet, 6-gun, rockets, napalm, 4000-pounds bombs, 605-mph

P-82	Twin Mustang (F-82)	North American	51'3"	39'1"	'45

Ordered during WWII—Delivered too late for combat in WWII
Two 1600-hp engines, 6-gun, 4 drop tanks, 4000-pounds bombs, 460-mph

R—RACER: 1920-1925

R-1	(VCP-R)	Engineering Division	32'	22'7"	'21

As VCP-R won 1920 Pulitzer Race—Redesignated R-1 in 1921
Standard biplane, open seat, 660-hp 12-cylinder engine, 156-mph

R-2	(MB-6)	Thomas-Morse	25'	20'	'21

Modification of MB-6 Pursuit—Entered 1921 Pulitzer Race
Standard biplane, tailskid, 400-hp engine, 175-mph

R-3		Engineering Division-Sperry	29'3"	22'5"	'22

Designed specifically as a Racer—Won 1924 Pulitzer Trophy
Low-wing monoplane, 500-hp engine, tailskid, 217-mph

NUMBER	NAME-VARIANTS	DESIGNER	SPAN	LENGTH	YEAR
R-5	(MB-9)	Thomas-Morse	27'	20'	'22

Unsuccessful version of MB-9 Pursuit for 1922 Pulitzer Race
All-metal, parasol high-wing monoplane, 600-hp engine

R-6		Curtiss	19'10"	18'11"	'22

Won 1st and 2nd place in 1922 Pulitzer—206-mph and 189-mph
Standard streamlined souped-up biplane, 550-hp engine, 240-mph

R-8	(Navy R2C-1)	Curtiss	21'	19'	'23

Built as Navy R2C-1 in 1923—Purchased by Army for one dollar
Crashed in 1924—Last of the ''R'' for Racer planes

R3C-1	Land Racer	Curtiss	22'	19'8"	'25

Won 1925 Pulitzer Trophy with Lieutenant Cy Bettis at controls
Standard biplane, enlarged fin, 590-hp engine, 249-mph

R3C-2	Sea Racer	Curtiss	22'	20'	'25

Won 1925 Schneider Cup piloted by Lt. Jimmy Doolittle—233-mph
Only Army pontooned biplane Racer, 610-hp engine, 246-mph

SR—STRATEGIC RECONNAISSANCE: 1962-today

SR-71	Blackbird (YF-12)	Lockheed	55'8"	107'4"	'64

World's fastest unarmed strategic aircraft—Strategic Air Command
Two 34,000-pound thrust turbojets, 85,100-foot ceiling, 2190-mph

TR—TACTICAL RECONNAISSANCE: 1978-today

TR-1		Lockheed	103'	63'	'80's

Tactical Reconnaissance successor to U-2 surveillance spyship
Single turbojet, 72,000-foot ceiling, 3300-mile range, 430-mph

TRAINERS: 1919-1925

TA-3	Chummy (TA-5)	Dayton-Wright	31'	22'7"	'21

Trainer-Aircooled successor to WWI Curtiss JN-4D Jenny
Standard biplane, side-by-side seats, 180-hp engine, 105-mph

TW-3	Trusty (TA-5, PT-1)	Dayton-Wright-Consolidated	34'9"	26'9"	'21

Trainer-Watercooled variant of TA-5 with solid-rim wheels
Standard biplane, side-by-side seats, 180-hp engine, 105-mph

TW-5	(TA-6, AT-1, AT-2)	Huff-Daland	31'1"	24'8"	'23

Trainer-Watercooled derived from TA-6 Trainer—Became AT-1
Tandem seats, biplane, spoked wheels, 180-hp engine, 110-mph

NUMBER	NAME-VARIANTS	DESIGNER	SPAN	LENGTH	YEAR

PT—PRIMARY TRAINER: 1924-1948

PT-1	Trusty (TW-3, PT-3)	Consolidated	34'9"	28'	'24

First official Primary Trainer—Refined version of TW-3
Standard biplane, tandem seats, 180-hp engine, 105-mph

PT-3	Trusty (PT-1, XPT-2, O-17)	Consolidated	34'6"	28'	'28

Like the PT-1 and XPT-2 except for uncowled engine
Standard biplane, spoked wheels, 220-hp engine, 105-mph

YPT-6	Husky Junior	Fleet	33'	26'	'30

Offshoot from the Consolidated family of primary trainers
Standard biplane, tandem seats, spoked wheels, 100-hp engine

YPT-9	Cloudboy (XBT-3, XBT-5)	Stearman	32'	24'8"	'31

Primary trainers became Basic Trainers when engines had 300-hp
Standard biplane, tandem seats, tailskid, 165-hp engine, 125-mph

PT-11	(YPT-12, XBT-6, YBT-7)	Consolidated	31'7"	26'11"	'32

Upgraded to basic trainer when fitted with 300-hp engine
Standard biplane, tandem seats, 200-hp engine, 120-mph

PT-13	Kaydet (PT-17, PT-18, PT-27)	Stearman	32'2"	24'9"	'36

PT-13B was highest powered of all Primary Trainers with 280-hp
Standard biplane, open seats, 220-hp, 120-mph

YPT-14		Waco	30'	23'1"	'40

Only Waco Primary Trainer tested by Air Corps
Standard biplane, 220-hp engine, open seats, 3-wheel, 125-mph

YPT-15		St. Louis	33'10"	26'5"	'40

Only St. Louis Primary Trainer tested by Air Corps
Standard biplane, 225-hp engine, open seats, 3-wheel, 130-mph

YPT-16	(PT-20)	Ryan	30'1"	21'6"	'39

First monoplane primary trainer—First Ryan primary trainer
125-hp engine, low-wing, 3 wheels, open seats, fancy pants, 130-mph

PT-17	Kaydet (PT-13, PT-18, PT-27)	Stearman-Boeing	32'2"	25'	'40

Most popular WWII biplane primary trainer
Standard biplane, tandem seats, 220-hp engine, 120-mph

NUMBER	NAME-VARIANTS	DESIGNER	SPAN	LENGTH	YEAR
PT-18	Kaydet (PT-13, PT-17, PT-27)	Stearman-Boeing	32'2"	25'	'40

Same as PT-17 except for engine change
Standard biplane, tandem seats, 220-hp engine, 120-mph

PT-19	Cornell (PT-23, PT-26)	Fairchild	36'	27'8"	'41

First of the Fairchild family of monoplane primary trainers
175-hp engine, tandem open seats, low-wing, 3-wheel, 135-mph

PT-20	(PT-16)	Ryan	29'11"	21'5"	'40

Slightly modified version of PT-16 with larger cockpits
125-hp engine, low-wing monoplane, 3-wheel, 130-mph

PT-21	Recruit (PT-20, PT-22)	Ryan	30'1"	22'5"	'41

Like PT-20 with new nose, uncowled engine, no fancy pants
135-hp engine, low-wing monoplane, open seats, 135-mph

PT-22	Recruit (PT-21)	Ryan	30'1"	22'5"	'42

Most popular Ryan all-metal monoplane primary trainer
160-hp engine, low-wing, tandem open seats, 3-wheel, 140-mph

PT-23	Cornell (PT-19, PT-26)	Fairchild	36'	27'8"	'42

Like PT-19 with uncowled engine and hood for front seat
220-hp engine, low-wing monoplane, 3-wheel, 130-mph

YPT-25		Ryan	32'11"	24'3"	'42

Experiment in building trainers out of plastic bonded wood
185-hp engine, standard biplane, tandem open seats

PT-26	Cornell (PT-19, PT-23)	Fairchild	36'	27'8"	'42

First monoplane primary trainer with enclosed cockpit
200-hp engine, low-wing, 3-wheel, blind flying, 130-mph

PT-27	Kaydet (PT-13, PT-17, PT-18)	Boeing	32'2"	25'	'42

Lend-Lease trainer—Highest number assigned in the ''PT'' category
Standard biplane, tandem seats, 220-hp engine, 120-mph

BT—BASIC TRAINER: 1930-1948

BT-1	(O-2)	Douglas	39'8"	29'6"	'30

First Basic Trainer—Last to fly with WWI Liberty engine
Standard biplane, 450-hp engine, 3-gun, 3-wheel, 135-mph

NUMBER	NAME-VARIANTS	DESIGNER	SPAN	LENGTH	YEAR
BT-2	(O-32)	Douglas	40′	31′2″	'31

Direct conversion from O-32 observation plane
Standard biplane, 450-hp engine, tandem seats, 3-wheel, 135-mph

YBT-7	(YPT-12)	Consolidated	31′7″	26′11″	'32

Primary trainers became basic trainers when engines had 300-hp
Standard biplane, 300-hp engine, open seats, 3-wheel, 135-mph

BT-8	(P-35)	Seversky	36′	24′4″	'34

First monoplane basic trainer—Enclosed cockpit
450-hp engine, low-wing, 3-wheel, fancy pants, 175-mph

BT-9	(BT-14, BC-1)	North American	42′	27′7″	'36

First North American military trainer—First combat trainer
Low-wing, 400-hp engine, 2-gun, one camera, closed cockpit, 170-mph

BT-12	Sophomore	Fleetwings	40′	29′2″	'38

First stainless steel military plane—Low-wing monoplane
450-hp engine, tandem seats, enclosed canopy, 195-mph

BT-13	Valiant (BT-15)	Vultee	42′	28′10″	'39

Most popular WWII basic trainer—Low-wing monoplane
450-hp engine, enclosed cockpit, retractable gear, 185-mph

BT-14	Yale (BT-9, BC-1)	North American	41′	28′5″	'41

Improved BT-9 with metal fuselage, new wing and engine change
450-hp engine, low-wing monoplane, enclosed cockpit, 175-mph

BT-15	Valiant (BT-13)	Vultee	42′	28′10″	'41

Like BT-13 except for engine change
450-hp engine, tandem seats, 3-wheel landing gear, 185-mph

AT—ADVANCED TRAINER: 1925-1948

AT-1	(TW-5)	Huff-Daland	31′1″	24′8″	'25

First advanced trainer—Spoked-wheel biplane
180-hp engine, tandem seats, tailskid, wing skids, 110-mph

AT-4	Hawk (P-1)	Curtiss	31′7″	22′8″	'27

First production advanced trainer—Standard biplane
180-hp engine, one seat, open cockpit, tailskid, 2-gun, 160-mph

AT-5	Hawk (P-1)	Curtiss	31′6″	23′3″	'27

Like AT-4 with engine change
220-hp uncowled engine, 2-gun, one open seat, 160-mph

NUMBER	NAME-VARIANTS	DESIGNER	SPAN	LENGTH	YEAR
BC-1	(BT-9, BT-14, AT-6)	North American	43'	27'9"	'38

First monoplane advanced-basic combat trainer
600-hp engine, low-wing, closed canopy, retractable gear, 205-mph

| AT-6 | Texan (BC-1, AT-16) | North American | 42' | 29'6" | '40 |

Famous WWII advanced trainer—Conversion from BC-1
600-hp engine, low-wing, 3-gun, 400-pounds bombs, 210-mph

| AT-7 | Navigator (UC-45, AT-11) | Beechcraft | 47'8" | 34'3" | '40 |

First WWII twin-engine navigator trainer
450-hp engines, one astrodome, 3-station, twin-fin, 225-mph

| AT-8 | Bobcat (AT-17, UC-78) | Cessna | 41'11" | 32'9" | '41 |

WWII advanced twin-engine pilot transition trainer
295-hp engines, low-wing monoplane, 5-seat, 200-mph

| AT-9 | Jeep | Curtiss | 40'3" | 31'8" | '40 |

WWII twin-engine advanced pilot transition trainer
280-hp engines, low-wing monoplane, 4-seat, 200-mph

| AT-10 | Wichita | Beechcraft | 47' | 34' | '41 |

First all-wood low-wing monoplane trainer accepted by USAAF
Two 295-hp engines, 4-seat, transition trainer, 200-mph

| AT-11 | Kansan (UC-45, AT-7) | Beechcraft | 47'7" | 34'3" | '41 |

Bombardier-gunner trainer—Bomb-bay, nose cone, hand-held guns
Two 450-hp engines, 4-station, 2-gun, 10-bomb, twin-fin, 215-mph

| AT-12 | Guardsman (P-35A) | Republic | 41' | 27'8" | '41 |

Enlarged P-35A for advanced transition combat trainer
1050-hp engine, 2-seat, 5-gun, 850-pounds bombs, 285-mph

| AT-16 | Harvard (AT-6) | Noorduyn | 42' | 29'6" | '42 |

Canadian-built version of the AT-6 for Lend-Lease
One 600-hp engine, 2-seat, closed cockpit, no arms, 210-mph

| AT-17 | Bobcat (AT-8, UC-78) | Cessna | 41'11" | 32'9" | '42 |

Popular WWII advanced twin-engine pilot transition trainer
245-hp engines, low-wing monoplane, 5-seat, 195-mph

| AT-18 | Hudson (A-29, C-63) | Lockheed-Vega | 65'6" | 44'4" | '42 |

WWII airborne bombcrew trainer—Nose cone, dorsal turret
1200-hp engines, mid-wing monoplane, 5-gun, twin-fin, 250-mph

| AT-19 | Reliant (UC-81) | Stinson-Vultee | 41'9" | 29'6" | '43 |

WWII Lend-Lease advanced pilot transition trainer
Single 285-hp engine, high gull-wing, 5-seat, 150-mph

NUMBER	NAME-VARIANTS	DESIGNER	SPAN	LENGTH	YEAR

AT-20 Avro Anson Federal 56'6" 42'3" '43
Canadian-built advanced twin-engine pilot transition trainer
330-hp engines, low-wing cabinplane, 4-seat, 180-mph

AT-21 Gunner Fairchild 53' 37'8" '43
WWII advanced airborne gunner trainer—Twin-engine—Top turret
450-hp engines, mid-wing monoplane, 5-seat, three-gun, 210-mph

AT-22 Liberator (C-87, TB-24) Consolidated 110' 67'2" '43
WWII flight engineer trainer—Flying classroom
Four 1200-hp engines, 5-crew, 2-instructor, 11-student, 305-mph

AT-23 Marauder (TB-26) Martin 71' 58'3" '43
WWII transition trainer—Target tug
Two 2000-hp engines, 7-seat, no arms, 285-mph

AT-24 Mitchell (TB-25) North American 67'7" 52'11" '43
WWII transition and aircrew trainer
Two 1700-hp engines, 6-seat, twin-fin, no arms, 285-mph

T—TRAINER: 1948-today

T-6 Texan (AT-6) North American 42' 29'6" '48
Advanced single-engine trainer—Korean ''Mosquito'' spotter
600-hp engine, all-around vision, full instrumentation, 210-mph

T-7 Navigator (AT-7, UC-45) Beechcraft 47'8" 34'3" '48
Advanced twin-engine navigational trainer
450-hp engines, astrodome, 3-station, twin-fin, 225-mph

T-11 Kansan (AT-11, UC-45, TC-45) Beechcraft 47'7" 34'3" '48
Advanced twin-engine bombardier-gunner trainer
450-hp engines, 4-station, 2-gun, 10-bomb, twin-fin, 215-mph

T-19 Cornell (PT-19) Fairchild 36' 27'8" '48
Monoplane primary trainer
200-hp engine, low-wing, open seats, 140-mph

TB-25 Mitchell (AT-24) North American 67'7" 52'11" '48
Advanced twin-engine transition and aircrew trainer
1700-hp engines, 6-seat, high-wing, twin-fin, 270-mph

T-28 Trojan North American 40' 33' '50
First new USAF trainer—First trainer with tricycle gear
800-hp engine, low-wing monoplane, 2-seat, 2-gun, 290-mph

NUMBER	NAME-VARIANTS	DESIGNER	SPAN	LENGTH	YEAR
T-29	Flying Classroom (C-131)	Consolidated-Vultee	91′9″	74′8″	'49

Navigator-bombardier trainer
Two 2500-hp engines, low-wing, 4-astrodome, 14-student, 300-mph

T-33	Silver Star— T-Bird (TF-80C)	Lockheed	38′11″	37′9″	'49

First USAF all-jet "T" for Trainer aircraft
One turbojet, low-wing, 2-seat, tip tanks, side intakes, 600-mph

T-34	Mentor	Beechcraft	32′10″	25′10″	'53

First new post-WWII low-powered primary trainer
One 225-hp engine, low-wing, 2-gun, 2-bomb, 6-rocket, 190-mph

T-37	Tweety Bird (A-37)	Cessna	33′9″	29′3″	'55

First trainer with side-by-side seats since the TW-3 of 1924
2-turbojet, low-wing, 2-seat, 2-gun, 8-bomb, 4-rocket, 425-mph

T-38	Talon (F-5)	Northrop	25′3″	44′2″	'61

First supersonic "T" for Trainer aircraft
2-turbo jet, low-wing, 2-seat, 1200-mile range, 840-mph

T-39	Sabreliner	North American	44′5″	43′9″	'60

UTX-Utility Trainer eXperimental aircraft
Two fuselage jets, side-by-side controls, 6-10 passenger, 565-mph

T-41	Mescalero	Cessna	36′2″	26′6″	'64

Low-powered monoplane primary trainer with side-by-side seats
150-hp engine, high-wing cabinplane, 4-place, tricycle gear, 135-mph

T-43	(Boeing 737)	Boeing	93′	96′11″	'73

Navigator-bombardier-radar trainer—Flying classroom
Two underwing turbofans, 2-crew, 19-station, 580-mph

TC-45	Expediter (T-11)	Beechcraft	47′8″	34′3″	'48

Advanced twin-engine aircrew trainer carried over from WWII
450-hp engines, 3-station, twin-fin, 220-mph

TF-80C	Shooting Star (T-33)	Lockheed	38′11″	37′9″	'48

Stretched F-80 fighter for jet transition training
One turbojet, low-wing, 2-seat, tip tanks, side intakes, 600-mph

U—UTILITY: 1952-today

U-2	Gray Ghost-Shady Lady	Lockheed	80′	49′4″	'55

Espionage surveillance photo-recon designated U-2 to hide identity
One turbojet, mid-wing, one-seat, 10-hour endurance, 600-mph

NUMBER	NAME-VARIANTS	DESIGNER	SPAN	LENGTH	YEAR
U-3	Administrator- Blue Bird (L-27)	Cessna	36'11"	29'3"	'58

Light administrative liaison and cargo transport cabinplane
Two 260-hp engines, low-wing, 6-seat, 600-pound payload, 240-mph

U-4	Commander (L-26)	Aero Design	49'6"	36'	'60

Presidential VIP staff transport cabinplane
Twin 350-hp engines, high-wing, 2-crew, 5-passenger, 250-mph

U-5	Twin Courier	Helio-GAC	41'	32'	'64

Light utility cargo transport VSTOL cabinplane
Two 250-hp engines, 2-crew, 4-passenger, 185-mph

U-6	Beaver (L-20)	DeHavilland	48'	30'4"	'62

Vietnam war light utility cargo transport and mercy ship
450-hp engine, high-wing, 8-seat, 1000-pound payload, 160-mph

U-7	Super Cub (L-21)	Piper	35'4"	22'7"	'62

Vietnam war utility aircraft for Army and Air Force
Single engine, high-wing, 2-seat, 750-mile range, 125-mph

U-10	Super Courier (L-28)	Helio-GAC	39'	31'	'61

Vietnam war STOL general utility plane for COIN operations
295-hp engine, high-wing, 6-seat, 1000-pound payload, 200-mph

HU-16	Albatross (SA-16)	Grumman	96'8"	62'10"	'62

High-wing monoplane search and rescue amphibian
Two 1425-hp engines, 7-crew, 12-litter, 15-survivor, 265-mph

U-17	Skywagon	Cessna	35'10"	25'8"	'65

Light cargo transport cabinplane for export under MAP
300-hp engine, 2-crew, 4-passenger, 300-pound cargo pack, 180-mph

AU-23	Peacemaker	Fairchild-Hiller	49'8"	36'10"	'71

Utility attack combat aircraft—"Credible Chase" program
Single engine, 2 miniguns, one cannon, 2310-pounds bombs, 175-mph

AU-24	Stallion	Helio-GAC	41'	39'7"	'72

Utility attack combat aircraft—"Credible Chase" program
Single engine, 10 seats, one cannon, 2240-pounds bombs, 215-mph

XS—SUPERSONIC EXPERIMENT: 1945-1948

XS-1	Rocketship (X-1)	Bell	28'	31'	'46

Forefather of supersonic research—First to exceed speed of sound
Rocket engine, B-29 mothership, 70,120-foot altitude, 964-mph

NUMBER	NAME-VARIANTS	DESIGNER	SPAN	LENGTH	YEAR

X—SPECIAL RESEARCH: 1948-today

X-1	Rocketship (XS-1, X-1A) Bell		28'	31'	'48

Redesignation of XS-1 for USAF—X-1A was first to exceed Mach-2
Rocket engine, B-29 mothership, 98,400-foot altitude, 1612-mph

X-2	Transonic	Bell	27'6"	35'8"	'52

First to exceed Mach-3 speed—First with escape capsule for pilot
Rocket engine, B-50 mothership, 126,200-foot altitude, 2148-mph

X-3	Stiletto	Douglas	22'8"	66'9"	'52

Subsonic aircraft for testing aerodynamic stresses on airframe
Twin-jet monoplane, 1200 pounds of research equipment, 650-mph

X-4	Bantam	Northrop	25'	20'	'48

Low-swept-wing "flying wing" with tailfin but no tailplane
Twin turbojets, one-seat, elevons, subsonic speed

X-5	Variable-Sweep	Bell	32'11"	32'11"	'51

World's first aircraft with variable-sweet wing
Single turbojet engine, sixty-degree sweepback, supersonic speed

X-13	Vertijet	Ryan	21'	24'	'55

First manned aircraft designed to take off from a launching pad
Single turbojet, delta-wing, VTOL, pivoting pilot's seat, subsonic

X-14	Thrust-Diverter	Bell	34'	25'	'56

VTOL plane with thrust-diverters that deflect exhaust downward
Twin turbojets, mid-wing monoplane, open cockpit, 160-mph

X-15	Hypersonic (X-15A-2)	North American	22'	50'	'59

World's fastest and highest flying hypersonic research aircraft
Rocket engine, B-52 mothership, 354,200-foot altitude, 4534-mph

X-18	Tilt-Wing (XC-122)	Hiller	48'	63'	'59

VTOL convertiplane with tilting wings for flying straight up
Two piston engines with 6-bladed contraprops, one turbojet, 250-mph

X-19	Tilt-Prop	Curtiss-Wright	34'6"	44'5"	'63

High-speed VSTOL plane with tilting props in wingtip nacelles
2-engine, 6-seat, 1200-pound payload, equal-span high-wings, 460-mph

X-21	Laminar-Flow	Northrop	93'6"	75'	'63

Laminar-flow system compresses air to generate more thrust
Twin fuselage jets, two underwing compressors, 5-man crew, 530-mph

NUMBER	NAME-VARIANTS	DESIGNER	SPAN	LENGTH	YEAR
X-22	Tilt-Duct	Bell	39'2"	36'4"	'66

Four props in circular ducts exhaust downward for VSTOL
4-engine, 2-crew, fan-type housings, rear wing, no tailplane, 345-mph

X-24	Lifting-Body (X-24B)	Martin-Marietta	13'8"	24'6"	'67

Wingless orbiter for soft reentry and landing studies
Rocket engine, B-52 mothership, 90,000-foot altitude, 1188-mph

X-25	Gyro-Glider	Bensen	20'	11'4"	'78

One-seat or two-seat flying toy for experimental rescue work
Powerless, 2-bladed rotor, looks like a miniature helicopter

SPACE VEHICLES: 1974-today

NASA-905	SCA (747-123)	Boeing	195'8"	231'4"	'76

Mothership piggy-back carrier for OV-101 "Enterprise"
Four 46,950-pounds thrust turbojets, pylon mountings, triple-fin

OV-101	Enterprise	Rockwell International	78'1"	122'3"	'76

Manned Orbital Vehicle for the Space Shuttle test program
Unpowered, low-wing, coke-bottle shape, single-fin, no tailplane

OV-102	Columbia	Rockwell International	78'1"	122'3"	'79

Reusable manned Space Shuttle—Rocket powered (45 engines)
2 solid-fuel boosters, 3 liquid-fuel engines, 7-man crew, 17,600-mph

ATTACK PLANES OF WWI FOREIGN-BUILT

During World War I, Attack planes were designed as two-seated light bombers for use as strafers against German infantrymen entrenched behind barbed wire entanglements along the Western Front. Although not highly effective in that role and having very little to do with the ultimate outcome of the war, that concept of air-to-ground close support operations was to become a potent weapon in later wars.

BREGUET MODEL-14

The Breguet 14A-2 was a French Corps d'Armée observation plane supplied to the Air Service of the American Expeditionary Force in 1917 for use in the Attack role as a trench strafer. It was a standard two-seated biplane with negatively staggered wings, a Renault 300-hp engine and a top combat speed of 130 mph. Two forward-firing machine guns were mounted on the fuselage and a flexible gun was operated by the observer-gunner in the back seat.

The Breguet 14B-2 light bomber Attack plane had 8 racks underneath, each carrying a 25-pound bomb for a total bomb load of 200 pounds. Otherwise it was similar to the Model 14A-2 in appearance and equipment. The first American bombing mission of WWI was made on June 12, 1918, by crews of the 96th Bomb Squadron flying Breguet 14B-2 bombers out of Amanty, France, for a raid on the German-held city of Metz.

The Air Service used Breguet-14 Attack planes while awaiting arrival of America's DH-4 Liberty Plane Attack-Bomber. Approximately 500 of the Breguet-14's were procured by the AEF. After the war a few were shipped back to McCook Field, Dayton, Ohio, where they were fitted with Liberty-12 engines for trials.

SALMSON MODEL-2

The Salmson 2A-2 was another French observation-attack plane acquired by the AEF in 1917 for use as a trench strafer. It was a standard biplane with a Salmson 270-hp engine and a top speed of 120 mph. Armament consisted of a 30-caliber forward-firing Vickers gun on the cowling and two Lewis flex guns in the back seat for defense. A total of 705 Salmson 2A-2 Attack planes were supplied to the AEF, partly for training and partly for combat.

ATTACK PLANES—WW I TO WW II

DH-4 LIBERTY PLANE

The De Havilland DH-4 prototype made its first flight on October 29, 1917, and the production model DH-4 Liberty Plane began rolling out of the factory in February 1918. The DH-4 was not an all-American aircraft—it was a successful mating of the British De Havilland airframe with an American-built Liberty-12 engine. The Dayton-Wright, Fisher Body, and Standard Aircraft Companies geared up for production, and the first shipload of combat-ready DH-4's arrived in France on May 11, 1918.

The DH-4 was a single-engine two-seated biplane with standard tail section, two fixed wheels up front and a tailskid at the rear. Pilot and observer-gunner sat in open tandem cockpits. The 400-hp engine drove a four-bladed prop to produce a top speed of 125 mph. Armament consisted of two forward-firing cowl guns and two flexible guns operated by the gunner in the second seat. A bomb load of 320 pounds was carried on racks under the wings for a combat range of 270 miles.

On August 2, 1918, the DH-4 became the first and only American combat plane to fly over enemy territory during World War I. According to advanced design theories gained from practical experience in combat, the DH-4 was considered obsolete even before making its first combat appearance. Several accidents caused by spontaneous bursting into flames while airborne resulted in European criticism of America for sending her young men over there to fight in "Flaming Coffins." The engine was located between the pilot and gunner, causing the two cockpits to be so far apart that the crewmen had difficulty talking to each other during combat. However, subsequent safety statistics showed the DH-4 to be no more hazardous than some of the other two-seated light bombers used by the Allies. The general opinion was one of disappointment—the Allies had expected more than that to come out of America.

The DH-4B in October 1918 was an improved version with the engine moved forward and the pilot's seat moved backward, bringing the two seats within shouting distance of each other. A few more safety features were added, but the DH-4B was too late for the war.

A total of 4,846 Liberty Planes were built in America, of which 1,215 were sent to France as standard equipment in 6 Bomb Squadrons and 7 Observation Squadrons. The remaining 3,731 planes remained in America for distribution to a variety of destinations. During WWI American Bomb units flew 150 missions and dropped 275,000 pounds of ordnance onto their targets.

DeHavilland DH-4 Liberty Plane, the only American-built Attack plane to enter combat in WWI. Pictured over Oahu, Hawaii on October 4, 1926. (USAF Photo 38593)

At the end of World War I American officials declared the DH-4 and many Trainer aircraft used by the American Expeditionary Force in France to be obsolete, justifying the action on the ground that it would cost more than they were worth to ship them home. Those planes were heaped in a pile near Paris and burned, causing some dissenters to dub it the "Billion Dollar Bonfire" in ridicule of the Army's aircraft procurement policies. It is noteworthy that during WWI procurement of military airplanes was placed under a government bureau, in keeping with the philosophy that it was best to have civilians temporarily running the show during a national emergency because they didn't have a military career at stake.

In the post-WWI Army, the 3,631 remaining DH-4's formed the core of available light Bombardment and Observation planes, being variously modified during the early 1920's. On September 4, 1922, Lt. Jimmy Doolittle made the first coast-to-coast flight from Pablo Beach, Florida, to San Diego, California, in a DH-4B; he made one refueling stop at Kelly Field, San Antonio, Texas, on the way. On June 27, 1923, a DH-4B flown by Capt. Lowell Smith and Lt. John Richter hooked up with another DH-4 using a 50-foot garden hose in the first airborne refueling operation. On August 27-28, 1923, the receiver remained airborne for a record 37 hours and 15 minutes while making fifteen inflight refueling contacts.

When the Army Air Corps replaced the Army Air Service on July 2, 1926, with Major General Mason Patrick commanding, approximately 1,000 DH-4's were still in service. Public pressure demanded that DH-4's be retained because the only way to get rid of them was to build another "Bonfire." Phasing out of the Liberty Planes began in 1927 with arrival of the Curtiss A-3 Falcon Attack plane. The last of the DH-4's were retired in 1932 after 14 years of service.

DB-1, GA-1, GA-2 ATTACK PLANES

During the Roaring Twenties and Depression Thirties the Army operated on tight budgetary limitations for development of airplanes. Money was scarce, taxes were low, and the public was disinclined to spend money on military airplanes during peacetime. The better part of the public funds appropriated for the Army planes was spent on Pursuit aircraft. Development of Bombers received second priority, and the remaining money was allocated for Attack planes. The role and size of an Attack plane was somewhere between that of a Pursuit and a Bomber. The Attack plane was fast, but not quite agile enough to participate in dogfights. It carried an internal bomb load, but didn't have nearly the capacity of a heavy Bomber.

The Gallaudet DB-1 in 1921 was the only Army plane to be designated in the category of "DB" for Day Bombardment. Designed as an air-to-ground Attack plane for use in anti-trench warfare, the DB-1 had a single 700-hp engine. It was a low-wing Monoplane having standard single-finned tail assembly and conventional under-carriage with two fixed wheels up front and a landing skid at the rear. Pilot and observer-gunner sat in open tandem cockpits. Armament consisted of four 30-caliber machine guns, located one in the nose, one in the belly, and two in the rear seat. A bomb load of 600 pounds was carried internally in the bomb-bay. The DB-1 was a good idea, but the Monoplane design was a little ahead of its time—structural knowledge as regards Bombers was yet in its infancy. Tailskids were used on most airplanes in the days when they operated on dirt landing strips. Tailwheels began to dominate on the late Biplanes and Monoplanes of the Thirties as concrete runways began to appear. Only one DB-1 was delivered to the Army Air Service for tests.

The Boeing GA-1 in 1921 was the first Air Service Attack plane to carry the designation "GA" for Ground Attack. The GA-1 was developed by the Army Engineering Division, McCook Field, Dayton, Ohio, as the GAX Ground Attack eXperimental, and Boeing was dealt the production order under the system of spreading the contracts out equally among all the aircraft manufacturers, regardless of who the designer was. The GA-1 was a huge twin-engine box-like Triplane with oversized tail section, double tires up front and a tailskid at the rear. The 435-hp pusher-type engines were mounted in the middle wing to produce a top speed of 105 mph. Two downward-firing guns were mounted in the fuselage and one flexible gun was located at the front of each engine nacelle. A bomb load of 250 pounds was carried underneath. The crew of four consisted of pilot and observer-gunner in the fuselage and two wing gunners, one in each nest at the front of the engine nacelles. Ten GA-1's were delivered for service trials, but the Triplane concept proved unsatisfactory.

The Boeing GA-2 in 1922 was the last of the "GA" for Ground Attack airplanes. It was a standard Biplane with a single 700-hp engine. Armament consisted of a 37mm cannon, three 50-caliber flexible machine guns, and two Lewis guns mounted on the top wing for firing in an arc over the propeller. Armor plating of 1,600 pounds of quarter-inch metal had been planned, but that was too heavy and too expensive for operational purposes. Only two GA-2's were built for experimental work.

A-3 FALCON

The "A" for Attack planes was introduced into the Army Air Service scheme of aircraft designations in 1924 when plans were being formulated for the forthcoming Air Corps

Act. The Douglas XA-2 in 1926, from the Douglas O-2 Observation biplane, was the first experimental aircraft assigned in the "A" category—there is no record of an A-1 having been assigned. Only one example of the XA-2 was delivered to the new Army Air Corps for evaluation testing.

The Curtiss A-3 Falcon in 1927 was the first successful aircraft in the "A" category for Attack planes. It was a modification of the Curtiss O-1B Observation plane into a much needed replacement for obsolete DH-4 Liberty Planes left over from WWI. The A-3 was a single-engine biplane with swept-back wings and clover-shaped tailfin that in profile resembled the letter "M." The landing gear had two wheels up front and a tailskid at the rear. Pilot and observer-gunner sat in open tandem cockpits. The 435-hp engine drove a 2-bladed prop to produce a top speed of 140 mph. Six 30-caliber guns were located two in the lower wing, two on the fuselage and two on a Scarff ring around the gunner's seat. Underwing racks were provided for a maximum bomb capacity of 300 pounds that could be taken on a combat radius of 315 miles. One version of the A-3 had dual controls in the two seats for use as an Advanced Trainer. Production totaled 153 A-3 Falcons for the Air Corps.

Curtiss A-3 Falcons of 3rd Attack Group. First successful Air Corps Attack plane to carry an "A" designation. (USAF Photo 16856)

YA-8, YA-10, A-12 SHRIKE

The Fokker-Atlantic-General Aviation XA-7 in 1931 was the Army's first experimental Monoplane Attack plane. Only one example was delivered to the Air Corps for testing. The Fokker Company was also known as Atlantic Aviation, which later became General Aviation and finally blended in with North American Aircraft.

The Curtiss XA-8 prototype first flew in 1931, and delivery of the limited production YA-8 Shrike began in 1932. The YA-8 was the Army's first successful all-metal Attack Monoplane and introduced the use of flaps at the wing's trailing edge. It was a low-wing monoplane with standard tail section and 3-wheeled undercarriage having streamlined housing on the fixed front legs. The pilot sat in an open cockpit at the wing's leading edge and the gunner's cockpit was remotely located about twelve feet away from the pilot. A single 600-hp engine drove a 3-bladed prop to generate a top speed of 160 mph. Four 30-caliber fixed guns were mounted in the front wheel housings and the gunner operated a flexible defense gun from the rear seat. Bomb load could be varied between four 100-pounders or ten 30-pounders on underwing racks for a combat range of 510 miles. Production totaled nine YA-8 Shrikes for the Air Corps.

The Curtiss YA-10 Shrike in 1933 was exactly like the YA-8 except for a 625-hp engine. Only five YA-10's were delivered for trials to Air Corps fields.

The Curtiss A-12 Shrike was the only Shrike to reach production. Deliveries began to arrive at the 3rd Attack Group (the Army's only Attack unit), Fort Crockett, Texas, in November 1933. (Fort Crockett was named in honor of Davy Crockett, the frontiersman and hero of the Alamo.) The A-12 was an improved version of its predecessors, with a 690-hp engine, a prominently rounded cowling and a top speed of 175 mph. The partially enclosed gunner's seat was moved forward closer to the pilot at a position just above the wing's trailing edge. Otherwise, the airframe and weapons placement was the same as on the other two Shrikes. Production totaled 46 A-12 Shrikes for the Air Corps. In those days the ''Y'' prefix indicated a limited production order, which generally meant less than fourteen airplanes. Horace Hickam was killed on November 5, 1934, in the crash of an A-12 at Fort Crockett.

Twenty export versions of the A-12 Shrike were shipped to China where they immediately went into combat when the Japanese invasion began on July 7, 1937. The last of those A-12's was destroyed by the Japanese in June 1938. At the time of the Japanese attacks on Pearl Harbor in December 1941, twelve A-12's were stationed at Hickam Field. Eight of them survived the raids and were shipped back home for minor roles until the last one was retired in September 1942.

YA-11 ATTACK PLANE

The Consolidated limited production YA-11 was delivered in January 1934 on a contract substituted for the Detroit-Lockheed YA-9, which was cancelled when Detroit went out of the airplane business. The YA-11, a spinoff from the P-30 (PB-2) project, was the first Attack plane equipped with retractable landing gear and a completely enclosed cockpit. It was an all-metal low-wing monoplane with a standard single-finned tail section. Pilot and observer-gunner sat in tandem under the sliding canopy. The single 675-hp water-cooled engine, without turbo-supercharger, drove a 2-bladed prop for top speed of 230 mph. Two guns were mounted on the fuselage, two in the wings and one at the gunner's station. A bomb load of 300 pounds could be carried underneath on a combat range of 500 miles. Only five YA-11's were delivered because Air Corps officials disapproved of water-cooled engines in their air-to-ground Attack planes.

A-17 NOMAD

The Northrop XA-13 prototype was delivered in June 1934 and the follow-up XA-16 first flew in March 1935. The production model A-17 Attack plane, a development from those two prototypes, was delivered beginning in August 1935 to the 3rd and 17th Attack Groups. Bill McChord was killed when his A-17 cracked up near Richmond, Virginia, on August 18, 1937, while on a mission for Air Corps Headquarters out of Bolling Field, DC.

The A-17 was an all-metal low-wing monoplane with standard tail section and 3-wheeled landing gear. The A-17 had fixed front wheels with streamlined fairings, which were changed on the A-17A to retractable front wheels with bare legs. Pilot and gunner sat in tandem with separate canopies inside an elongated cockpit. The single 825-hp engine drove a 3-bladed prop to give it a top speed of 220 mph. Four guns were mounted in the wings and a flex gun for tail defense was operated from the rear seat. A 600-pound bomb load could be stowed internally on twenty racks, each holding a 30-pounder. Combat range was 700 miles. One version was modified to have three seats and designated A-17AS, which stood for Staff Transport. After Douglas absorbed Northrop in August 1937, the A-17's in service were referenced with the double-designer name of Northrop-Douglas.

A shipment of the similar Northrop Model-2E Export version went to China in 1937 for use against the Japanese invaders. In June 1940 a batch of A-17A's were transferred from the Air Corps to Britain where they were shipped to South Africa for use as target tugs and operational Trainers when the United Kingdom, in dire need of training bases, used South Africa as a training ground for thousands of aircrewmen. For leading such a wandering life, those A-17's were nicknamed ''Nomads.'' One shipment of A-17's was sent to Casablanca for use by the Free French Forces in North Africa. There is no record of the A-17 having been flown in combat by Americans.

Production totaled 241 A-17 Nomads, including 148 for the Air Corps, 61 for the British and 32 for the Free French.

YA-18 SHRIKE

The Curtiss XA-14 prototype made its initial flight in 1934 as the first twin-engined Attack plane in Air Corps history. The limited production YA-18 Shrike, a development from that prototype, was delivered for service trials in December 1937. The only difference in the two planes was an engine change.

The YA-18 was a twin-engine low-wing monoplane with standard tail section and conventional 3-wheeled non-retractable landing gear. Pilot and gunner sat tandem style under an enclosed cockpit. Two 600-hp engines driving 3-bladed propellers produced a top speed of 238 mph. Four fixed forward-firing guns were fitted in the nose and a flexible defense gun was operated from the rear seat. A bomb load of 650 pounds could be carried internally on a combat range of 800 miles. The name of Shrike was used even though the YA-18 was an entirely different aircraft from its predecessors with the same name.

Only 13 YA-18's were delivered to the Air Corps and there is no record of their still being on hand when WWII started.

Curtiss A-18 Shrike. First successful Air Corps twin-engined Attack plane. (USAF Photo 978-AC)

YA-19 LIGHT BOMBER

The limited production Vultee YA-19 was a derivative of the model V-11 Light Bomber built for export to China, Brazil, Russia and Turkey. First delivery of the YA-19 to the Air Corps was made in July 1939 for service trials.

The YA-19 was a single-engine low-wing monoplane with retractable wheels up front and a tailwheel in the rear. The tailplane was unusually located about six inches up on the tailfin, protruding forward prominently in a V-shape. The pilot and one or two other crewmen sat in a very long enclosed cockpit. The single 1,200-hp engine produced a top speed of 230 mph—one example substituted a 1,800-hp engine for tests. Armament consisted of six 30-caliber forward-firing machine guns. A bomb load of 1,080 pounds could be carried internally on a combat range of 675 miles.

Only seven YA-19's were delivered to the Air Corps, three of which were used as flying test beds for experiments with an assortment of engines. The Light Bombers sent to China were used against the Japanese in Burma and Indo-China (Southeast Asia).

There is no record of Americans having flown the A-19 in combat.

ATTACK PLANES OF WWII

A-20 HAVOC

The Douglas A-20 Attack plane was developed from the Model DB-7 Douglas Bomber that made its first flight on August 17, 1939. Deliveries of the production model A-20 Havoc for service evaluation began in April 1940 as the first Air Corps combat plane to have a tricycle landing gear. Derived from export models named Boston already tested in battle by France and England, the combat-ready A-20A began to arrive at Air Corps bases in December 1940.

The A-20 was a twin-engine mid-wing monoplane with standard tail section and retractable tricycle landing gear. The 1,600-hp engines drove 3-bladed propellers to produce a top speed of 340 mph. Armament on the most-produced model, A-20G, consisted of six guns in the nose, two in the top turret and one in the belly position. An internal bomb load of 2,600 pounds plus 1,400 pounds underwing added up to a capacity load of 4,000 pounds that could be hauled on a combat range of 675 miles. A 374-gallon drop tank increased the combat range to 1,100 miles with an endurance of ten hours. Interchangeable nose sections came in a variety of configurations. The solid cone model had a pilot in the forward cockpit and gunner in the turret position; the glassed-in version added a navigator-bombardier position in the nose; one French model had 8 guns in the nose, and an English version had 12 nose guns; another British Nightfighter had a Turbin-lite (searchlight) nose for illuminating enemy aircraft—that was in the days before radar. Russian models carried a 2,000-pound torpedo and later American models could be fitted with 8 underwing rockets. A Navy Patrol variant carried anti-sub depth charges. The F-3 version was a Photo-Recon plane; the P-70 was a radar-equipped Night Havoc; and the TA-20 Havoc was a 3-seated Advanced Operational Trainer.

Hitler ordered the blitzkrieg against France on May 10, 1940, and on May 31, 1940, Havocs were used for the first time in a strike against the Germans at Saint Quentin. After France fell on June 21, 1940—less than six weeks after the assault began—shipments already enroute were diverted to England, where they were named ''Boston,'' and used as day Bombers and Nightfighters. Some were immediately diverted to North Africa for use as Fighter-Bombers against Rommel's Afrika Korps.

On July 4, 1942, members of the 15th Bomb Squadron, flying out of their English base in six A-20's borrowed from the RAF, had the distinction of being the first American military men to enter combat over Western Europe. That attack, aimed at German-held airfields in Holland, marked the beginning of American daylight bombing raids over the continent. Two of the Havocs were shot down. For his actions on that raid Charles Kegelman was awarded the Distinguished Service Cross. The 9th Air Force used Havocs in tactical sorties to disrupt enemy communications as a prelude to D-Day, the invasion of

Douglas A-20A Havoc, first USAAF plane to enter combat over Western Europe in WWII. Photo snapped above Oahu, Hawaii, May 29, 1941. (USAF Photo 38365-AC)

Europe, originally planned for June 5, 1944, but delayed one day due to bad weather over the English Channel. On D-Day Havocs laid down smoke screens to cover the troops by releasing chemicals stored in bomb-bay tanks. Final statistics from the European theater showed that A-20 Havocs flew 39,500 sorties, dropped 57 million pounds of bombs, shot down twelve enemy aircraft, and lost 275 planes.

During 1941 a Squadron of A-20A Havocs was sent to Hawaii where they were caught on the ground and almost completely destroyed during the Japanese raids on December 7, 1941. On August 4, 1942, when General George Kenney assumed command of the Southwest Pacific Air Force, A-20's immediately began operations on skip-bombing raids against Japanese shipping. The term "skip" bombing meant dropping a bomb like a torpedo so that it skimmed over the water to delay detonation in order that the pilot could avoid flying objects from the explosion. On September 12, 1942, Havocs of the 89th Attack Squadron were led by Don Hall on a low level raid at Buna, New Guinea. Modified by Paul "Pappy" Gunn to carry two extra fuel tanks and 8 guns, those A-20's were the first to drop parafrag bombs. A "parafrag" bomb was a parachute-attached fragmentation bomb that was timed to explode just above the ground and scatter its metal fragments over a wide area. They were highly effective on Japanese airfields.

On July 14, 1945, Havocs started a holocaust in the oilfields at Boela, Ceram, on the first attack in which rockets were used against Southwest Pacific targets. The Soviets used their A-20's effectively as torpedo bombers in actions against enemy shipping. A-20's had been in at the war's beginning and served right up until the last blow was struck.

Production ended on September 20, 1944, after 7,385 Havocs and English Bostons had rolled out of the factory, including 1,644 for the USAAF, eight for the Navy, 3,125 for Russia, 1,469 for Britain, 108 for France and 31 for Brazil. All surviving A-20's were withdrawn from service after the war ended. None of them were around when the "A" category was dropped by the United States Air Force at the time it was made an autonomous arm of the military service in 1947, and became operational in 1948.

A-24 DAUNTLESS

The Douglas A-24 Attack bomber grew out of the Navy SBD-3 Dive Bomber that was itself a development of the Northrop XBT-1 Dive Bomber of 1934. In 1940, General George Marshall turned to the Navy for a couple of their best aircraft to be adapted to the air-to-ground Attack role. An order was submitted for SBD-3A's already under construction to be delivered as A-24's without the carrier arrester hook and with a new tailwheel. Deliveries of the A-24 Dauntless began to arrive in June 1941—the same month in which the Army Air Corps was reorganized into the United States Army Air Forces (USAAF).

The A-24 was a single-engine low-wing monoplane with up-curled wings, standard tail section and conventional landing gear having retractable front wheels. Pilot and gunner rode tandem style in a long canopy-covered cockpit. The 1,000-hp engine with a three-bladed prop generated a top speed of 250 mph. Offensive armament consisted of two 50-caliber machine guns on top of the fuselage; defense relied on two flex guns operated by the gunner from the rear seat. A bomb load of 1,200 pounds could be carried on a combat range of 780 miles. The A-24B had a 1,200-hp engine and was equivalent to the Navy SBD-5.

A shipload of A-24's was enroute to the Philippines at the time of the Japanese attack on Clark Field, December 8, 1941. The A-24's were rerouted to Australia for lack of a place to disembark. Assembled at Brisbane and flown across to Darwin, those A-24's crossed the Indian Ocean to enter combat in February 1942 with the 91st Bomb Squadron at Java in the Dutch East Indies (Indonesia). Other A-24's were flown in actions around Port Moresby, New Guinea, by the 8th Bomb Group. The A-24A's were withdrawn from combat in July 1942 after those initial operations proved unsuccessful. In November 1943 the higher-powered A-24B joined the battle in the Gilbert Islands, being flown out of Makin Island by the 531st Bomb Squadron while supporting ground forces at Tarawa before moving on to Kwajalein in the Marshalls. The A-24B enjoyed more success than the earlier A-24A.

During the battle of Midway at the western tip of the Hawaiian Islands in June 1942, Navy SBD-5's participated in the dive bombing attacks that sank the Japanese aircraft carriers *Akagi, Kaga* and *Soryu,* and badly damaged the *Hiryu* in which admiral Nagumo limped back into Tokyo Bay.

In 1943, fifty A-24B's went to the French Air Forces in North Africa for operations out of Syria. Those aircraft were shipped to France after D-Day; they entered the war in Europe at Toulouse in September 1944 against the retreating Germans.

After the war some of the A-24B's remained in service for experiments in Fighter-Bomber tactics, and a couple were used in tests with remotely controlled drones. When the "A" category was abolished in 1948, the Dauntless was reclassified as a Fighter and became the F-24B until the last ones were retired in 1950. Production ended in 1944 after 863 A-24 variants had been delivered to the USAAF.

A-25 SHRIKE-HELLDIVER

The Curtiss A-25 Attack bomber was a direct development from the Navy SB2C-1 Dive Bomber. Deliveries of the A-25A Shrike began to arrive at USAAF units on September 29, 1942, in response to a call for help from Army officials needing anything that could be used in the air-to-ground Attack role. The A-25A differed from the SB2C-1 by having stiff wings, no hooking mechanism, and landing gear suitable for land-based rather than shipboard operations. The name was later changed from Shrike to Helldiver.

The A-25 was a single-engine low-wing monoplane with standard tail assembly and 3-wheeled undercarriage having retractable front legs. The crew of two consisted of pilot and gunner seated one behind the other under a long canopy with sliding panels. The 1,750-hp engine drove a 3-bladed prop to develop a top speed of 275 mph. Four 50-caliber machine guns or two 20mm cannon could be mounted in the wings, and the gunner fired a swing gun from the rear seat. An internal bomb load of 1,000 pounds, supplemented by 400 pounds under the wings, could be carried on a combat range of 700 miles.

Experience gained from the A-24 Dauntless taught USAAF officials that Navy Dive Bombers should be left to Navy duties and that land-based Attack planes would have to be designed using different specifications. Trials with the A-25 were only mildly successful and it was decided to abandon the dive-bombing concept altogether.

Production totaled 900 A-25's for the USAAF. They were put to use as Operational Trainers and Target Tugs. In October 1943 the USAAF redesignated 492 A-25 Helldivers as RA-25 Helldivers with the new "R" for Restricted, signifying that they were being used out of category. The other 408 Helldivers were turned over to the Marines for use in the Southwest Pacific, where they reverted back to the original designation of SB2C-1A. There is no record of the A-25 Helldiver having been flown in combat by USAAF crews.

A-26 INVADER

The Douglas XA-26 prototype made its initial flight on July 10, 1942, and delivery of the production model A-26B began in August 1943. (There was no A-26A production version.) Military requirements specified an aircraft to pack a powerful punch at high speeds on extended combat ranges in a double role. The A-26B filled the need for a fast and agile low-level Attack plane. The A-26C satisfied the specifications for a Medium Bomber operating in the middle altitudes. Both versions had the distinction of being the fastest aircraft developed in the Attack category.

The A-26B Attack version was a twin-engine shoulder-wing monoplane with retractable tricycle landing gear and a standard tail section featuring an enlarged fin. The 2,000-hp engines with 3-bladed props generated a top speed of 355 mph. Offensive armament consisted of six nose guns, four blister guns on the fuselage sides and eight guns in four optional underwing pods. Defensive armament had four guns, two in the top turret and two in the belly. Total armament was 22 guns and 6,000 rounds of ammunition. A bomb load of 4,000 pounds was carried internally and an additional 2,000 pounds could be slung at underwing hardpoints for a combat range of 1,100 miles. Additional capabilities included eight rockets and two 165-gallon drop tanks to increase the range by 300 miles. Eight more rockets could be substituted for the drop tanks on short range missions. The 3-man crew consisted of pilot and navigator-radioman in the front cockpit forward of the wing's leading edge, and gunner in the top turret to the rear of the wing's trailing edge.

Douglas A-26 (B-26) Invader, fastest American Attack plane of WWII. Redesignated B-26 Invader in 1948 and fought in the Korean war. (USAF Photo 19814-AC)

The A-26C Medium Bomber version had a glassed-in nose and a total of only six guns, two in the nose and two in each turret. The top speed was increased to 370 mph and a copilot-bombardier was added as the fourth crewman. Otherwise, the A-26C was like the A-26B Attack version. The FA-26C was a Photo-Recon plane and the XA-26F experimented with a turbojet engine in the rear fuselage.

Douglas Invaders arrived in the United Kingdom during September 1944 for assignment with the 9th Air Force, and first entered combat on November 19, 1944. The A-26's proved to be especially deadly on low level air-to-ground attack missions and performed brilliantly on intrusion forays. Their job was to knock out ground installations, support troop advances and secure the sky against German counter-attacks.

Invaders joined the war in the Pacific during January 1945, and their presence provided a big boost to the morale of war-weary ground troops—the A-26 was a beautiful airplane. Spectacular actions against the Japanese forces established the A-26 as a superior warrior, summed up by one G.I. in the words, ''I wish it could have got here sooner.''

Combat life for the A-26 Invader was short, preventing the accumulation of impressive battle statistics during mopping-up actions in Europe and the Pacific at a time when the USAAF had more combat planes than could be effectively used.

Production was halted, with cancellation of all outstanding contracts, at war's end after 2,450 A-26 Invaders had been delivered. Most of the WWII A-26 veterans were assigned to the Tactical Air Command (TAC) when that organization was formed in March 1946.

In July 1948 the newly formed U.S. Air Force did away with the ''A'' for Attack designation, and all surviving A-26 Invaders were redesignated B-26 Invaders with no modification to the aircraft. Those Light to Medium Bombers went on to serve in Korea and Vietnam to round out a triple-war career. Other roles performed by Invaders were: trainer, staff transport, experimental drag-chute braking action, missile guidance research, and drone director for remotely controlled targets.

A-27 ATTACK BOMBER

The North American A-27 Attack Bomber came into the USAAF strictly by accident in an emergency move. The Model NA-69 combat plane, designed for export to friendly nations during the late 1930's, was a souped-up member of the AT-6 Texan Advanced Trainer family. In October 1941 a shipment of Model NA-69's on the way to Siam (later Thailand) were pulled off the ship while it was anchored in the Philippines for fear that if delivered they might fall into Japanese hands. Hastily assembled, and designated A-27 Attack Bombers, they were not to wait long before getting into a fight. When the Japanese invaded the Philippines on December 8, 1941, A-27's took to the air in counter-attacks against enemy troops to achieve the distinction of being the first Attack planes to enter combat in the Pacific. (Note: A-20 Havocs were present in Hawaii during the attacks there on the previous day, but they never got airborne.) After the Philippines were overrun by the Japanese, those A-27's disappeared from the record.

The A-27 was a single-engine low-wing monoplane with standard tail section and conventional undercarriage having retractable front wheels. Pilot and gunner sat tandem style within an enclosed cockpit with sliding panels. The 775-hp engine drove a 3-bladed prop to produce a top speed of 250 mph. Armament consisted of four forward-firing guns and one defensive swing gun in the rear seat. A bomb load of 400 pounds was carried underneath for a combat range of 575 miles.

That shipment of ten planes short-stopped in the Philippines was the only batch of A-27's used by the USAAF. No production order was ever submitted for the aircraft and it was not given a fighting name.

A-28, A-29 HUDSON

The Lockheed A-28 and A-29 were members of the famous Lockheed family of Hudsons that were developed from the Model-14 Super Electra civil airliner. The Hudson I made its first flight on December 10, 1938, and export shipments began to arrive in the British Isles during February 1939. Hudsons subsequently evolved through six Roman numerals for use by the Air Corps, England and Australia. The Hudsons I, II and V were built strictly for export; the Hudsons IV and VI were the A-28 for Lend Lease export after the Lend Lease Bill was enacted on March 11, 1941; and the Hudson III was the A-29 Attack version. Offshoots with basically the same airframe were the A-28A Troop Transport, A-29A Bombcrew Trainer, A-29B Photo-Recon version, AT-18 Gunner Trainer, AT-18A Navigational Trainer, C-63 Cargo Transport, C-111 Cargo Transport and the Navy PBO-1 Patrol Bomber. Deliveries of the A-29 Hudson III to the Air Corps began in May 1941.

The Hudson was a twin-engine mid-wing monoplane with conventional landing gear having retractable front wheels. The classic Lockheed tail section had a long tailplane with twin fins and rudders. The 1,200-hp engines had 3-bladed props generating a top speed of 250 mph. Five guns could be carried, two in the nose, one in the belly and two in an optional top turret near the tail. A bomb load of 1,600 pounds could be carried internally for a combat range of 1,500 miles. Racks were provided for depth charges and later models had provision for underwing rockets. The four-man crew included pilot, copilot, navigator-bombardier and gunner.

When Great Britain declared war against Germany on September 3, 1939, the Hudson I immediately entered the war, and on October 8, 1939, while flown by an RAF pilot,

Lockheed-Vega A-29 Hudson-III. Outstanding American Attack plane of WWII nicknamed "Old Boomerang" by the British. (USAF Photo 21675-AC)

became the first American-built plane to destroy an enemy aircraft in WWII. Hudson I's on patrol duty over the North Sea and Atlantic Ocean were credited with sinking many German U-boats. In February 1940 a Hudson II led the British Navy to the German prison ship *Altmark*. In December 1941 Australian Hudson V's operating out of the Netherlands East Indies (Indonesia) attacked Japanese transports and barges that were landing troops at Kota Bharu, Malaya, losing two airplanes in the action. Hudsons flew attack sorties during the evacuation at Dunkirk in May-June of 1940. British RAF Commanders affectionately referred to the Hudson as "Old Boomerang" because, regardless of the kind of mission, they almost invariably came home.

When America declared war against Germany and Italy on December 11, 1941, A-29 Hudson III's were modified to carry depth charges and were assigned in 1942 to the 13th Bomb Group at Westover Field, Massachusetts, for anti-submarine patrol duty in the Atlantic. A Navy PBO-1 Hudson armed with four 325-pound depth charges flying out of Newfoundland on March 1, 1942, sank the first German U-boat of the war to be destroyed by Americans. In May 1943 a USAAF A-29 became the first aircraft in history to sink a German submarine with rocket bombs.

When production was halted in May 1943 a total of 2,822 Hudsons had been built, distributed as follows: 306 A-29's, 300 AT-18's, 364 C-63's, three C-111's, 20 PBO-1's, 1,237 for England and 592 for the Australians.

A-30 BALTIMORE—XA-22 MARYLAND

The family of Martin Attack planes out of Baltimore, Maryland, began with the XA-22 Maryland delivered to Wright Field, Dayton, Ohio, for tests on March 14, 1939. First deliveries of the production Model-167 export medium bomber were made to the French in October 1939, who first used them in action against the Germans on May 22, 1940. When France fell to the Blitzkrieg on June 21, 1940, seventy-five undelivered

Model-167's were diverted to Britain where they were given the name of Maryland I. Some of the Model-167's fell into Vichy French hands and were turned against the Allies. The Maryland II was delivered to the British in December 1940 for use in England, North Africa and the Middle East as a medium bomber and long-range reconnaissance plane. Marylands on Recon missions spotted the Italian fleet at Taranto in 1940 and discovered the battleship *Bismarck* hidden at Trondheim in Norway. The XA-23 in 1940 would have been a modified variant of the Maryland, but it never got off the drawing board. Only one XA-22 was delivered to the Air Corps for experimental purposes. A total of 400 Model-167's were built for our Allies to help hold the line against Axis forces during the crucial early days of World War II.

The Martin Model-187 medium bomber, an improved version of the Maryland, was delivered in 1941 as the Baltimore I, II and III to English and Australian forces holding out in North Africa and the Mediterranean. When America joined the battle in December of 1941, a batch of Model-187's was stopped on the way to England and given the designation of A-30 Baltimore IV for delivery to the USAAF. The A-30A Baltimore V in December 1942 was exactly like its predecessor Baltimores except for more efficient engines.

The A-30 was a mid-wing monoplane with retractable landing gear and a standard tail assembly. Two Wright R-2600 1,200-hp engines drove 3-bladed props to produce a top speed of 300 mph and give it a maximum range of 2,100 miles. The crew of four included pilot and radio operator in the main cockpit, bombardier-navigator in the plastic nose cone and a gunner in the upper turret protruding from the fuselage behind the cockpit. Armament consisted of four fixed guns in the wings, four mounted in the fuselage sides, two swing guns in the upper turret and two guns jutting from the underside of the fuselage. Internal stowage was provided for a ton of bombs carried on a combat radius of 600 miles.

Production ended in May 1944 after 1,575 Baltimores had been built, including 300 for the USAAF as A-30's, which were used only in secondary roles and never flown in combat.

A-31 VENGEANCE—A-35 VENGEANCE

The Vultee A-31 was another of the dive-bomber type aircraft procured by the USAAF in the early days of WWII. Originally built for export to Britain as the Model V-72 combat bomber, it was redesignated A-31 when the Lend-Lease Act was signed by President Franklin D. Roosevelt in March 1941. First deliveries to the USAAF were a batch stopped before shipment to England in December 1941 by authority of the Faddis Act, which authorized military officials to repossess any war materials deemed vital to our nation's security. Those A-31 Vengeances were used as Operational Trainers and Target Tugs.

The A-31 was a single-engine monoplane with standard tail section, conventional retractable undercarriage and a low wing that curled upward on each side of the fuselage. A 1,600-hp engine with a three-bladed propeller produced a top speed of 275 mph. Pilot and gunner sat in tandem in a long enclosed cockpit. Two 50-caliber guns were fitted in the wings, two on the cowling and one flex gun was operated from the rear seat. A bomb load of 2,000 pounds was carried internally for a combat range of 1,050 miles.

The Vultee A-35A Vengeance delivered in 1942 was like the A-31 except for four wing guns and one gun in the rear seat. The A-35B had a 1,700-hp engine, top speed of 280 mph and six wing guns to go with the one in the back seat. A-35B's were flown by the British RAF in Burma from 1942 until the war ended. There is no record of either the A-31 or A-35 having been flown in combat by Americans.

Production ended in 1943 after 1,931 Vengeances had been built, including 583 for the USAAF, 1,319 for Britain and 29 for Brazil.

A-33 NOMAD

The Douglas A-33 Attack plane was a direct conversion from the Model DB-8A Douglas Bomber built for export to small Allied countries. The DB-8A was itself an outgrowth of the Northrop-Douglas A-17 Nomad. In January 1942 the USAAF invoked the Faddis Act to repossess a shipment of DB-8A-5's on the way to Peru. They were designated A-33 and put to work as Operational Trainers.

The A-33 was a single-engine low-wing monoplane with a standard tail assembly and conventional landing gear having retractable front wheels. Pilot and gunner sat in separate cockpits under a long canopy with sliding panels above each seat. The 1,200-hp engine drove a 3-bladed prop to reach a top speed of 250 mph. Four forward-firing machine guns were fitted in the wings and a flexible defense gun was operated from the back seat. Internal stowage was provided for 1,800 pounds of bombs that could be carried on a combat range of 900 miles.

Production was discontinued in 1942 after only 121 A-33's had been built, including 31 in that USAAF batch stopped on its way to Peru, 36 to Canada for use in combat training of Norwegian pilots who managed to escape the German blitzkrieg, and 54 for small friendly nations. There is no record of A-33's having served in combat during WWII. The term ''blitzkrieg'' means a fast-hitting attack conducted simultaneously by ground and air forces, and naval forces if required by the geographical location.

A-36 MUSTANG

The North American A-36 Ground Attack plane was a modified version of the P-51 Mustang. Urgently needing aircraft for use on low level attack missions in the European Theater of Operations (ETO), in August 1942 the USAAF ordered a batch of 500 P-51's already under construction to be completed as A-36A Mustangs. Delivery of the A-36A began in September 1942, and the last of that order arrived in March 1943.

The A-36A was a single-engine low-wing monoplane with a prominently protruding intake underneath the fuselage at about the center of the plane. The tail section was standard and the landing gear had conventional retractable wheels. The enclosed pilot's cockpit, centered on the wing, blended into the fuselage back, thereby hampering the rear view. A 1,325-hp engine drove the 3-bladed prop to produce a top speed of 310 mph— sluggish compared to the P-51 Pursuit versions. Four 50-caliber machine guns were mounted in the wings and two guns were fitted on the forward fuselage sides. Two underwing bomb racks carried a 500-pounder each for a combat range of 580 miles.

The A-36A arrived on the scene just in time to participate in the mopping-up actions of the war in North Africa that ended when Field Marshall Erwin Rommel's forces were defeated in Tunisia on May 12, 1943. On the subsequent invasion of Sicily, A-36A's flew low-level missions in close support of the glider troops that were able to land on the

North American A-36 Attack-Mustangs in formation over Southern California. The A-36 was the first USAAF combat variant of the P-51 Pursuit-Mustang —in 1942 it was rated as the world's fastest Divebomber. (USAF Photo 23244-AC)

island. After Sicily was secured, A-36A's backed up the landings at Salerno and Anzio to establish beach-heads on the Italian coast near Rome. One British and two American combat units flew the A-36A during the year-long struggle before Rome was captured by the Allies on June 4, 1944—two days before D-Day on the coast of Normandy. During their combat tour in Europe A-36A Mustangs flew 23,400 sorties, dropped 16 million pounds of bombs, shot down 84 German and Italian fighters, and destroyed 17 enemy aircraft on the ground. The USAAF lost 177 A-36's in combat, establishing a comparatively low loss rate of less than one percent per sortie.

Production of the A-36A totaled only that one order of 500 planes, but later P-51's were similarly modified after receipt in the field. Although numbers were assigned all the way up to A-45, the A-36 turned out to be the highest numbered successful aircraft in the "A" for Attack category. None of them were carried forward into the USAF when that organization became operational in 1948.

ATTACK PLANES — 1962 TO THE PRESENT

A-1 SKYRAIDER

In the 1962 tri-service scheme of aircraft designations, the "A" for Attack category was reinstated by the Air Force after having been dropped in 1948.

The Douglas BT2D-1 Dauntless II was designed for the Navy's use in close support operations from aircraft carriers in WWII. The first flight was made on March 18, 1945, and the plane began to appear in dribbling quantities at Naval Air Stations during 1946— too late for the war. After WWII the BT2D's were redesignated AD-1 Skyraiders and sent into the Korean conflict on July 3, 1950, with Squadron VA-55. During and after Korea Skyraiders evolved thru the AD-7 variant. When older models were withdrawn from service they were pickled for future use if needed. In 1962 the Navy redesignated their AD-5's as A-1 Skyraiders to fit in the new tri-service Attack category.

Realizing that jet Fighters were not suited for some operations in Vietnam, the Air Force began resurrecting that reliable old warrior from graveyards all over the country. After a trip to the factory for reconditioning and for increasing the weapons capability, they were designated Douglas A-1E Skyraiders and delivered to the Tactical Air Command (TAC). Immediately rushed into war with the 1st Air Commando Group, Skyraiders were used on COunter-INsurgency (COIN) operations backing up the South Vietnamese ground forces. Under the code name of "Spad," the A-1E was operated by Air Force crews in the highlands around Pleiku to become famed as one of the real workhorses of the war. Affectionately referred to as the "Flying Dump Truck" because of its heavy ordnance load, the A-1E earned the respect of Army "grunts" for providing close support tirelessly until running out of fuel. Accuracy was its forte, placing ordnance exactly where the FAC (Forward Air Control) boys called for it.

Skyraiders were also flown in Vietnam by the Navy and the South Vietnamese Air Force. Pilots of the latter learned to handle their heavily-laden Dump Trucks almost overnight. Navy Squadron VA-25 brought fame to their shipboard variant when two Skyraiders combined to shoot down an attacking Russian MIG Fighter.

The Skyraider was a single-engine low-wing monoplane with standard tail section and conventional retractable landing gear. The cockpit came as either a single-seater or a double-seater with full dual controls. A cockpit cabin was available on one version for carrying 8 passengers or 4 stretchers. The 2,700-hp engine drove a 4-bladed prop to achieve a top speed of 300 mph. Armament consisted of four wing-mounted 20mm cannon. A weapons load of 7,000 pounds carried underneath could be varied among

Douglas A-1 Skyraider of 56th Special Operations Wing at Danang, South Vietnam, April 6, 1970.
(USAF Photo 107430)

bombs, gun pods, napalm canisters and up to twelve rockets for a maximum combat range of 1,500 miles.

Production of the AD-5 ended in February 1957 after 3,180 had been delivered to the Navy. The Air Force received only 50 modified A-1E Skyraiders. BT2D-1, AD-1 thru AD-7, A-1, Dauntless II, Skyraider, Dump Truck, Spad, Sandy, Able One and Able Dog all refer to the same airplane.

A-7 CORSAIR II

The Ling-Temco-Vought (LTV) A-7D was a modified version of the Navy A-7 Corsair II. The A-7D made its first flight on April 5, 1968, and was accepted by the Air Force in December 1968. Deliveries began to arrive in March 1969 at the 54th Tactical Fighter Wing, Luke AFB, Phoenix, Arizona.

The A-7D Tactical Fighter is a single-jet high-swept-wing monoplane with retractable tricycle landing gear. The tailfin is unusually large, and twin half-tailplanes are located on each side near the rear bottom of the fuselage. The pilot's cockpit is located on top of the nose section. A single 15,000 pounds thrust turbofan engine has a fish-mouth intake under the nose—only a small part of the nose cone protrudes forward from the cockpit and intake. Forward-firing armament consists of a Vulcan M-61 20mm cannon mounted in the lower left side of the fuselage and carrying 1,000 rounds of ammo with a firing rate of 6,000 rounds per minute if desired. Two Sidewinder air-to-air missiles can be hung on the fuselage sides under the wing's leading edge; there is no rear-firing armament. A 13,000-pound weapons load carried on six underwing pylons can be varied among 18 bombs, eight napalm canisters, air-to-ground rockets, multi-purpose missiles or special gun pods. A late edition was modified to carry the Pave Penny Laser target pod. Maximum speed is

680 mph, and 1,200 gallons of fuel can be carried in drop tanks to stretch the range beyond 1,400 miles.

Combat experience in Korea and Vietnam demonstrated that speed isn't everything in jungle warfare against guerrillas, and that some operations require slower, more maneuverable aircraft that can provide close support to ground troops for long periods of time. For that purpose the A-7D came along as the first sub-sonic Jet Attack-Fighter to be accepted by the Air Force in the fifteen years since the North American F-100 Super Sabre began breaking speed records back in 1953.

In the latter part of the 1960's aircraft built for tactical raids were no longer referred to as Bombers but were called Strike planes. Specifications called for a quick, sleek, light aircraft with the nimble agility of a Fighter and the heavy duty potential of a Bomber. The Strike role calls for a plane that can penetrate enemy territory flying below radar's protective screen and hit the target with a load of either conventional or thermonuclear explosives. As a Fighter the A-7D can out-maneuver any of its speedier opponents. As a Bomber it can skim over the ground, zig-zag to locate the target, drop its bombs, leap quickly to altitude and run for safety, using its air-to-air weapons to fight off enemy Interceptors if necessary.

The A-7D got its first taste of conflict in the Vietnam war during October 1972 with the 354th Tactical Fighter Wing that had deployed to Southeast Asia from Myrtle Beach AFB, South Carolina. American bombing of North Vietnam ceased on January 15, 1973, but in that brief combat tour the A-7D proved to be highly effective.

Production ended in December 1976 after 460 A-7D's had been delivered to the Air Force. The Vought A-7K in 1979 was a 2-seated Trainer copied from the USAF A-7D for delivery to the Air National Guard.

LTV A-7A Corsair-II (Navy), 1968, forerunner of the USAF A-7D jet-powered Attack plane. (USAF Photo 178729)

A-10 THUNDERBOLT-II

The Fairchild-Republic YA-10A prototype first flew on May 10, 1972, and was accepted for tests in February 1975 over its competitor, the Northrop XA-9A Attack prototype. Deliveries of the production model A-10A began to arrive at the 333rd Tactical Fighter Training Wing at Davis Monthan AFB, Tucson, Arizona in March 1976. On June 3, 1977, just one week before the Air Force planned to make the A-10A operational, the Director of Flight Operations for Fairchild-Republic, Sam Nelson, was killed during the Paris Air Show when his $4 million A-10 crashed into the same runway on which Charles Lindbergh landed at Le Bourget Airport fifty years before. Lindbergh's widow, Anne Morrow Lindbergh, recoiled in shock as she witnessed the accident from her seat in the stands.

The A-10 is a twin-jet low-straight-wing monoplane with twin-fin tail section and retractable tricycle landing gear. The titanium armored bubble type cockpit located above the front wheels shields the pilot from ground fire and provides a clear all-around view. The two turbofan engines generating 9,065 pounds of thrust each are installed in large external pods on top of the fuselage rear. Armament consists of a GAU-8A 30mm 7-barrel rapidfire cannon in the nose carrying 1,350 rounds of ammunition, and six Guided Air Missiles (GAM) on underwing pylons at the wing's outer sections. A maximum bomb load of 16,000 pounds can be carried for a combat radius of 300 miles when using all 18 hardpoints on the four underwing pylons at the wing's inner sections. A load of 12,000 pounds of ordnance can be accommodated when all internal tanks are full. When a 600-gallon belly drop tank is fitted, six bombs can be carried 600 miles if the mission calls for a deeper strike. Depending upon any specific requirement, the weapons load can be

Fairchild A-10A Thunderbolt-II of the 57th Tactical Fighter Wing, 1978. (USAF Photo 114327)

varied to carry either six MK-84 Laser guided bombs or six AGM-65 Maverick air-to-ground missiles on the outer pylons. The weapons delivery system uses an optical sight in conjunction with the Pave Penny Laser target seeker pod mounted on the lower right side of the fuselage below the cockpit. The barrels on the nose cannon rotate like a Gatling gun in a coordinated manner such that the one firing is always on the center line—otherwise the aircraft would vibrate uncontrollably. The top speed is 520 mph and the ferry range extends out to 2,900 miles when three 600-gallon drop tanks are attached under the wings and fuselage.

A two-seated version of the A-10 Thunderbolt-II (named after the WWII P-47 Thunderbolt) made its first flight on May 4, 1979, carrying a Weapons Systems Officer in the rear seat. Designed for adverse weather operations, nocturnal combat missions and training flights, the 2-seater has larger tailfins, relocated landing gear, modernized radar and a new cathode ray tube. As of July 1, 1979, production was still in progress and a total of 339 A-10A's had been budgeted for. Future plans call for a total of 740 A-10 variants by the end of fiscal year 1983.

A-37 DRAGONFLY

The Cessna A-37 Attack plane was a conversion from the T-37B Tweety Bird Trainer by way of the YAT-37D Combat Trainer that made its first flight on October 22, 1963. Examples of the A-37A began to arrive at Air Force Attack units in August 1966. Back in 1962, two T-37B Tweety Birds were equipped with machine guns, rockets and practice bombs for tests by the Special Air Warfare Center (SAWC) at Eglin AFB, Florida, to determine their adaptability for COunter-INsurgency (COIN) operations in Vietnam. Then, in October 1963, two YAT-37D Combat Trainers were delivered for trials. Because of the success of those tests, plans were changed to the ''A'' for Attack category and the designation became A-37A Dragonfly.

The A-37A was a twin-jet, low-straight-wing monoplane with retractable tricycle landing gear and a tailplane mounted one-third the way up the single tailfin. Pilot and co-pilot sat side by side with dual controls under a jettisonable clamshell canopy centered above the wing's leading edge. Two turbojet engines producing 2,400 pounds of thrust each were mounted at the juncture of the fuselage and wings, with intakes at the wing's forward edge. Armament consisted of a GAU-2 7.62mm minigun mounted in the nose. A bomb load of 4,855 pounds, including 38 rockets, could be carried on 8 hardpoints under the wings. The top speed was 505 mph and the ferry range extended to 1,200 miles when the aircraft was fitted with four 100-gallon underwing fuel tanks.

The A-37A had the distinction of being the smallest aircraft of its type in the Air Force—it was about the same size as the Bell P-39 Airacobra of World War II. Reporting for combat in Vietnam on May 2, 1967, Dragonflies were used on COIN operations out of Tan Son Nhut and Bien Hoa in the Saigon area—Saigon has since been renamed Ho Chi Minh City. COIN Fighters were equipped with lots of firepower, and plenty of fuel capacity to allow a substantial loiter time in close support of ground forces; they were not equipped for air-to-air dogfights. Serving in combat with the 604th Air Commando Squadron, A-37's completed their 10,000th sortie in May 1968—quite an accomplishment for that little workhorse in only a year of front-line service. The A-37B delivered in January 1968 was an advanced version built mostly for supply to friendly nations.

Cessna A-37 Dragonfly, COIN-operated Attack plane of the Vietnam war, pictured inflight during 1968. (USAF Photo 178467)

A total of 39 A-37A Dragonfly conversions were delivered to the Air Force for use in Vietnam; the last of them was retired in 1974. Production of the A-37B ended on July 1, 1977, after 560 had been built for Air Force distribution under the Military Assistance Program (MAP).

THE PROP-DRIVEN BOMBERS—WWI TO WWII

America participated in World War I without owning a single Heavy Bomber. The French Breguet 14B-2 Light Bomber was flown by AEF military aviators while awaiting arrival of the British-American DeHavilland DH-4 Light Attack Bomber for use on trench-strafing missions. The DH-4 Liberty Plane—so named because it was powered by an American Liberty-12 engine—arrived at the front in August 1918, just three months before the war ended. The DH-4 was nicknamed the ''Flaming Coffin'' because it sometimes caught fire spontaneously while in flight and was obsolete before it was introduced.

When WWI ended, many Army warplanes were shipped back home to form the nucleus of the postwar Air Service. But many hundreds of obsolete aircraft were declared to be worth less in America than it would cost to ship them home, including the surviving DH-4's. Those obsolete planes were heaped in a pile and burned, causing dissenters to refer to the occasion as the ''Billion Dollar Bonfire'' in ridicule of the Army's aircraft procurement policies.

Martin MB-2 (NBS-1) ''Martin Bomber.'' Air Service Day Bomber and Night Bomber, Short-Range; sank the captive German battleship Ostfriesland, *proving Billy Mitchell's contention that planes could sink ships.* (USAF Photo 28184-AC)

Witteman-Lewis NBL-1 "Barling Bomber," the Army's pioneer experiment with heavy Bombers. Pictured here in 1923 with four of its six engines. (USAF Photo 8359-AS)

During World War I the Martin Company blueprinted, designed, developed and assembled its GMB Bomber, which was purchased by the Army and took to the air in August 1918 as the MB-1 Martin Bomber. That was the first American conceived and constructed combat airplane in history—too late for the war! Fourteen MB-1's were built.

Huff-Daland-Keystone LB-5 "Pirate," Light Bombardment biplane of the "Roaring Twenties," shown here over Pope Field, Ft. Bragg, North Carolina. (USAF Photo 7200AS)

The Martin MB-2 Bomber in 1920 was similar to the MB-1 except for folding wings that swung backwards on hinges about halfway to the wingtips for stowage in hangars with narrow access doors. When the aircraft was ready for flight the wings were extended and locked into place. In the course of criticizing higher military officials, Billy Mitchell boasted that an Army bomber could easily sink a battleship. When the suggestion was ridiculed, he arranged a demonstration in July 1921, during which an MB-2 Bomber sank the captured German Battleship *Ostfriesland* that had been sailed back to America among the spoils of war. Five of the MB-2's were delivered to the Air Service.

The Martin NBS-1 Night Bomber-Short Range in 1920 was a redesignation of the MB-2 at the factory to put it in the new "NBS" category recently established as a part of the Army's first aircraft designation scheme. The NBS-1 was America's first mass-produced Bomber and also the only successful aircraft in the "NBS" classification. A total of 125 NBS-1's were built.

The Witteman-Lewis NBL-1 Barling Bomber was first flown by Lt. Hal Harris on August 22, 1923, as the pioneer Night Bomber-Long Range. The NBL-1 was a huge six-engine triplane with a 120-foot wingspan and eight landing wheels. The Martin NBL-2 in 1923 was the only other plane in that category—it had a span of 98 feet. Both of those "NBL" experiments were unsuccessful, but served as milestones during pioneer experiments with very large airplanes.

The Huff-Daland XHB-1 in 1926 was the first Air Corps Bomber to carry the new "X" prefix indicating that it was an experimental prototype. It was the only aircraft in the "HB" (Heavy Bomber) category, and also the only single-engine Heavy Bomber ever built for the Army. With a wingspan of 85 feet and a length of nearly 60 feet, the one-eyed aircraft was nicknamed "Cyclops" after the one-eyed giant in Homer's *Odyssey*.

The Huff-Daland-Keystone series of "LB" (Light Bombardment) biplanes began in 1926 as the Huff-Daland Company which was absorbed by Keystone in 1927. The Huff-Daland LB-1 in 1926 was the only single-engine Bomber in the "LB" category. The LB-5 in 1927 was delivered as a Huff-Daland and the LB-5A was known as a Keystone. The Keystone LB-6 through LB-14 in 1928-29 were the rest of the "LB" designations. They were advances upon the LB-5, with larger airframes, twin engines hung between the wings, twin or single tailfins, five 30-caliber machine guns and a 2,000-pound bomb load.

Curtiss B-2 Condor twin-engined biplane. First successful Air Corps Bomber to carry a "B" designation. (USAF Photo 28915-AC)

A total of 153 Light Bombers were delivered to the Air Corps. The Light Bombardment classification was abolished in 1930, and all surviving "LB" aircraft were redesignated into the "B" for Bombardment category.

The Curtiss B-2 Condor in 1928 was developed from the Curtiss NBS-4 (Night Bomber-Short Range) that first flew at Mitchell Field, NY, on August 1, 1925. The first successful aircraft in the new "B" for Bombardment category had twin 2-gun positions, one at the back of each engine nacelle mounted on top of the lower wing. A 4,000-pound bomb load could be carried underneath. Only twelve of the Condors were built because the wide span of 90 feet necessitated parking in the open—they were too big for existing hangars!

The Keystone B-3 in 1929 was a direct conversion from the LB-10 at the factory when the "LB" category was abolished. The B-3 was modified to have three guns, a 2,500-pound bomb load, a four-man crew and a single tailfin. The B-4, B-5 and B-6 were also Keystone Bombers, differing from the B-3's only in minor changes and late model engines. Approximately 130 of those biplanes formed the core of first-line Bombers for the Air Corps until the arrival of the Martin B-10 monoplane in 1934.

The Douglas YB-7 in 1932 was the first monoplane Bomber in the Army's history. It had a high birdlike wing, twin engines mounted on wing struts, a standard tail section and a 3-wheeled undercarriage with backward retracting front legs. The 4-man crew sat in open cockpits, armament consisted of two 30-caliber machine guns, and 1,200 pounds of bombs could be carried underneath. The top speed of 180 mph was 50 mph faster than that of the old Keystone biplanes; however, only eight of the YB-7's were built because more sophisticated all-metal monoplanes were already on the way.

The Boeing YB-9 in 1932 was the first all-metal Bomber in the Army's history. It was a twin-engine midwing monoplane with three open cockpits, two guns and a 2,200-pound bomb load. The supercharged Hornet engine gave it a top speed of 190 mph, but that wasn't adequate: only seven of the YB-9's were delivered to the Air Corps for trials. Like the Douglas YB-7, the YB-9 could hear the engines of more advanced monoplane designs warming up on the horizon.

Douglas YB-7 Bomber, the Air Corps, first monoplane Bomber. Seen here flying along Air Mail Route 18 from Salt Lake City, Utah, to Oakland, California. (USAF Photo 18285-AC)

Martin B-10 "Martin Bomber." First production model twin-engine monoplane Bomber in Air Corps history. (USAF Photo A-23570-AC)

The Martin B-10 in 1934 was the real forerunner of more advanced WWII Bombers yet to be designed. It was an all-metal, twin-engine, mid-wing monoplane with retractable landing gear and a standard tail section. The pilot rode in the cockpit, the radio operator and one gunner sat in the dorsal (back) turret, and one other gunner position was in the rotatable nose turret. Winner of the 1932 Collier Trophy for achievement in aviation, the B-10 featured such refinements as internal bomb racks, Norden bomb-sight, completely enclosed crew stations and controllable pitch props. The YB-10A had a turbo-super-charger that gave it a higher ceiling and a top speed of 236 mph, faster than any Pursuit then in the Air Corps arsenal—*something had to be done about the pursuits, quick!* During July 19-August 20, 1934, Hap Arnold led a flight of ten B-10 Bombers to set a roundtrip record from Bolling Field, DC, to Fairbanks, Alaska and back, a distance of 8,290 miles. Arnold was awarded the Mackay Trophy for that feat. The Martin B-12 Bomber in 1934, similar to the B-10, was fitted with twin floats and used on coastal patrols. A Martin B-12 Float-Bomber was flown by Frank Andrews between Langley Field, Virginia, and New York City on August 24, 1935, to set 3 world payload-speed records. A total of 152 B-10 and B-12 Bombers were delivered to the Air Corps and 190 examples were exported to our allies. China flew their B-10's on bomb raids against Japan in 1938, and Dutch pilots used their version on war missions in the Netherlands East Indies in 1939. The Air Corps modified all surviving B-10's into target tugs for use in training exercises during WWII.

The Boeing XB-15 in 1937 was developed as a prelude to long-range Hemispheric Bombers. Coming on the scene as the largest American Bomber of its time, the XB-15 boasted a 149-foot span, four engines, 35-ton gross weight, a ten-man crew, six machine guns and a 12,000-pound bomb load. Only one example was built, and after the experiments were finished it was converted to the XC-105 Cargo Transport, hauling 64 passengers or 21,000 pounds of cargo at a top speed of 197 mph.

The Douglas B-18 Bolo in 1937 was developed from the DC-2 Douglas Commercial as a replacement for the aging B-10 Martin Bomber. The B-18 was a twin-engine, low-wing monoplane carrying a crew of six, three machine guns and 6,500 pounds of bombs. Two units of B-18's were wiped out at Hickam Field during the Japanese attacks on December 7, 1941. Bolos stationed in the States were immediately modified for anti-submarine duties in the Atlantic and Caribbean by installing a dorsal radome and adding magnetic anomaly-detection equipment. On July 7, 1942, a B-18 from the 396th Bomb Squadron became the first USAAF aircraft to sink a German submarine in WWII, sending a U-boat to the bottom off Cherry Point, North Carolina. A total of 350 B-18 Bombers were delivered to the Air Corps. They were replaced in first-line service by Boeing B-17 Flying Fortresses, which later came rolling out of the factories at the rate of 500 planes per month.

The Douglas XB-19 in 1941 was the next step up in size from the Boeing XB-15. Developed strictly as an experiment for providing scientific information on very large aircraft, the XB-19 was the first American plane to have a wingspan greater than 200 feet. It was a four-engine low-wing monoplane with a crew of ten, fourteen defensive guns, a 37,000-pound bomb load and a maximum range of 7,750 miles. Only one example was built, and after tests were completed it was used as a Cargo Transport with provision for 123 fully equipped combat troops or 56,000 pounds of freight. Knowledge gained from the XB-19 was applied to subsequent development of the Consolidated-Vultee B-36 Global Bomber as a response to possibly having to bomb targets in Europe and Japan from bases within the continental limits of the United States.

The Douglas B-23 Dragon in 1939 was the first military Bomber to have a tail gun turret. Similar to the Douglas B-18, the B-23 had more powerful engines and a top speed of 280 mph. The 6-man crew was retained and one of them manned the fourth gun in the

Boeing XB-15 "Super-Flying Fortress" Bomber; first Air Corps Bomber to have a wingspan exceeding 100 feet. Photographed here on the ground at Langley Field, Virginia, March 9, 1939. (USAF Photo 173021)

Douglas XB-19 Heavy Bomber. First American Bomber to have a wingspan wider than 200 feet. (USAF Photo 21019)

tail turret. The B-23 was used briefly on coastal patrols before being used as a trainer; then it was converted to the UC-67 Cargo Transport with 21 seats or an 8,000-pound cargo capacity. Only forty of the B-23 Dragons were built for the Air Corps.

Production of all those significant Bombers delivered between WWI and WWII totaled approximately 900 planes.

PROP-DRIVEN BOMBERS OF WW II

B-17 FLYING FORTRESS

When America entered WWII the Air Corps inventory of Bombers included a few Martin B-10 and B-12 Mediums, 155 Douglas B-18 Bolos and about 40 Douglas B-23 Dragons, all designed for housekeeping duties and coastal patrol. From the time of its first flight on July 28, 1935, until December 7, 1941, only 131 Boeing B-17 Flying Fortress Bombers had been delivered to the Air Corps. During trials of the first American offensive Bomber to have a turbo-supercharged engine, the prototype XB-17 crashed on October 30, 1935, killing Major Pete Hill and Boeing test pilot Les Tower.

The B-17 carried 6,000 pounds of bombs on a 2,000-mile mission or could be loaded with 10,800 pounds of explosives for short runs. Twelve to 13 defensive 50-caliber machine guns, carrying 6,380 rounds of ammunition, were fitted in five or six gun turrets. Four Wright R-1820 1,200-hp engines produced a maximum speed of 318 mph, and fuel tanks holding 2,490 gallons of gas gave it a ferry range of 3,700 miles. The crew of 10-11 men included pilot, co-pilot, navigator, bombardier, radio operator, flight engineer and four or five gunners.

Variants of the B-17 included the B-40, a de-bombed version rigged with 15 guns and carrying 11,200 rounds of ammo for flying as an Escort defender alongside the Bombers. The B-40 was not successful because it was too heavily armored and could not keep up with the formations. The YC-108 in 1941 was a Cargo Transport used as General MacArthur's personal VIP Staff Transport, carrying up to 38 passengers. The F-9 in 1942 was a long-range Photo-Recon plane equipped with trimetrogon cameras (the term means snapping three pictures simultaneously, one straight down and two at oblique angles, over the area being photographed). The PB-1 Patrol Bomber was for the Navy's use on anti-submarine patrols.

B/Gen. Frank Andrews was a staunch supporter of the B-17 Bomber. Because of his fights with the War Department to step up production of the B-17 he was demoted to Colonel in 1939—but when he was subsequently proved to have been right, his rank was restored. General Andrews died in a crash in Iceland on May 3, 1943.

During the surprise Japanese raids of December 7-8, 1941, three units of B-17's in Hawaii and the Philippines were almost completely destroyed. In the words of Admiral Isoroku Yamamoto, the Japanese Naval Chief, the attack at Pearl Harbor "awakened a sleeping Giant." And within a year B-17 Bombers were rolling out of American factories

Boeing B-17 Flying Fortress. Renowned WWII Heavy Bomber used on daylight raids over Germany's fortress Europe. (USAF Photo 12134-AC)

at the rate of 500 aircraft per month. Flying Fortress crews of Lewis Brereton's 19th Bomb Group in the Philippines quickly regained their composure and flew the first offensive strike of WWII on December 9, 1941, hitting Japanese ships off the coast of Luzon.

B-17 Bombers flew softening-up raids in the Solomons just prior to the American invasion at Guadalcanal on August 7, 1942, that marked the beginning of the Allied offensive in the Southwest Pacific. In the Battle of the Bismarck Sea of March 1-4, 1943, Fifth Air Force B-17's from the 43rd Bomb Group out of Port Moresby, New Guinea, participated in sinking an entire Japanese convoy carrying supplies and reenforcements from Rabaul, New Britain, to Lae, New Guinea. Only one B-17 was shot down in that action; Woody Moore and his entire crew from the aircraft were strafed to death while floating down in their parachutes.

During the Northeast African offensive that began on October 23, 1942, at El Alamein, Egypt, B-17's of Lewis Brereton's hastily put together 9th Air Force flew bombing sorties against Germany's crack Afrika Korps. Flying Fortresses of Jimmy Doolittle's 12th Air Force arrived in Northwest Africa during operation Torch, the Allied invasion of Morocco and Algeria on November 8, 1942, and participated in the annihilation of Erwin Rommel's desert Army.

Ben Warmer, waist gunner on a 12th Air Force B-17 out of Algeria, became an instant Gunner-Ace on July 5, 1943, when he shot down seven German Fighters on a bombing raid over Gerbini airfield near Mount Etna during softening-up operations prior to the Allied invasion of Sicily. In July 1943, B-17 Bombers of the 12th Air Force participated in the first raid on Rome, Italy. In March 1944, Flying Fortresses of the 15th Air Force out of Foggia—the largest base in Italy—helped demolish the town of Cassino in a prelude to the spring offensive against the Germans along the Gustav Line (41st parallel).

On August 17, 1942, twelve B-17E's of Frank Armstrong's 97th Bomb Group led by Ira Eaker, Commanding General of the 8th Air Force, took off from Polebrook, England, for the first American daylight, heavy Bomber raid over the Continent, hitting railway yards at Rouen in German-occupied France. In January 1943, Forts made the first raid on Germany, bombing U-boat havens in the northern coast city of Wilhelmshaven. On August 17, 1943, 8th Air Force Bombers inflicted heavy damage on the Messerschmitt aircraft works at Regensburg and the ball-bearing factories in Schweinfurt; of the original 376-plane force, sixty B-17's were shot down.

The first Flying Fortress to complete 25 combat missions over Europe was Bob Morgan's "Memphis Belle," which was presented to the city of Memphis, Tennessee, for display as a lasting memorial to the WWII Bombers and the courageous men who flew in them.

During March 3-8, 1944, B-17's of the 13th Bomb Wing led by Al Elton participated in the initial raids over Berlin, thus establishing the capital of Germany as a part of the Western Front, no longer sheltered from the onslaught of American daylight bombing raids. On the night of June 21-22, 1944, during Operation Frantic, German Bombers destroyed 47 Flying Fortresses on the ground at the Ukranian airfield near Poltava, into which they had been shuttled after losing 44 planes in a raid over Berlin. Those 91 Bombers that failed to return home were the highest 24-hour loss for the 8th Air Force in WWII. During February 13-15, 1945, B-17's of the 8th Air Force took part in the raids that turned the city of Dresden into ashes and killed over 100,000 of its inhabitants in support of Russian troops approaching from the east. In March 1945, Fortresses flew in the massive raid on Essen where 1,079 planes dumped nine and a half million pounds of explosives onto targets in and around the city.

During the war in Europe B-17 Flying Fortress Bombers shot down 6,660 enemy aircraft and dropped 640,000 tons of bombs onto their targets.

When production ended in July 1945, a total of 12,731 B-17 Bombers had been delivered to the USAAF.

B-24 LIBERATOR

The Consolidated B-24 Liberator was delivered in June 1941—the same month in which the Air Corps was reorganized as part of the USAAF. Distinguishing features of the B-24 were a twin-fin, twin-rudder tail section and the fact that it was the first large aircraft with tricycle landing gear. The C-87 was a Cargo Transport variant with a crew of five, twenty passenger seats, ten sleeper berths and a 12,000-pound cargo capacity. The C-109 Liberator was a flying filling station for carrying fuel over the Himalayan "Hump" to B-29 Superfortresses operating out of Chengtu, China. The F-7 Liberator was a long-range Photo-Recon version with up to twenty cameras, and the AT-22 (TB-24) was a flying classroom for training Flight Engineers (crew chiefs). The B-24N was to have been changed to a single-finned tail section, but contracts were cancelled before it went into production. The PB4Y Privateer was a single-fin Navy Patrol Bomber with twelve guns, and the RY-3 was a single-fin Navy Cargo Transport.

The B-24 carried 5,000 pounds of bombs on a normal range of 2,850 miles, or 12,800 pounds on shorter missions. Ten 50-caliber guns carrying 4,700 rounds of ammunition were fitted in six gun positions. Four Pratt & Whitney R-1830 1,200-hp engines gave it a top speed of 303 mph. The ten-man crew consisted of four officers and six enlisted men, including pilot, co-pilot, navigator, bombardier, radio operator, flight engineer and four gunners.

The first combat version was the B-24D that reported to the Pacific theater in April 1942. On August 17, 1943, Fifth Air Force Liberators from the 90th Bomb Group out of Port Moresby opened the air campaign in New Guinea with a raid on landing strips at Wewak, Boram and Dagua. On October 12, 1943, B-24's from the 43rd and 90th Bomb Groups out of Port Moresby destroyed shipping in Simpson Harbor at Rabaul. During the air battle that followed, gunners in the Liberators shot down a fourth of the attacking force of Japanese Interceptors. In September 1944, 13th Air Force B-24's out of Morotai in the Moluccas began bombing Japanese oil refineries in Borneo, the Celebes and Ceram. Flying out of Gaya, India, Liberators of the 10th Air Force made the first raid on Japanese-held installations at Bangkok, Siam (Thailand). On July 18, 1943, B-24's of the 36th Bomb Squadron, 11th Air Force, led by Bob "Pappy" Speer, made the first raid on Paramushiro in the Kurile Islands.

On November 13, 1943, B-24's of the 7th Air Force, flying with the 431st Bomb Squadron out of Funafuti in the Ellice Islands, launched the Central Pacific air campaign with a raid on the Japanese stronghold at Tarawa in the Gilbert Islands. In October 1944 Kelly's Cobras of the 494th Bomb Group, 7th Air Force, moved into the Palau Islands and immediately began bombing Japanese targets on Leyte in a prelude to General MacArthur's return to the Philippine Islands on October 21, 1944.

On June 11, 1942, B-24's of Harry Halverson's "Halpro" detachment with the Middle East Air Force, conducted the first American bombing of Europe in a raid on the oil refineries in Ploesti, Rumania. Flying with the 9th Air Force out of Northeast Africa in December 1942, Liberators made the first raid on Italy, demolishing targets at Naples. In July 1943, B-24's were the first wave of Bombers to hit German targets in Rome.

Consolidated B-24 Liberator, famous WWII twin-finned Heavy Bomber. Most-produced combat plane in American history. (USAF Photo 25767-AC)

On August 1, 1943, five Bomb Groups consisting of 177 Liberators and 1,726 men departed Benghazi, Libya, to participate in Operation Tidal Wave, a suicidal treetop bombing raid on the heavily fortified oil refineries at Ploesti, Rumania. John Kane's 98th Bomb Group Pyramiders, Leon Johnson's 44th Eight Balls, Addison Baker's 93rd Traveling Circus, Kay-Kay Compton's 376th Liberandos and Jack Wood's 389th Sky Scorpions made up the force. Fifty-seven Bombers and 532 crewmen failed to return home from that mission. For actions above and beyond the call of duty the Congressional Medal of Honor was awarded to five brave airmen, one from each Group. John Kane of the 98th and Leon Johnson of the 44th survived to receive their awards. Addison Baker of the 93rd, John Jerstad of the 376th and Lloyd Hughes of the 389th were decorated posthumously.

In September 1943, B-24's from the 12th Air Force out of Northwest Africa participated in bombing raids over Italy prior to American landings at Salerno. On July 30, 1944, Liberators of the 15th Air Force bombed oil refineries at Budapest, Hungary.

The B-24J named "Blue Streak" received a Presidential Citation for its 110 missions under fire as the "first-est" Liberator of them all—first to bomb Europe, first over Ploesti, first to bomb the Italian fleet, first over Rome and first over the aircraft factories at Wiener Neustadt, Austria. The "Blue Streak" was originally named "Florine Jo Jo" when it was flown by Ken Butler of Halverson's detachment out of Fayid, Egypt, in 1942.

B-24J Liberators joined the B-17 Flying Fortresses in Britain during the autumn of 1942 to carry out daylight raids over the continent. On March 6, 1944, B-24's made their first run on Berlin under cover provided by P-47 Thunderbolts and P-51 Mustangs. On March 18, 1945, Liberators of the 8th Air Force participated in the largest daylight raid on Berlin when 1,250 aircraft hit targets in and around the city.

During the war B-24 Liberators dropped 635,000 tons of ordnance on their targets and shot down 4,200 attacking enemy Interceptors.

Counting all variants, the Consolidated B-24 airframe was produced in greater quantity than any combat plane in American history. Production ended with the B-24M in October 1944 after 18,188 aircraft had been built for the war.

B-25 MITCHELL

The North American B-25 Mitchell delivered in 1941 was named after America's greatest military martyr—General Billy Mitchell. Because he loudly and steadfastly criticized policies governing control of military aircraft, Mitchell was convicted in December 1925 of "conduct of a nature to bring discredit on the military service." He resigned his commission and returned to civilian life where he died of a heart attack in February 1936. Then, in July 1946, after the experience of World War II, Congress awarded him the Nation's highest honor, posthumously—*for having been right in the first place!*

Designed as a Medium Bomber, the B-25 was distinguished by a twin-fin, twin-rudder tail section and tricycle landing gear. The B-25G was armed with a 75mm cannon in the nose—the largest gun ever carried in an aircraft up to that time. The B-25H was the most lethal of all the versions, carrying 3,000 pounds of bombs on a normal combat range of 1,560 miles, 14 machine guns with 2,300 rounds of ammunition, eight 5-inch rockets underwing and a 75mm cannon with twenty-one 15-pound shells. Some of the field-modified examples had as many as 18 guns. Two Wright R-2600 1,700-hp engines gave it a top speed of 280 mph. The 5-6 man crew included pilot, copilot, navigator-bombardier,

North American B-25 Mitchell. Twin-finned Medium Bomber that achieved eminence on Jimmy Doolittle's "Tokyo Raid" in WWII. (USAF Photo 20664)

radio operator and 1-2 gunners. Variants of the B-25 included the F-10 Photo-Recon version, the AT-24 (TB-25) Advanced Trainer and the Navy-Marines PBJ-1 Patrol Bomber.

When America entered WWII in December 1941, the USAAF arsenal contained 183 B-25 Mitchells, which were immediately readied for deployment overseas. On December 24, 1941, a B-25 became the first combat plane to sink a Japanese submarine. By the end of 1942, Mitchells were fighting on virtually every front in the war.

On April 18, 1942, 16 B-25's, each loaded with 2000 pounds of bombs, led by Jimmy Doolittle, took off from the deck of the aircraft carrier *Hornet*—Admiral Halsey commanding—on the "Tokyo Raid," a daring low-level mission to bomb Japan for the first time. That raid broke the morale of Japanese civilians in Tokyo who had been propagandized into believing that their defenses were impregnable. Three crewmen on that fateful flight—Dean Hallmark, Bill Farrow and Hal Spatz—were captured by the Japanese and publicly executed for their part in the action. Lt. Col. Doolittle received a jump-promotion to Brigadier General the following day and was subsequently awarded the Congressional Medal of Honor. Ted Lawson's plane, the "Ruptured Duck," crash landed in the coastal waters near Hanchang, China. As a result of being thrown through the windshield on impact, Lawson sustained injuries that necessitated amputation of his left leg, but he survived the ordeal.

On August 4, 1942, when General George Kenney assumed Command of the Southwest Pacific Air Force from General George Brett, he inherited only one unit of B-25's, the 90th Bomb Squadron of the 3rd Bomb Group. On March 3, 1943, Mitchells of the 90th Bomb Squadron, modified by Paul "Pappy" Gunn to carry ten machine guns and 500-pound bombs, were led into the Battle of the Bismarck Sea by Ed Larner. Flying out of Port Moresby, New Guinea, the B-25's sank or damaged eleven Japanese ships; all twelve of the Mitchells returned to home base. On August 17, 1943, B-25's from the 3rd Bomb Group out of Port Moresby climbed over the Owen Stanley mountains and swooped down on landing strips at Dagua, Wewak and Boram to cripple Japanese air power at the outset of the Allied offensive in New Guinea. During that action Ralph Cheli was awarded the Congressional Medal of Honor for intrepidity over Dagua.

On September 11, 1943, B-25's out of Attu in the Aleutians made a raid on Paramushiro in the Kuriles to bomb Japanese soil. On January 28, 1944, 7th Air Force B-25's of the 41st Bomb Group, flying out of Tarawa in the Gilberts, conducted softening-up raids prior to American invasions in the Marshall Islands.

In July 1942 Mitchells of the 12th Bomb Group joined General Lewis Brereton's Middle East Air Force (later the 9th Air Force) at Fayid, Egypt, in Northeast Africa to participate in the battle of El Alamein near Cairo that began on October 23, 1942, under the Command of British General Bernard Montgomery. On May 11, 1944, Tactical B-25's of the 12th Air Force began "Operation Strangle" in Central Italy, strafing and bombing supply routes along the Gustav Line, culminating in the liberation of Rome on July 4, 1944.

But the B-25 is also remembered as the airplane that crashed into the 79th story of the Empire State Building in New York City on July 28, 1945, during dense fog, killing its crew of six and thirteen occupants of the building.

Production totaled 11,655 B-25 variants, including 9,209 for the USAAF, 706 transferred to the Navy and 1,840 exported to allies. After the war B-25's served as Staff Transports and TB-25's were used as twin-engine Advanced Trainers until the last Mitchell was eliminated from the Air Force inventory at Eglin AFB, Florida, on May 21, 1960, after 18 years and 7 months of honorable service.

B-26 MARAUDER (MARTIN)

The Martin B-26 Marauder led a controversial early life. The fast speed was due in part to a comparatively small wing that made it skittish to handle and resulted in a high loss rate during check rides and transition training. The pilots who flew that "widow maker" at Tampa, Florida, originated the phrase, "One a day in Tampa Bay." A six-foot increase in the wingspan cured the problem—but the stigma was to linger throughout the war.

The B-26 carried 3,000 pounds of bombs on a range of 1,300 miles, or up to 4,000 pounds on shorter missions. Up to 12 machine guns, carrying 4,400 rounds of ammo, were fitted at five gun positions. Two Pratt & Whitney R-2800 2,000-hp engines produced a top speed of 280 mph. The 6-7 man crew included pilot, co-pilot, navigator-bombardier, radio operator and 2-3 gunners. The only variant of the B-26 was the AT-23 (TB-26) Marauder in 1943 that served briefly as an Operational Trainer and was used mostly as a Target Tug. A few were transferred to the Navy as JM-1's.

On June 4, 1942, during the naval Battle of Midway, four 7th Air Force B-26's, each loaded with a torpedo, took off from Midway under the leadership of Jim Collins to attack the Japanese aircraft carrier *Akagi*. The torpedoes missed, and two of the Marauders were shot down. That turned out to be the only time USAAF Bombers were fitted with torpedoes for combat.

Marauders began combat in the Southwest Pacific in April 1942, flying out of New Guinea on a raid against the Japanese at Rabaul in New Britain. On September 12, 1942, B-26's out of Port Moresby, New Guinea, attacked airstrips at Buna to cripple air support for Japanese ground troops coming through the Owen Stanley mountains and getting perilously close to overrunning the Allied camp at Port Moresby itself. Soon replaced in the Pacific by longer ranging B-25 Mitchells, the Marauders were moved to Europe and the Mediterranean, where they flew a majority of the short-range sorties for the rest of the war.

In July 1942, B-26's went to Northeast Africa where they flew close support operations with British and Australian ground troops against Rommel's Afrika Korps. In May 1943 Marauders proved highly effective on strafing runs against the Germans attempting to evacuate out of Tunisia across the Mediterranean Sea. On March 15, 1944, B-26's of the 12th Air Force loaded with 1,000-pound bombs participated in a devastating raid on the town of Cassino, Italy. During the hard winter of 1944-45, when the Allied push was bogged down due to weather, Marauders of the 12th Air Force struck transportation arteries in Northern Italy, cutting supply routes to the Germans holding out along the Gothic Line (roughly, the 43rd parallel). In April 1945, B-26's joined the spring offensive into the Po Valley of Northern Italy that ended with unconditional surrender of all Axis forces in Italy on May 2, 1945.

The first appearance of B-26's operating out of England was a tragedy. On May 14, 1943, a small formation of 8th Air Force B-26B's attacked flak-infested coastal installations at Ijmuiden in Holland and every plane on the raid was destroyed. Prior to D-Day on June 6, 1944, Marauders from the 386th Bomb Group, 9th Air Force, flying interdiction missions out of the British Isles, helped destroy all bridges across the Seine leading into Normandy. On June 8, 1944, B-26's of the 387th Bomb Group, 9th Air Force, led by Rollin Childress in adverse weather, completely destroyed a German Panzer fuel dump near Caen, France. On August 15, 1944, Marauders of the 12th Air Force participated in

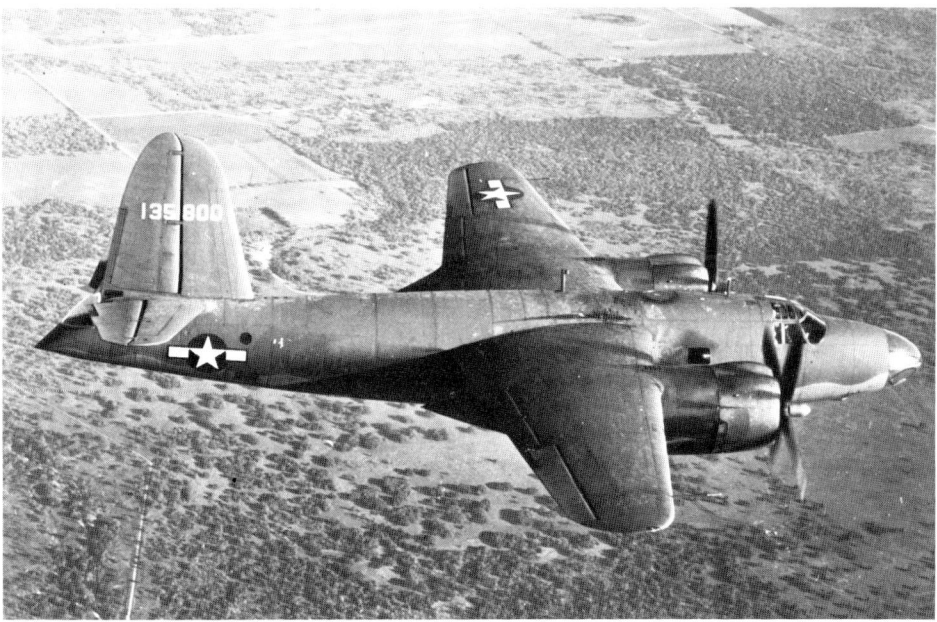

Martin B-26 Marauder. Distinguished WWII twin-engined Medium Bomber snapped here in 1944. (USAF Photo 30947-AC)

the invasion of France from the south at Nice, Toulon and Marseilles to become the 4th and final Front in the European war—the other three Fronts were that of the Russians on the east, the Allies advancing from Normandy in the west and the Allies coming up through Italy.

One of the most famous Marauders of all was the one named "Flak Bait" from the 322nd Bomb Group, 9th Air Force, which was once hit in the face by a 20mm cannon and limped home, absorbed a thousand hits in combat thereafter, flew its 101st and 102nd missions on D-Day, completed 202 combat missions without relief, accumulated 717 hours of combat flying time, and still managed to hobble back to the Smithsonian Institute for display in the airplane museum.

Production was halted in March 1945 after 5,157 B-26 Marauders had been delivered to the USAAF. The last were retired in 1948 and did not figure in the inventory of the new U.S. Air Force.

B-29 SUPERFORTRESS

The Boeing B-29 Superfortress is one of the most famous aircraft in military history. Early deliveries of the first Bomber in USAAF history to have completely pressurized crew stations began to arrive at American bases in July 1943. The F-13 was a de-bombed Photo-Recon version of the B-29 with long-range fuel tanks in the bomb-bay and K-18 and/or K-22 cameras in the nose, fuselage and tail. B-29 Super Dumbos equipped with floats and loaded down with all types of survival gear accompanied the Bombers over Japan as air-sea rescuers for fishing crewmen out of the water. The B-29 Escort variant was armed with 19 machine guns, two 20mm cannon and one 37mm cannon to help fight off the Interceptors over Tokyo.

Boeing B-29 Superfortress. Famed for raids over the Japanese Islands in WWII and for its role in Korea. (USAF Photo 28044)

The B-29 carried 18,000 pounds of bombs for a distance of 4,200 miles, or 20,000 pounds on shorter missions. Defensive armament consisted of 12 machine guns with 11,500 rounds of ammo and a 20mm tail cannon with 100 shells. Four Wright R-3350, 2,220-hp engines produced a top speed of 360 mph. The 11-man crew included pilot, co-pilot, navigator, bombardier, radar operator, flight engineer, radio operator and four gunners.

Because the B-17 Flying Fortresses and B-24 Liberators were doing an outstanding job in Europe, Superfortresses were never sent to that theater. B-29's in the Pacific operated out of three main staging areas: Kharagpur near Calcutta in India, Chengtu near Chungking in China, and the Mariana Islands in the Pacific Ocean, about 1,400 miles south of Tokyo.

The first B-29 deployed overseas arrived at Kharagpur on April 2, 1944, and in June the 58th Bomb Wing became the first unit formed under General Ken Wolfe's 20th Bomber Command. On June 5, 1944, the 58th Bomb Wing flew its first mission, hitting targets at Bangkok, Siam, to cut Japanese supplies and reenforcements flowing into Burma. In August of 1944 General Curtis LeMay arrived in India and took over the 20th Bomber Command from General Wolfe.

On May 10, 1944, the runway at Chengtu was finished, but it took about a month to "truck" supplies and equipment over the "air bridge" spanning the Himalayan "Hump." On June 15, 1944, B-29's out of Chengtu flew their first mission over the Japanese mainland, bombing iron and steel factories at Yawata. General LaVerne Saunders flew in the lead plane piloted by Howard Engler, and dropped the first bombs to land on Japan since the Doolittle raid of April 18, 1942. On November 5, 1944, Superfortresses of the 468th Bomb Group hit Singapore in full force, knocking the shipping docks out of commission. Ted Faulkner, leader of the 468th, was killed on that raid.

The Mariana Islands were liberated in August of 1944, and work began immediately on construction of five runways to accommodate 900 B-29 Bombers. On October 12, 1944, General Haywood Hansell landed his "Joltin' Josie" at Isley Field on Saipan to begin operating the 21st Bomber Command. On October 28, 1944, B-29's flew their first mission out of the Marianas, hitting the Japanese Naval staging base at Truk in the Carolines. On November 1, 1944, an F-13 Photo-Recon plane named "Tokyo Rose," piloted by Ralph Steakley out of Tinian, became the first American land-based plane to fly over the Ginza in downtown Tokyo.

On November 24, 1944, B-29's of the 21st Bomber Command, led by Emmett O'Donnell in "Dauntless Dotty," made the first bombing raid over the capital city—the beginning of the Battle of Tokyo. The target on that mission was the Musashino aircraft engine plant, which was later converted by American occupation forces into a family housing complex named "Green Park."

On January 19, 1945, General LeMay moved over from the 20th Bomber Command and replaced General Hansell as head of the 21st at Guam. Curtis LeMay believed incendiary bombs would be more effective than explosives against Japanese civilian and military targets on the mainland. On March 10, 1945, B-29's from the 314th Bomb Wing out of Guam, stripped of their guns and gunners, but loaded with napalm and cluster incendiaries, carried out the first low-level nighttime fire-bombing of Tokyo. Led by Tom Powers, those 325 aircraft created a holocaust that wiped out about fifteen square miles of the city and killed 83,800 of its people. On March 29, 1945, the last low-level fire-bombing raid was conducted over Tokyo.

One of the toughest assignments of the war was completed successfully by Colonel Paul W. Tibbets when he molded a group of men into the 509th Composite Group, an outfit trained specifically for the purpose of dropping atomic bombs out of airplanes—though they didn't know it at the time. After the explosion of the first experimental nuclear bomb in the Rocky Mountains on July 16, 1945, the 509th Composite Group was quietly deployed to Tinian. On August 6, 1945, three B-29's of the 393rd Bomb Squadron, 509th Composite Group, lifted off from Tinian on a top secret mission. The lead plane, named ''Enola Gay'' and flown by Paul Tibbets, was loaded with ''Little Boy,'' an atomic bomb ten feet long and two feet four inches in diameter. The support aircraft were flown by Charles Sweeney and George Marquardt. At 8:15 that morning, from 31,600 feet above the island of Honshu, Tom Ferebee, the bombardier, toggled the switch, releasing the bomb that flattened the city of Hiroshima and killed 86,530 of the inhabitants.

But Japan insultingly refused to surrender!

On August 9, 1945, another formation of three B-29's departed from Tinian. Flown as the lead plane by Charles Sweeney, the ''Bock's Car'' was loaded with ''Fat Man,'' a nuclear bomb ten feet eight inches long and five feet in diameter. Support aircraft were flown by Fred Bock and Jim Hopkins. At 10:58 AM Kermit Beahan dropped the second atomic bomb, annihilating the city of Nagasaki on Kyushu Island and killing approximately 40,000 of its people. It had already been decided that, if necessary, Tokyo would be the third target.

From the first raid on Bangkok of June 5, 1944, until the last and largest conventional bombing raid on the Japanese Islands by 754 Superfortresses and 169 Pursuits just before VJ-Day, B-29 Bombers dropped 177,000 tons of explosives on their targets and flew 100 million miles. During the entire ground war on the islands in the Pacific, Allied forces inflicted 760,000 combat casualties on Japanese troops. By comparison, during the nine-month air war over the Japanese mainland, B-29 Bombers out of the Marianas inflicted 806,000 civilian casualties, of which 330,000 were deaths.

On August 15, 1945, Emperor Hirohito accepted an unconditional surrender ultimatum from the United States War Department as his warlords committed suicide around him. On September 2, 1945, aboard the Battleship *Missouri,* anchored off Yokohama, the final surrender agreement was signed by representatives of all the nations involved. Thus, the war that began when warplanes of the Rising Sun attacked an American battleship at Pearl Harbor, ended on an American battleship in Tokyo Bay.

When contracts were cancelled in August 1945, a total of 4,281 B-29 Superfortresses had been delivered. Most survivors of the war remained on active duty to serve in Korea, where they flew 42,000 combat hours while dropping fifty-six and a half million pounds of explosives on enemy installations. The last B-29 was retired in November 1954, after eleven years and four months of honorable service.

When Yugloslavia's Marshal Tito died in 1980, Japan's 79-year old Emperor Hirohito was the last surviving leader of major nations involved in World War Two.

B-32 DOMINATOR

The Consolidated B-32 Dominator was that Company's entry into competition with the Boeing B-29 as America's first-line long range Heavy Bomber. Deliveries to the USAAF began to arrive in November 1944; by that time the B-29's were already dropping bombs on Japan. Contracts for the B-32 were cancelled after 115 aircraft had been built. Fifteen of that total saw combat on low-level missions in the Western Pacific. Forty B-32's were converted to TB-32 Bombcrew Trainers during the twilight days of the war. All of the Dominators were retired in 1946.

The first B-32's were designed to carry 14 guns, two cannon and a 20,000-pound bomb load. The production model carried 8,000 pounds of bombs for a distance of 2,500 miles and was fitted with ten guns at five gun positions. Four R-3350 2,200-hp engines gave it a top speed of 355 mph. The crew of nine consisted of pilot, copilot, navigator, bombardier, radio operator and four gunners.

B-34 and B-37 VENTURA

The Lockheed B-34 Ventura Medium Bomber was delivered in 1942 for use on coastal patrols and overwater missions. Akin to the Lockheed C-60 Lodestar Cargo Transport, the B-34 was a speedy low-wing monoplane having the classic twin-fin tail section originated with the Lockheed Electra and Hudson families. The B-34A carried 3,000 pounds of bombs on a normal range of 950 miles. Ten guns were mounted: four in the nose, two each in a dorsal (back) turret and ventral (belly) turret, and one each in the fuselage sides. Two Pratt & Whitney R-2800 2,000-hp engines gave it a top speed of 320 mph. The crew of four consisted of pilot, copilot, bombardier-navigator and gunner. The B-34B was a modified version used for Navigator Training. Production totaled 463 B-34 Venturas built for the USAAF.

The Lockheed B-37 Ventura was similar to the B-34, retaining accommodations for the four-man crew but having only 8 guns. The B-37 carried 2,000 pounds of bombs on a range of 1,300 miles or 2,500 pounds for shorter distances. Two Wright R-2600 1,700-hp engines produced a top speed of 300 mph. Only 18 B-37's were delivered for use on coastal patrols. The USAAF relinquished its rights to production of the Venturas in favor of the Navy, which subsequently produced 1,627 aircraft as the PV-1 Patrol Bomber. Production of all the Ventura variants added up to 3,350 aircraft for the USAAF, Navy and Lend Lease.

On May 3, 1943, twelve British Venturas attacked a German power plant in Amsterdam, Holland. The lead plane, flown by Leonard Trent, shot down a German BF-109 Fighter, was the only Ventura to reach the target, got shot down just after the bombs were released and had its entire crew captured by the enemy. For heroism in action on that daring raid, Squadron Leader Len Trent was awarded the Victoria Cross.

Final tabulations showed that American factories produced a total of 50,768 Bombers discussed in this section for use in WWII by the United States Army Air Forces (USAAF).

PROP-DRIVEN BOMBERS SINCE WW II

B-17 FLYING FORTRESS

The Boeing B-17 Flying Fortress, after a busy three years and nine months of combat duty in WWII, suddenly found itself with nothing to do when the unconditional surrender agreement was signed aboard the Battleship Missouri in Tokyo Bay on September 2, 1945. When SAC—the Strategic Air Command—was formed on March 21, 1946, all surviving B-17's were reconditioned and joined B-29 Superfortresses to make up America's pioneer Strategic Heavy Bomber force. Those few B-17's served briefly until phasing out began in June 1947 with arrival of Boeing B-50 Superfortresses in sufficient quantities as replacements. All of the B-17's were gone about the time combat-ready Consolidated-Vultee B-36B's began to arrive in July 1948, thus ending a career of hard work and meritorious service that spanned eight years and four months.

B-25 MITCHELL

The North American B-25 Mitchell was one of the Medium Bombers carried over into the post-WWII USAAF and on into the autonomous United States Air Force (USAF) that was formed on September 18, 1947, and became fully operational by July 1, 1948. A few B-25's were integrated into the SAC plan in 1946, but later became Staff Transports. When the Tactical Air Command (TAC) was formed on March 21, 1946, B-25's joined that organization to supplement Douglas A-26 Invaders as Tactical Light Bombers. The TB-25 Advanced Trainer version served as a twin-engine Transition and Aircrew Trainer until the last one was retired at Reese AFB, Texas, in 1959. The last B-25 aircraft was eliminated from inventory on May 21, 1960, at Eglin AFB, Florida, after a long career of 18 years and 7 months.

B-26 (A-26) INVADER (DOUGLAS)

The Douglas A-26 Invader Attack Medium Bomber delivered in 1943 was one of the finest aircraft developed in WWII. When TAC was formed in March 1946 the A-26 was selected as that organization's standard Tactical Light Bomber. The new USAF abolished the "A" for Attack designation as a separate category in 1948, and all A-26 Invaders were redesignated B-26 Invaders with no modification to the aircraft. By that time all of the WWII Martin B-26 Marauders had been retired from service to eliminate any mixup.

When North Korean soldiers crossed the 38th parallel into South Korea on June 25, 1950, and Russian-built YAK Fighters attacked the Kimpo airfield at Seoul, the Air Force had about 1,100 B-26 Bombers in stock. Rushed into action on June 27, 1950, B-26's were quickly modified to carry 18 machine guns, an internal bomb load of 4,000 pounds and either four bombs or fourteen 5-inch rockets underwing. Invaders had enough speed, firepower and heavy payloads to exact a heavy toll of enemy ground equipment and installations. Assigned primarily on night raids against enemy supply lines and on close-support missions backing up Marines and Infantrymen, B-26's flew 12,000 sorties in three years and one month of Korean combat duty—a total of 226 B-26's were lost in the process. On September 14, 1951, John Walmsley earned the Congressional Medal of Honor when his B-26 Night Intruder crashed while on a nocturnal raid near Yangdok, killing the entire crew. A B-26 made the last bomb drop of the war within an hour before the cease-fire agreement was signed at Panmunjom on July 27, 1953. Invaders joined B-29 Superfortresses and B-45 Tornado Jets to form America's Korean Bomber force.

Phasing out of the B-26 began after the Korean War; it was gradually replaced in service by Martin B-57 Intruder Jets and Douglas B-66 Destroyer Jets in the new all-Jet scheme of Tactical bombing.

When the USAF was ordered by President John Kennedy in 1962 to reenforce Southeast Asia in the face of Communist aggression, B-26's were recommissioned for active duty. Used on COunter-INsurgency (COIN) operations in the Vietnam War, Invaders served honorably until being retired in 1963 with the distinction of having been one of only a few aircraft to fight in three wars—World War II, Korea and Vietnam—although it was known as the A-26 in WWII; in all, the B-26 had 20 years of active duty. The Douglas C-47 Gooney Bird (Gunship) and the North American P-51 (F-51) Mustang were two other notable warriors that served in those three conflicts.

B-26K COUNTER-INVADER

But the B-26 refused to succumb to old age without one last gasp. During 1963-64 the On Mark Engineering Company developed a total of forty B-26K Counter-Invaders for use on COIN operations in the Southeast Asia War. To satisfy requirements of the Special Air Warfare Center (SAWC) located at the Proving Grounds test area on Eglin AFB, Florida, B-26K's were loaded with 11,000 pounds of assorted weapons, permanent wing-tip fuel tanks, sophisticated electronics equipment and a removable weapons-bay camera. The B-26K reported for combat in May 1964 as the most beautiful plane in the Vietnam War, flying close support missions with the "grunts" down on the ground until being withdrawn from combat a few years thereafter.

B-29 SUPERFORTRESS

The Strategic Air Command originally began operations on March 21, 1946, with a Heavy Bomber force of about 254 B-17 Flying Fortresses and B-29 Superfortresses left over from World War II. Versatility of that giant among giants allowed the B-29 to be converted for a variety of roles. The KB-29 Tanker was an airborne refueler from which the KC-97 Cargo Transport/Tanker was developed. The B-29D was converted to the B-50 Bomber, from which the KB-50 Tanker was developed. A few B-29's were converted to P2B-1's for the Navy to use in their Airborne Early Warning Service. Other variants included the RB-29 Photo-Recon, WB-29 Weather Recon and HB-29 Air Sea Rescue versions. In 1950 seventy Superfortresses were refurbished for sending to Britain under the Military Assistance Program (MAP).

Immediately pressed into action when the Korean Conflict began in June 1950, approximately 130 B-29 Superfortresses flew 42,000 combat hours while dropping fifty-six and a half million pounds of explosives on enemy installations.

One version of the B-29 had served as a "mothership" carrying the Bell XS-1 (X-1) Special Research vehicle that first broke through the sound barrier in October 1947 with Charles Yeager at the controls.

Phasing out of the B-29 began shortly after the Korean War ended. The last of the fleet was retired on November 1, 1954, after eleven years and four months of service.

YB-35 FLYING WING

The Burgess-Dunne Tailless aeroplane in 1912, with a span of 46′6″ and a length of 23 feet, was the original pioneer in development of Flying Wing aircraft. It was a single-engine pusher biplane with two landing wheels supported by ski-type skids for stability.

The first Northrop Flying Wing mock-up with a span of 38 feet and two Franklin 120-hp engines took to the air in July 1940. It was designated the N1M in the Northrop mock-up series. Every portion of the aircraft contributed directly to lift. The wingtips were turned down noticeably to compensate for the lack of normal vertical control mechanisms.

The Northrop YB-35 Flying Wing Heavy Bomber with a wingspan of 172 feet and a length of 53 feet first flew in June 1946 with four Pratt & Whitney R-4360 engines generating 3,000 horsepower each. It carried a maximum bomb load of 51,000 pounds and was fitted with 20 defensive machine guns. The crew of six consisted of pilot, copilot, navigator, bombardier, flight engineer and tail gunner. A single YB-35 was fitted with six Allison J-35 Jet engines to become the YRB-49A, distinguished as the world's first jet-powered Flying Wing Bomber. Production totaled 15 of the YB-35 prop-driven Flying Wings.

B-36 PEACEMAKER

Faced with the possibility of having to bomb European and Japanese targets from bases inside the continental limits of the United States, the USAAF ordered the Consolidated-Vultee B-36 long range Heavy Bomber in 1941; but development was slow and the B-36 didn't make its first flight until August 1946, one year after the end of World War II. The first unarmed B-36A was delivered to SAC in August 1947 for aircrew transition training. The combat-ready B-36B with six 3,000-hp radial engines arrived in July 1948 to join the B-29 and B-50, giving the Strategic Air Command a three-pronged attack force. The B-36D added four J-47 turbojet engines to earn the distinction of being the only ten-engine Bomber in Air Force history. The RB-36D Photo-Recon version carried fourteen cameras and a 22-man crew. One B-36 was converted into the XC-99 Cargo Transport for delivery in May 1949 as the world's largest prop-driven Carrier with accommodations for 400 combat troops, 300 stretchers or 101,000 pounds of freight. Two Peacemakers were equipped with eight jet engines each to become the experimental all-Jet YB-60

The Consolidated B-36 Peacemaker, with a bomb-bay volume equal to that of four railroad freight cars, was the largest Bomber in American military history. Photo taken during its first flight in August of 1946. (USAF Photo 32254-AC)

Hemispheric Bomber. One B-36 was redesigned into the X-6 Special Research vehicle for trials in the Vertical/Short Take-Off and Landing (VSTOL) program.

Capable of delivering nuclear weapons to any target on earth, the B-36 was dubbed ''Peacemaker'' because of its deterrent influence on Communist nations during the turbulent 1950's.

In 1952 the B-36 was used in experimental work on the feasibility of carrying its own parasite McDonnell XF-85 Fighter underwing for protection when enemy Interceptors were encountered. During May 1953 trials under the Fighter Conveyor (FICON) project the GRB-36D ''mothership'' carried a Republic RF-84F Thunderflash underwing to the last leg of a mission where it was launched for high-speed penetration to the target. Subsequent advancements in airborne refueling techniques eliminated the need for that type of mating.

The B-36 could carry 10,000 pounds of bombs for a distance of 10,000 miles or a 72,000-pound load on shorter hauls. The B-36D could carry two 42,000-pound Grand-Slam bombs. Defensive armament consisted of sixteen 20mm cannon carried two each in the nose, tail and six remotely controlled turrets. The normal crew consisted of 15 men plus four relief—the crew moved back and forth on a wheeled trolley through an 85-foot tunnel. Six Pratt & Whitney R-4360 3,000-hp engines driving pusher type airscrews produced a top speed of 410 mph; the jet-assisted B-36D had a max speed of 440 mph when all ten engines were pushing at the same time.

Production totaled 382 B-36's of all variants. The final one was retired from SAC in February 1959 with the distinction of being the last piston-engined Heavy Bomber to serve in the Air Force. Although the B-36 was combat-ready during the Korean War,.it was never used in combat.

B-50 SUPERFORTRESS

Work began on the Boeing B-50 Superfortress in 1944 without a prototype as such. It was a development from the B-29D which was, itself, originally designated the XB-44. The production model B-50A was delivered to the Strategic Air Command and flew its first mission on June 25, 1947, to begin the phasing out of the aging WWII fleet of B-17 Flying Fortresses.

On February 26, 1949, a B-50 named "Lucky Lady II" departed from Carswell AFB, Texas, on the first non-stop flight around the world. With Jim Gallagher as its Aircraft Commander, the Lucky Lady II, aided by four inflight air refuelings from KB-29 Tankers, landed back at Carswell on the morning of March 2, 1949, after covering 23,452 miles in 94 hours at an average speed of 249 mph. That feat caused nations around the globe to take notice of the fact that America had become the world's leader in aeronautical technology.

The B-50A and B-50B were adapted for airborne refueling by the KB-29 Tanker, using the British-developed hose method. The B-50D in 1949 introduced the Flying Boom technique of inflight refueling, and also had two extra 700-gallon underwing fuel tanks to stretch the range about 300 miles. The RB-50 Photo-Recon version was modified to accommodate radomes, sophisticated equipment and a 16-man crew. The TB-50 was an Advanced Aircrew Trainer, and the WB-50 Weather Recon version was used as a Hurricane and Typhoon chaser. (A hurricane and a typhoon are the same kind of storms; if they form in the Atlantic Ocean they are called hurricanes, and if they form in the Pacific

Boeing B-50 Superfortress, post-WWII Heavy Bomber that made history's first non-stop flight around the world aided by inflight refueling. (USAF Photo 34743-AC)

Ocean they are called typhoons.) The Boeing B-54 would have been an enlarged Bomber version, but that contract was cancelled.

The KB-50 Tanker version was never used by the Strategic Air Command. Equipped with four radial engines and carrying a 6-man crew, the KB-50 was delivered to the Tactical Air Command as an airborne refueler for TAC's jet-powered Medium Bombers and Tactical Fighters. When those aircraft got too speedy, the KB-50J added two J-47 turbojet boosters underwing to increase the speed during actual hookups.

One version of the B-50 acted as a "mothership" to the Bell X-2 Special Research vehicle that crashed into the heat barrier, killing the pilot, Capt. Milburn Apt.

The B-50 could carry 10,000 pounds of bombs for 6,000 miles or a maximum load of 28,000 pounds on shorter missions. Thirteen 12.7mm guns were carried in a tail turret and four remotely controlled dorsal (back) and ventral (belly) turrets; a 20mm cannon could be substituted for one of the tail guns. The normal crew of ten consisted of pilot, copilot, navigator-radar operator, bombardier, flight engineer, radioman and four gunners. The crew in the Photo-Recon version was augmented by six electronics experts. The crew in the KB-50 Tanker was composed of pilot, copilot, navigator-radar operator, flight engineer and two boom operators. Four Pratt & Whitney R-4360 3,500-hp engines produced a top speed of 380 mph, which was increased to 444 mph on the jet-boosted Tankers.

Including all variants, production totaled 370 aircraft, of which 222 were the most-produced B-50D model. Conversions from Bombers to Tankers totaled 126 examples. Gradual phasing out of the B-50 began in June 1955 with arrival of the Boeing B-52 Stratofortress. The last B-50 departed SAC in 1958 with the distinction of having been the highest numbered prop-driven Bomber for the Air Force. The B-50 was never needed in combat, but was ready if the call had come. Production of prop-driven Bombers at Boeing ended with the B-50.

JET BOMBERS OF THE AIR FORCE—1946 TO 1962

XB-43—FIRST JET BOMBER

The Douglas XB-43 took to the air on May 17, 1946, as America's first all-jet Bomber prototype. The XB-43 was a development from the Douglas XB-42 Mixmaster, so-named because the power package was a mixture of two pusher-prop engines and two jet boosters. The XB-43 had twin General Electric J35-GE turbojet engines delivering 3,750 pounds of thrust each to produce a top speed of 500 mph. The engines were mounted one above the other inside the fuselage with intakes at the shoulders. The bomb load of 8,000 pounds was carried for a range of 1,800 miles. It had no defensive armament, but relied on speed for protection. The 3-man crew consisted of pilot and copilot in separate side by side cockpits and a bombardier in the clear-view nose cone. Only two XB-43's were built, strictly for experimental purposes—they were an important link between already successful turbojets for the Fighters and development of more powerful jet engines with sufficient thrust to get the heavier Bombers off the ground.

The Douglas XB-43 ''Flying Museum Piece'' was the first jet-propelled Bomber in Air Corps history. Picutred while flying out of Edwards AFB over the San Gabriel Mountains of California. (USAF Photo 45697)

North American B-45 Tornado, America's first operational all-jet Bomber. Shown here on January 28, 1948, just before delivery to the USAAF. (USAF Photo 34477-AC)

B-45 TORNADO

The North American XB-45 prototype medium Bomber made its initial flight in March 1947 out of Muroc (later Edwards) Field, California. Deliveries of the production model B-45A Tornado began in February 1948 as America's first operational Jet Bomber. By November 1948 the 47th Bombardment Group (TAC), Barksdale AFB, Louisiana, was flying the B-45 on training missions as the first Jet Bomber unit in Air Force history. The B-45B was ordered and cancelled before any were built. The B-45C in 1949 was an improved version with two 1,200-gallon wingtip drop tanks that increased the span by seven feet. The TB-45 was used as a Trainer and Target Tug, the DB-45 was a Drone director, and the JB-45 was an engine test bed.

The RB-45C joined the 91st Strategic Reconnaissance Wing in June 1950 equipped with twelve elaborate cameras, including a trimetrogon K-17C system for taking wide-angle photographs. Easily convertible for either the Bomber or Photo-Recon role, the RB-45C joined the Douglas B-26 Invader and Boeing B-29 Superfortress in Korea to become America's first Jet Bomber used in combat. On long-range Photo-Recon missions the bay was loaded with photoflash bombs and extra fuel tanks, thereby increasing the range to 2,500 miles.

Another milestone was the first inflight refueling of a Jet Bomber when an RB-45C hooked up with a KB-29P on February 2, 1951, using the Flying Boom technique. In 1952 the B-45A had its first overseas tour, reporting for duty with USAFE in the United Kingdom.

The Bomber version carried 10,000 pounds of bombs for a range of 1,900 miles or 22,000 pounds on short missions. Two 12.7mm defensive guns were fitted in the tail—the B-45 relied on speed as its best forward defense. Four General Electric J47-GE engines supplied 5,200 pounds of thrust each to produce a top speed of 580 mph. The four-man crew consisted of pilot and copilot under a bubble fighter-type canopy, bombardier in the nose cone and gunner in the tail turret.

Production totaled 142 aircraft, including 3 prototypes, 106 of the Bombers and 33 Photo-Recon examples. The final B-45 rolled out of the factory in 1952 and the last one was retired in 1958, having been phased out with arrival of the Douglas B-66 Destroyer.

B-47 STRATOJET

Designed as the first full weapons system bomber in Air Force history, the XB-47 prototype made its initial flight on December 17, 1947, and the production B-47A took to the air on June 25, 1950, with acceleration rockets to assist at takeoff time and a 16-foot drag chute for braking action when landing. Deliveries began to arrive at SAC bases in October 1951 as that organization's first long range Heavy Jet Bomber. The Stratojet was the first swept-wing Jet Bomber ever produced; it was capable of operating with atomic bombs or conventional loads over extended distances after aerial refueling. Cockpit arrangement was awkward and crew fatigue on long missions presented a slight problem. The B-47A through B-47E were Bomber versions and the KB-47B was an experimental Tanker. Other variants were the RB-47 Photo-Recon, WB-47 Weathership, TB-47 Bomb-crew Trainer, DB-47 Missile Launcher and QB-47 Target Drone. The YDB-47E served as a "mothership" carrying the Bell GAM-63 Rascal (X-9 Shrike) air-to-surface guided missile.

Milestones established by the B-47 were many. In June 1953 a Stratojet set the trans-Atlantic speed record of 618 mph from Maine to England. In September 1953 a B-47 Bomber was refueled by a KB-47 Tanker in the first jet-to-jet inflight refueling operation ever made. In 1955 CWO Andy Waters served as the first Weather Liaison Officer during tryouts for operation "Reflex" with the 308th Bomb Wing Commanded by Col. Paul W. Tibbets, who also ramrodded the dropping of the first atomic bomb on Japan. Operating out of Hunter AFB, Georgia, "Reflex" action was an exercise in shuttling B-47 units back and forth between America and the United Kingdom in support of worldwide peace-

Boeing B-47 Stratojet —first USAF long-range jet-powered Heavy Bomber. Caught here landing at Limestone AFB, Maine, after a flight with the 305th Bomb Wing out of MacDill AFB, Florida. (USAF Photo 49138-AC)

keeping responsibilities. In 1959 Stratojets participated in operation "Oil Burner," conducting low level strike runs in attempts at penetrating below enemy radar detection while delivering nuclear weapons to strategic targets. In November 1959 a B-47 established the world endurance record for a jet aircraft by remaining aloft 80 hours and 36 minutes, covering a distance of 39,200 miles. The ERB-47H was specially equipped for flying surveillance missions near the Russian border; international tension mounted in 1960 when an ERB-47H was shot down over the Bering Sea by Russian MIG Fighters.

The B-47 carried a 12,000-pound bomb load for 4,000 miles or 20,000 pounds on shorter hauls. Two radar-controlled 12.7mm guns in the tail with 350 rounds per gun were standard defensive armament, but two guns could be fitted in the nose also. Six General Electric J47-GE turbojet engines supplied 6,000 pounds of thrust each to produce a top speed of 620 mph. The three-man crew consisted of pilot, copilot and navigator-bombardier-gunner with room for a Standboard inspector. On the Photo-Recon version three additional electronics operators were accommodated in a bomb-bay pod.

A total of 2,041 B-47's of all variants had been built when production ended in February 1957, of which 1,600 were Bombers and 441 were the other versions. The last of the Stratojet fleet were retired in 1966, being replaced in service by Boeing B-52 Stratofortresses.

YB-49 FLYING WING

The Northrop YB-49 Flying Wing long range Heavy Bomber made its initial flight on October 21, 1947, as a jet-powered conversion from the prop-driven XB-35 prototype,

The Northrop YB-49 Flying Wing was the first USAAF experimental tailless jet-powered Heavy Bomber. (USAF Photo 44172-AC)

with a wingspan of 172 feet and a length of 53 feet. The one-piece wing swept back slightly on either side and the landing gear was of the 3-wheeled retractable type—it had no tail section. The crew of six was composed of pilot and copilot in the forward bubble cockpit offset to the left of centerline, bombardier at the right front window, navigator and engineer in the center and a gunner in the rear top turret. Folding bunks were provided for a full 6-man backup crew in the pressurized central nacelle. The YB-49 carried no defensive armament; it relied entirely on speed for protection. A 16,000-pound bomb load was carried on a range of 3,550 miles. The eight Allison J-35 gas turbine engines with 3,750 pounds of thrust each, produced a top speed of 510 mph. The eight engines actually appeared as four because of their twin grouping on either side of the central nacelle.

Too much pitching and yawing resulted in destruction of two prototypes during tests. On June 5, 1948, Glen Edwards, Dan Forbes and three crewmen were killed when their YB-49 Bomber prototype crashed in the desert near Muroc AFB, California. In April 1949 the idea of developing the YB-49 as a Bomber was cancelled. The reconnaissance portion of the program was continued, but in 1950 the entire project died a natural death with delivery of only one YRB-49A Photo-Recon example with photography equipment fitted in a bulge below the center section.

Only three experimental aircraft were delivered on the contract that called for a jet-powered B-49 Tailless Flying Wing Recon-Bomber.

B-52 STRATOFORTRESS

The Boeing B-52 Stratofortress is distinguished as having remained operational longer than any Bomber in U.S. Military history. The YB-52 prototype first flew on April 15, 1952, with Tex Johnson at the controls. The production model B-52A flew in August 1954 and first deliveries of the B-52B began to arrive at the 93rd Bombardment Wing (SAC), Castle AFB, California, in June 1955. The thin flexible wings of the B-52 have a natural sag that is supported by outrigger struts near the tips while on the ground—when airborne the sag straightens out.

On May 21, 1956, a B-52 flown by Dave Critchlow and his crew while participating in operation "Redwing" dropped the first hydrogen bomb in history on Bikini Atoll. On January 16, 1957, five B-52 Stratofortresses departed Castle AFB, California, on the first non-stop trip around the world by a jet aircraft. Led by Archie Old, three of those B-52's completed the trip, landing back at March AFB on January 18, 1957, after covering 24,325 miles in 45 hours and 19 minutes at an average speed of 530 mph. That was less than half the time it took the B-50 Lucky Lady II on the first-ever non-stop flight around the world by a prop-driven Bomber in March 1949. A Stratofortress took off in 1958 with a total weight of 488,000 pounds, the heaviest load ever carried aloft by an airplane of any type. In January 1962 a B-52 established a non-stop, non-refueled world distance record by flying 12,520 miles from Okinawa, Japan to Madrid, Spain, in 22 hours and 10 minutes at an average speed of 565 mph.

When B-52's joined the war in Vietnam on June 18, 1965, as conventional 108-iron Bombers, the normal bomb load of 14 tons was more than doubled, and a 6-barrel ASG-21 Gatling gun firing 6,000 rounds per minute was used as the defensive armament. During the siege by 20,000 Viet Cong soldiers of 6,000 Marines and South Vietnamese at the Khe Sanh firebase in 1968, Stratofortresses flying three-plane cells at 90-minute

Boeing B-52 Stratofortress. Distinguished as the most durable Bomber in American military history. Seen in flight above Eglin AFB, Florida.154854-AC)

intervals around the clock, completed 2,600 sorties and dropped 75,000 tons of bombs. During operation "Linebacker II" in December 1972, B-52's flying night missions over North Vietnam hit military targets at Hanoi and Haiphong from sundown to sunup for eleven days, hoping that the saturation bombing would bring an end to American involvement in Vietnam. Ground-based Russian surface-to-air missiles (SAM) shot down fifteen B-52's and damaged nine others.

The B-52 underwent eight major design changes that evolved from the B-52A through the B-52H. Sophisticated equipment installed in various models included jet-powered GAM-77 Hound Dog missiles carried underwing near the fuselage sides, GAM-87 Skybolt ballistics missiles that were cancelled, SRAM attack missiles, jet-powered GAM-72 Quail decoy missiles and SCAD decoy-attack missiles that were cancelled. Stratofortresses have been used as "mothership" to carry the North American X-15 Hypersonic Special Research vehicle that flew at a speed of 4,534 mph and reached an altitude of 354,200 feet with Joe Walker at the controls, and the Martin-Marietta X-24 Wingless Orbiter that paved the way for the Space Shuttle "Enterprise." Should the occasion arise, the B-52 can be rigged to carry up to twenty nuclear-headed Cruise Missiles. A normal bomb load of 28,250 pounds can be carried by the B-52 on an extended range of 10,000 miles. In Vietnam a heavy load of 60,000 pounds included eighty-four 500-pound bombs in the bay and twenty-four 750-pounders underwing carried on a combat range of 3,000 miles. Defensive armament mounted in the tail turret consisted of either four 50-cal machine guns or two 20mm cannon or one multi-barreled 20mm gun. Eight Pratt & Whitney J57-P turbojets, producing 12,500 pounds of thrust each, give it a top speed of 660 mph. The B-52H substituted TF-33 turbofans to increase the range. The crew of six consists of pilot, copilot, bombardier, navigator-radar operator, electronics engineer and tail gunner seated in the flight deck, remote from the tail section.

The last of 744 Stratofortresses built rolled out of the factory in June 1962 and reached its destination with SAC in October 1962.

B-57 INTRUDER

The Martin B-57 Intruder Medium Bomber was adapted from the English Electric Canberra as the first foreign aircraft accepted by the Air Force since the Reverse Lend Lease transactions of WWII. Development began in 1951 and the production model B-57A began to arrive at Tactical Air Command bases in July 1953. On the outside the Intruder looked almost exactly like the British Canberra, but the interior structure was greatly strengthened for low-level, close-support operations. An American feature introduced was a pre-loaded revolving bomb-bay door that rotated through 180° just before the bombs were released.

The B-57A Tactical Bomber carried a 2-man crew of pilot and bombardier-navigator-radarman seated in side-by-side seats. The 8,000-pound weapons load was distributed 5,000 pounds in the bomb-bay door and 3,000 pounds underwing consisting of 16 small rockets, eight 5-inch HVAR rockets, or four napalm tanks. Armament consisted of eight 50-caliber forward-firing wing-mounted machine guns. Two Wright J-65 engines with 7,200 pounds thrust each produced a top speed of 580 mph and a range of 2,300 miles. The RB-57A carried a camera in the bomb-bay for quick conversion in a dual role of either Bomber or Photo-Recon. The B-57B Night Intruder was delivered to TAC in 1955 with two tandem seats. By September 1956 B-57B's had replaced Douglas B-26 Invaders in four of the TAC Wings. The B-57C was used as a transition Aircrew Trainer. The B-57E was a Target Tug.

The Martin B-57 Intruder was the first foreign-designed jet Bomber accepted by the Air Force. Shown here is a "Night Intruder" during a trial flight above the Chesapeake Bay bridge near the coast of Maryland. (USAF Photo 47478-AC)

The RB-57D Strategic Reconnaissance version was an entirely different airplane with a wingspan of 106 feet and a top speed of 630 mph. It came in either a one-seat or two-seat variety and had twin J57-P turbojets.

Production ended in 1959 after 403 Intruders of all variants had been delivered to the Air Force.

The General Dynamics-Martin RB-57F Strategic Reconnaissance Intruder in 1964 was yet another entirely different airplane having a span of 122 feet and an endurance of ten hours. Two TF33-P turbofans were fitted in the engine nacelles to increase the range, and two J60-P boosters were added underwing for use on takeoff. The RB-57F had a cruising altitude of 93,000 feet. Twelve aircraft were converted from B-57B Bombers in this manner.

In retrospect the Martin B-57 was really three different aircraft, the first such occurrence since the Republic P-84 (F-84) Pursuit-Fighter-Reconnaissance family of Thunderjet, Thunderstreak and Thunderflash in the late 1940's and early 1950's.

B-58 HUSTLER

The Convair B-58 Hustler was the world's first supersonic Strategic Bomber. The XB-58 prototype made its initial flight on November 11, 1956, and deliveries of the production model B-58A began in December 1959 to the 43rd Bomb Wing (SAC), Carswell AFB, Texas. In 1961 the 305th Bomb Wing at Bunker Hill AFB, Indiana, became the only other SAC unit to be equipped with B-58's, the first aircraft to have adjustable air intakes.

Convair B-58 Hustler, the world's first supersonic Strategic Bomber. (USAF Photo 160404-AC)

The B-58A conformed to the Delta wing design developed on the XF-92 Dart, F-102 Delta Dagger and F-106 Delta Dart, three Convair Fighter-Interceptor aircraft. Strong wings allowed it to fly at top speed for approximately an hour before easing off on the throttle. The Hustler could fly non-stop missions for an endurance of 18 hours, giving it the capability of delivering a nuclear weapon to any target on the globe. Designed under the Weapons System concept, the B-58 featured sophisticated refinements such as a two-component droppable weapons/fuel pod under the fuselage that could house any one of three different Weapons Systems: an air to surface missile system, an electronics counter-measures system, or a nuclear bombing system. Advances included individual escape capsules for each crew member, a Bendix electronic auto-pilot control system and a Sperry bombardier-navigator automatic internal guidance system.

Low level strike capability was demonstrated in 1960 when a Hustler made a 1,400-mile flight at 700 mph and never got above 500 feet. In January 1961, six closed circuit speed records were set by B-58's flying out of Edwards AFB, California. In March 1962 a live bear came flying out of a B-58 to become the first living creature that made an emergency exit from a supersonic aircraft when he was ejected from the plane flying 870 mph at 35,000 feet—and survived. In September 1962 a B-58A climbed to 85,361 feet and by October 1963 the Hustler's string of records totaled 19: notable among them was the accomplishing of an 8,028-mile non-stop flight from Tokyo to London in 8 hours and 35 minutes at an average speed of 935 mph.

The B-58 carried a secret bomb load of thermo-nuclear weapons for an unrefueled range of 2,400 miles. Defensive armament was one General Electric Vulcan multi-barreled, radar-aimed tail gun. Four General Electric J-79 turbojet engines delivered 10,300 pounds of thrust each to produce a top speed of 1,380 mph. The 3-man crew consisted of pilot, bombardier-navigator and systems operator seated tandem style in three separate cockpits. A total of 116 B-58 Bombers were delivered, production ending in autumn of 1962.

YB-60 STRATEGIC BOMBER

The Consolidated-Vultee YB-60, first flown on April 18, 1952, with an all-jet configuration, was a modified version of the B-36 Heavy Bomber. Designed in competition with the Boeing B-52 Stratofortress as America's supreme Strategic Bomber, the YB-60 maintained about three-fourths of the parts and features of the B-36. Distinguishing differences were the eight Pratt & Whitney J57-P turbojet engines mounted at four installations, two on each wing. Each engine generated 8,700 pounds of thrust to push the huge aircraft forward at a top speed of 508 mph. In comparison with the B-36, the wings on the YB-60 swept back notably, as did the tailplane. The maximum designed bomb load of 72,000 pounds was to be carried on a range of 6,200 miles. Defensive armament would have consisted of ten 20mm cannon installed in retractable tail turrets and carrying 3,600 rounds of ammunition. Accommodations were provided for an eleven-man crew.

Only two YB-60 prototypes were built, since the aircraft was rejected in favor of the much-faster B-52 that had a top speed of 660 mph. Accordingly, the YB-60 was never loaded with its bombs, armed with its weapons, or manned with its full crew. Nevertheless, that giant of the airways contributed valuable scientific information about jet propulsion in extremely large aircraft.

The YB-60 represented a concluding stage in distinguishing the B-36 family as possessing the sturdiest airframe ever designed, having originally supported six piston engines in the B-36B, accommodated ten engines in the B-36D mixture of six piston engines and four jet boosters, and been fitted with eight turbojets in the all-jet YB-60.

B-61 MATADOR

When the pilotless missile program began at Holloman AFB, New Mexico, in the late 1940's, it was decided that missiles should fall into the same category as manned Bombers and designated by the ''B'' for Bomber classification. The first of those unmanned vehicles was the Martin B-61 Matador Tactical Guided Missile that was first flown in December 1950 and began tactical tests on June 21, 1951. The Matador was equipped with both a rocket booster and an Allison turbojet engine. It blasted off under rocket power from a mobile launcher; after it was airborne and a critical speed had been reached, the rocket automatically fell off and the jet engine was activated. With a span of 28'7" and a length of 39'6" the B-61 looked more like a flying tube than an airplane. The small, high wing swept back, hugging the body, and the rear assembly had a tiny tailplane mounted at the top of the single fin. The Matador had a speed of 600 mph with a range of 620 miles and carried either a conventional or nuclear warhead.

The first Pilotless Bomber Squadron of the Tactical Air Command was formed in October 1951 at Patrick AFB, Florida, equipped with B-61 Matadors for training personnel in that revolutionary new means of warfare. In March 1954 two Pilotless Bomber Squadrons were sent to West Germany to bolster the deterrent forces in the United States Armed Forces Europe (USAFE). By 1955 the B-61 and its contemporary unmanned missiles had been redesignated into their proper classifications as ground-launched medium-range Tactical Guided Missiles.

Other unmanned vehicles that received Bomber designations before being reclassified were: the Northrop B-62 Snark, Bell B-63 Rascal, North American B-64 Navaho, Convair B-65 Atlas, Radioplane B-67 Crossbow, Martin SB-68 Titan, Lockheed RB-69 Neptune, McDonnell B-72 Quail, Douglas B-75 Thor, Martin B-76 Mace, North American B-77 Hound Dog, Chrysler B-78 Jupiter, Boeing B-80 Minuteman, Martin B-83 Bullpup, and the Douglas B-87 Skybolt that turned out to have the highest number assigned in the ''B'' category.

B-66 DESTROYER

The Douglas B-66 Destroyer Tactical Bomber didn't have a prototype as such because it was a direct modification from the Navy A-3D Skywarrior program. The first production model was the RB-66A Photo-Recon version that was assigned to the Tactical Air Command in June 1954 for purposes of familiarization and training.

The B-66B was the only variant built strictly as a Bomber. The first B-66B took to the air in January 1955, and deliveries to TAC units began in March 1956 as replacements for the North American B-45 Tornado. The B-66B was equipped with two Allison J-71 turbojet engines delivering 10,200 pounds of thrust each to produce a top speed of 630 mph. A bomb load of 15,000 pounds could be carried on a combat range of 1,500 miles, or 10,000 pounds for a distance of 1,800 miles. Defensive armament consisted of two

Douglas B-66 Destroyer, a tactical jet Bomber of the Vietnam War era. (USAF Photo 155149-AC)

M-24A 20mm tail guns. The crew of three was composed of pilot, navigator-bombardier and defensive systems operator. The radar system provided instant warning of approaching aircraft, tracked the enemy, took aim and automatically fired the two cannon remotely located in the tail turret. A B-66B participated in operation "Redwing," the dropping of a hydrogen bomb at Bikini Atoll in May 1956. During July-August 1958 Destroyers joined a 150-plane Tactical Composite Strike Force sent to Adana, Turkey, as support for Marines flown to Beirut, Lebanon, by C-130 Hercules Cargo Transports to protect Middle East interests threatened by war in Lebanon.

The RB-66B was designed for the Photo-Recon role, but was also convertible to a Bomber when needed. It was equipped with the very best photographic and detection devices of the day—the costs of some items were staggering. RB-66B's served with TAC in Europe, supporting the peace-keeping force by flying day and night surveillance missions along the Communist borders. The RB-66C was a specialized Photo-Recon version with radomes on the wingtips, a K-38 day camera, three K-46 night cameras and photo-flash bombs. The regular crew of three was augmented by a 4-man team of electronics experts housed in a compartment where the bomb-bay was normally located. Joining the war in Southeast Asia, RB-66C's operated in the role of Electronics Reconnaissance Escort on strike missions over North Vietnam. The WB-66D was the first aircraft in history designed specifically for Weather Recce duty. Its five-man crew included the three regular members plus two airborne weather observers.

A total of 295 Destroyers were delivered to the Air Force, seventy-two of which were B-66B Bombers.

XB-70 VALKYRIE

Development of the North American XB-70 Valkyrie began in the late 1950's when Pentagon officials submitted specifications for an aircraft to replace the Boeing B-52

Stratofortress. Designed as a high-flying triple-sonic Heavy Bomber capable of easily cruising higher than Russian Interceptors could climb, and safely above the range of Soviet Surface-to-Air Missiles (SAM), the Valkyrie was one of the most impressive aircraft ever built. On May 1, 1960, while all the hard work and research on the XB-70 were going at a feverish pitch, Gary Powers and his U-2 Strategic Reconnaissance plane were shot down from the Stratosphere over Communist territory. That unanticipated event shattered our dreams that the XB-70 could fly unmolested over enemy terrain. A conflict erupted among the Department of Defense, the Department of the Air Force, Congress and the President concerning the feasibility of continuing the B-70 program, especially in the face of rapidly accelerating construction costs.

Pentagon research director Harold Brown recommended cancellation of the B-70 project and President John Kennedy was forced into the same kind of decision that President Jimmy Carter would have to face with the Rockwell International B-1 Bomber sixteen years later. In March 1961, Mr. Kennedy stated that America's missile capability at that time and in the foreseeable future made it impractical and economically unjustifiable to develop the B-70 as a complete Weapons System. Sound familiar?

In March 1964 the decision was made to build only two XB-70 prototypes, the first of which took to the air on September 21, 1964. On October 14, 1964, the Valkyrie exceeded Mach-3 at an altitude of 70,000 feet. On June 8, 1965, one of the prototypes crashed during a test flight, killing the entire crew. The surviving XB-70 was subsequently retired and placed on display at the Wright-Patterson Aviation Museum.

Powered by six General Electric YJ93-GE turbojet engines generating 19,500 pounds of thrust each, the XB-70 had a top speed of 2,300 mph. The crew of four was composed of pilot, copilot, bombardier-navigator and defensive systems operator. It was often referred to as "tailfirst" because the elevons were installed on a foreplane up by the cockpit and the delta wings were at the rear of the fuselage. Remember the Wright Brothers' first aeroplane at Kitty Hawk? It had the elevators (empennage) forward of the wings also. The XB-70 was distinguished as having the highest number assigned to a manned Bomber in the "B" category before the tri-service scheme of designating aircraft started over at the beginning of the list in 1962.

The name "Valkyrie" translates literally into "chooser of the slain," and was taken from the mythological Norse maiden of Odin named Valkyrie, who chose the heroes to be slain in battle and conducted them to Valhalla. The Epilogue to this Saga is, "the modern XB-70 Valkyrie did not live long enough to choose her heroes to be slain in battle."

JET BOMBERS — 1962 TO THE PRESENT

B-1 STRATEGIC BOMBER

The Rockwell International B-1 Strategic Bomber was conceived in 1971 as a replacement for the Boeing B-52 Stratofortress that had already been on active duty for over fifteen years. The XB-1 prototype made its first flight on December 23, 1974, in answer to requests for a Bomber with variable-sweep wings that could fly in excess of Mach-2 at optimum altitude and fly faster than Mach-1 while avoiding radar detection on low-level Strike missions. The XB-1 first exceeded the speed of sound on April 26, 1975, and proceded to justify its claims as the best Bomber in history. Airframe comparisons showed the B-1 to be about two-thirds the size of a B-52, while being able to carry a conventional or nuclear bomb load nearly double that of the Stratofortress. Engine comparisons revealed that one of the B-1 turbofans generated approximately twice the thrust of a B-58 Hustler turbojet, with a twenty percent fuel economy to boot. Wings on the B-1 had a forward sweep of fifteen degrees at full extension during takeoff or landing, and a backward sweep of 68 degrees to form a delta shape while cruising at altitude.

Almost from the beginning the B-1 program was a political football. Opponents won the first round by succeeding in temporarily shelving the project, using as their strongest argument the fact that a five-inch crack developed in a section of the prototype. Rockwell designers defended themselves by showing that the crack appeared under abnormally extreme stress during an intentionally severe laboratory test; engineers later proved that they had permanently solved the problem.

In May 1976 the ''nays'' pushed through an amendment delaying decision on the B-1's destiny until after the winner of the coming Presidential election would be inaugurated— the bill also included a proviso that allowed extension of the final decision until July 1, 1977. Then, in December 1976, the outgoing lame-duck ''ayes'' attempted to force the issue by awarding $704.9 million in contracts spread out among Rockwell who constructed the airframes, General Electric who built the engines, and Boeing who supplied the electronics equipment. Options were proffered to build eight more aircraft, and promises were made that it might be possible to contemplate full production. Air Force dreams at that time called for 244 B-1 Bombers at a cost of roughly $100 million per aircraft. Every time the program was deferred, the cost-per-plane took a big jump.

On June 30, 1977, President Jimmy Carter, fulfilling a major campaign promise, put an end to the B-1 project. Despite the fact that Strategic Arms Limitations Treaty (SALT) talks with Russia were going badly, Mr. Carter came to the conclusion that the $24.8

North American-Rockwell B-1 Strategic Bomber of the 1970's that was deferred as a replacement for aging B-52's. Caught here over Edwards AFB, California, in 1976.

billion price tag was too much of a burden for the taxpayers to bear. Presenting an alternative reason for his decision, the President revealed that Cruise Missiles, which can be directed to targets at low altitudes through their own computerized guidance systems, had become effective, cheaper alternatives. At the time, Cruise Missiles could be built for half a million dollars apiece. And too, at that time he knew of the existence of the "MX" Missile, a mighty weapon of the future.

The B-1 had a bomb-load capacity of 115,000 pounds and a maximum range of 6,100 miles. Four General Electric YF101-GE turbofan engines generated 17,000 pounds of thrust each, which could be boosted to 30,000 pounds each with afterburner. Top speed was 1,400 mph. The four-man crew included pilot, copilot, offensive weapons engineer and defensive weapons engineer. Two additional fold-down seats accommodated technical advisers or trainees. Defensive technology relied on electronics counter-measures and counter-countermeasures equipment such as warning, jamming, homing, chaff and infra-red devices. Only four prototypes had been built when the program was deferred. Authority was granted to continue research, using those four prototypes for experimental purposes.

FB-111 STRATEGIC BOMBER

The General Dynamics F-111 Tactical Fighter (TFX) was delivered to the Air Force in April 1965 as a response to a joint Air Force-Navy request for a multi-purpose aircraft with variable-geometry wings—meaning that the wings are extended straight out during takeoff or landing and sweep back against the fuselage to form a delta shape when airborne.

The General Dynamics XFB-111A prototype made its first flight in July 1967 as a direct development from the F-111 Tactical Fighter. Designed with funds for the war in Vietnam as a supplement to the Convair B-58 Hustler and as relief for the overworked and aging Boeing B-52 Stratofortress, the FB-111A Strategic Bomber production model was first delivered in October 1969 to the 340th Bombardment Training Group (SAC) at Carswell AFB, Texas. After sufficient crews had been trained at Carswell, deliveries were made to the 509th Bomb Wing, Pease AFB, New Hampshire, and the 380th Strategic Aerospace Wing at Plattsburg AFB, New York.

The FB-111A carries a conventional bomb load of 31,500 pounds in an internal bomb bay and on external racks underneath for an unrefueled range of 3,100 miles. Alternate ordnance loads can be varied among six nuclear weapons or six Short Range Attack Missiles (SRAM). Two Pratt & Whitney TF-30 turbofans supply 18,500 pounds of thrust each to produce a top speed of 1,450 mph. The crew of two is composed of pilot and systems engineer who sit side by side in a jettisonable cockpit which is a self-propelling capsule that acts as a survival shelter upon landing in water or on the ground. The FB-111A carries no defensive weapons, relying on its speed to escape enemy Interceptors.

Production was completed in 1971 after 76 FB-111A's had been delivered to the USAF. In 1979 it was proposed that the 76 active duty FB-111A's plus an additional 75 F-111D Fighters be converted to FB-111B configuration by widening the span, increasing the fuel capacity and installing new engines.

STEALTH AIRCRAFT

On August 22, 1980, after rumors of a new Fighter-Bomber-Reconnaissance-Strike aircraft leaked out during the Democratic Convention in New York City, Secretary of Defense Harold Brown more or less confirmed that a modernized combat plane was being developed under the code name of "Project Stealth." No details are available because Stealth technology is of a nature to have been considered for the new "Royal Secret" security classification, a category proposed in 1980 to designate material so sensitive that only a handful of high-level dignitaries are allowed access to it.

CARGO TRANSPORTS — 1919 TO 1962

T—TRANSPORT: 1919 TO 1925

The "T" for Transport category was introduced in the 1919 scheme of Army airplane designations proposed for use by the Army Air Service, which became operational on June 4, 1920, under the Command of Major General Charles Menoher.

The Martin T-1 cabinplane in 1919 was the Army's first officially designated Transport aircraft. It was a modification of the biplane MB-1 Martin Bomber with twin Liberty-12 400-hp engines and a top speed of 105 mph. Conversion from the Bomber version involved removing the armament, expanding the fuselage to make room for walking passengers, installing five windows on each side of the fuselage and mounting ten passenger seats in the cabin. Only one of the T-1's was delivered for service.

The Fokker T-2 personnel carrier in 1922 was purchased from the Fokker factory in Holland as the Army's first monoplane Transport. With a wingspan of almost 80 feet the T-2 was the world's largest passenger plane. A single Liberty-12 420-hp engine gave it a

The Fokker T-2 Transport was the Army's pioneer monoplane passenger carrier. It was flown on history's first non-stop cross-country flight in May 1923. (USAF Photo 18760-AC)

top speed of 110 mph. Only two of the T-2's were procured for the Army Air Service. One of them was modified to carry two litter patients and redesignated A-2 for use as an airborne hospital in the "A" for Ambulance category. Both planes were accepted in 1923.

On October 5-6, 1922, Lts. John MacReady and Oakley Kelly established a world endurance record of 35 hours and 19 minutes in a Fokker T-2 out of Rockwell Field, California. Extra fuel tanks were added to the T-2 for the first nonstop cross-country flight in history on May 2-3, 1923, piloted by Lts. MacReady and Kelly, who flew from Roosevelt Field, New York, to Rockwell Field, San Diego, California, a distance of 2,700 miles in 26 hours 51 minutes at an average speed of a little over 100 mph.

The Lowe-Willard-Fowler (LWF) T-3 Transport in 1924 was the last to carry a "T" designation. Only one LWF T-3 was built for Army trials. The project was cancelled because the favored Douglas C-1 Cargo Transport was already under construction and about to start rolling out of the factory.

C—CARGO TRANSPORT: 1925 TO 1962

In 1925 the "T" category was replaced by the "C" for Cargo Transport designation in the new scheme for the Army Air Corps that was to become operational on July 2, 1926.

The **Douglas C-1** in 1925 was the Army's first Cargo and Personnel Transport. Designed from the Douglas O-2C Observation biplane, the C-1 had a Liberty 400-hp engine and a top speed of 115 mph. Modification from the 0-2 involved swapping two cockpits

A Douglas C-1 biplane Cargo-Tanker, the first Army Transport to carry a "C" designation, is about to make a refueling hookup with a Fokker C-2 Trimotor, the Air Corps' first "C" designated monoplane, during an endurance flight above Bolling Field, DC, on December 17, 1928. (USAF Photo 7811-AS)

Ford C-4 Trimotor similar to Admiral Byrd's South Pole-crossing plane. Photographed here is Assistant Secretary of War for Air F. Trubee Davison's C-4A. (USAF Photo 17755-AC)

for one, enlarging the fuselage girth, and installing eight or nine seats in the cabin. Pilot and copilot occupied open side-by-side seats in the new cockpit located forward of the wings. Production totaled 27 C-1's for use by the Air Corps.

The **Fokker C-2 Trimotor Cargo Transport** was purchased in 1926 from the Fokker plant in Holland as the first 3-engined high-wing monoplane in the "C" category. Three Wright R-790 220-hp engines gave it a top speed of 110 mph. Accommodations were provided for pilot, copilot and ten passengers. Eleven C-2's were produced for the Army Air Corps.

On May 9, 1926, Commander Richard E. Byrd made the first crossing of the North Pole in a Fokker Trimotor piloted by Floyd Bennett. The trip was financed jointly by John D. Rockefeller, Jr., and Edsel Ford: the plane was named "Josephine Ford" after Edsel's daughter. On June 28, 1927, a C-2 named "Bird of Paradise" brought fame to the Air Corps by making the first flight in history from Oakland, California, to Hawaii, a distance of 2,400 miles in 25 hours 49 minutes at an average speed of 93 mph; Lts. Les Maitland and Al Hegenberger were at the controls. During January 1-7, 1929, a C-2 named "Question Mark," flown by Carl Spaatz, Ira Eaker, Elwood Quesada and Harry Halverson established a world endurance record of 150.67 hours, being aided by air refuelings from a Douglas C-1 Cargo-Tanker. On March 31, 1931, Knute Rockne, famous Notre Dame football coach, was killed when the Fokker Trimotor civilian airliner in which he was traveling crashed over southeastern Kansas.

The **Ford C-3 Trimotor Cargo Transport** in 1928 was designed from the "Tin Goose" commercial airliner, a pioneer in all-metal airframes in Detroit, Michigan.

The **Ford C-4 Trimotor** in 1929, developed directly from the C-3, was a high-wing cabinplane with three Wright R-1340 450-hp engines and a top speed of 145 mph. One engine was installed in the nose and two outer engines were mounted in the wings. The

3-man crew sat in a slightly elevated cockpit forward of the cabin. Five individual windows along each side provided a full view for ten passengers. In 1930 all of the older C-3's were given new engines and redesignated C-9 Cargo Transports. Production of the Ford Trimotors for the Air Corps totaled 13 aircraft designated C-3, C-4 and C-9 Cargo Transports.

Practically all of the early commercial airlines employed their Ford Trimotors as trail blazers in developing air routes around the Northern Hemisphere. At 8:55 on the morning of November 29, 1928, Admiral Richard E. Byrd made the first crossing of the South Pole in a Ford Trimotor piloted by Bernt Balchen. The plane was named "Floyd Bennett" in honor of Byrd's old friend and pilot who had died of pneumonia earlier that year.

The **Sikorsky C-6** in 1929 was the first Amphibian Cargo Transport in Air Corps history. It was a speedy twin-engine twin-fin high-wing monoplane with a top speed of 185 mph. Twelve seats were provided for crew and passengers. Eleven C-6's were delivered to the Air Corps for use in the Philippines and Panama on coastal patrols and in the roles of transport and target tug.

The **Fairchild C-8** was initially delivered as the **F-1** in 1930 for use as the first officially designated Photography plane in Air Corps history. In 1930 the "F" for Photography category was introduced into the Air Corps scheme of aircraft designations. The F-1 was a single-engine high-wing cabinplane with conventional 3-wheel landing gear. The Pratt & Whitney R-1340 410-hp engine produced a top speed of 140 mph. The wing was unique in that it folded backward on each side for stowage through narrow hangar doors. When used on Photography missions the F-1 carried a crew of three, including pilot, copilot and photographer. Later in 1930 all F-1's were redesignated C-8's, but continued to be used in

Sikorsky C-6 Amphibian, first Air Corps Amphibian Cargo Transport. (USAF Photo 157724-AC)

The Douglas C-21 (OA-3) Dolphin was Douglas' first Amphibian Cargo Transport. It was redesignated OA-3 Observation Amphibian in 1933. Photo shows General Drum's OA-3 over Oahu, Hawaii, on April 22, 1935. (USAF Photo 38594)

the Photography role in addition to performing as Cargo Transports with accommodations for a crew of two and five passengers. Production totaled 15 aircraft designated F-1 and C-8 for the Air Corps.

The **Fokker YC-14** in 1931 was used in both the roles of Cargo Transport and Airborne Ambulance. It was a single-engine high-parasol-wing monoplane with conventional landing gear and a standard tail section. The term "parasol" applied to a high-wing that was mounted on braces and elevated above the fuselage, resembling an opened umbrella, rather than being welded into the airframe. The Wright R-1750 525-hp engine gave it a top speed of 135 mph. Pilot and copilot sat in the forward cockpit and six passengers were seated in the cabin. A total of 20 YC-14's were delivered to the Air Corps. One example was converted into the XC-15 for use as an Airborne Ambulance. The prefix "Y" was added to the designation scheme in 1928 to indicate that the aircraft was ordered into Limited Production for service trials—generally, the number of aircraft ordered had to be 13 or less in order to qualify for the "Y" category.

In July 1931 responsibility for coastal patrol of the United States was transferred from the Navy to the Army Air Corps.

The **Douglas C-21 Dolphin**—the first Douglas Amphibian—was developed in 1932 when the Air Corps needed help with its new coastal defense responsibilities. The C-21 was a high-wing monoplane with two Wright R-975 300-hp engines and a top speed of 140 mph. The close-together engines were supported by studs protruding upward from the wing. Fixed wheels were installed in the long center float, and a smaller balancing skid-float was fitted under each wingtip. The tail assembly had a tailplane mounted about one-third the way up the fin. Seven seats were provided for crew and passengers. Eight C-21's were delivered to the Air Corps, and in 1933 all of them were redesignated OA-3 Observation Amphibians.

The **Douglas C-26 Dolphin** in 1932 was like the C-21 except for two Pratt & Whitney R-985 350-hp engines and eight seats for crew and passengers. Fourteen of the C-26's were built for the Air Corps, and in 1934 seven of them were redesignated OA-4 Amphibians.

The **Bellanca C-27 Airbus** in 1932 was a civilian airliner adapted to military standards. It was a single-engine high-wing monoplane with three fixed wheels and a standard tail assembly. Streamlined fairings on the front legs were uniquely mounted by V-type braces that had one end attached to the fuselage and the other end fitted underneath the wings. The Pratt & Whitney R-1860 550-hp or Wright R-1820 750-hp engines were equipped with a 3-bladed prop. Twelve seats were provided for crew and passengers, and a large cargo door allowed easy loading of freight packages. A total of 14 C-27's were built for the Air Corps.

On July 8, 1929, a civilian Bellanca named "Pathfinder," manned by Roger Williams and Lew Yancey, flew from Old Orchard Beach, Maine, to Rome, Italy, stopping over along the way at Santander, Spain, for fuel. On July 28-30, 1931, Russell Boardman and John Polando, flying a Bellanca named "Cape Cod" from Floyd Bennett Field, N.Y., to Istanbul, Turkey, set a non-stop distance record of 5,011 miles in 49 hours. On October 4-6, 1931, Clyde Pangborn and Hugh Herndon made the first non-stop crossing of the Pacific Ocean from Sabishiro, Japan, to Wenatchee, Washington, in a Bellanca named "Miss Veedol," completing the 4,465-mile flight in 41 hours and 13 minutes.

The **Douglas C-29 Dolphin** in 1933 was a direct modification from the C-26B Dolphin with twin Pratt & Whitney R-1340 575-hp engines and a top speed of 150 mph. Otherwise it was exactly like the C-26, which was, itself, a derivative from the C-21 Dolphin. Only two C-29's were built for the Air Corps. In July 1935 Hezekiah McClellan flew a C-29 Amphibian from Nome, Alaska, on a hop to Point Barrow, thus becoming the first Air Corps plane to land at that northernmost site.

The **Curtiss C-30 Condor** in 1933 was a development from the Curtiss B-2 Bomber. It was a standard biplane with conventional landing gear and a twin-finned tail section. Two Wright R-1820 675-hp engines gave it a top speed of 130 mph. The 3-man crew included pilot, copilot and crew chief. Fifteen passengers could be accommodated in the cabin. Two C-30's were delivered to the Air Corps.

In October 1933 Admiral Byrd used a commercial Condor Transport to explore 450,000 square miles of ice-covered terrain on his second Antarctic expedition. Harold June was Byrd's pilot on that survey trip.

The **Douglas C-32** in 1936 was the first Air Corps Cargo Transport from the famous "DC" series of civilian airliners. The "DC" family began with development of a single DC-1 Douglas Commercial that made its first flight on July 1, 1933. The DC-2 followed in April 1934, and when the Air Corps received a generous allocation of funds from Congress in Fiscal 1936, the first thing they did was purchase a DC-2 and designate it the C-32: thus was born the original forefather of the famous C-47 Gooney Bird. The C-32 was a twin-engine low-wing personnel Transport with a crew of three, including pilot, copilot and crew chief. The C-32 accommodated fourteen passengers or a cargo load of

Douglas C-32 Cargo Transport, first of the famous Douglas "DC" commercial transports adapted for the military. (USAF Photo G-1061-2-8526)

4,000 pounds. It had a range of 900 miles and a top speed of 180 mph. In 1942 the USAAF ordered 24 more aircraft as C-32A's for use in WWII.

the **Douglas C-33** on that same order in 1936 was a DC-2 with more powerful engines, a cargo capacity of 5,000 pounds and a top speed of 205 mph; it was the first Air Corps Cargo Transport that could fly faster than 200 mph. Otherwise it was like the C-32 in appearance and accommodations. Eighteen C-33's were delivered to the Air Corps.

The **Douglas YC-34**, also on the same 1936 order, was merely a C-32 with different internal arrangements providing for sixteen passengers. Only two of the YC-34's were built for the Air Corps.

The **Lockheed XC-35** in 1936 was a single example developed from the civilian Model-10A Electra strictly for experimental purposes. In 1937 the XC-35 was awarded the Collier Trophy for its incorporation of research work on high altitude pressurization and engine super-charging. The prefix "X" was introduced into the 1925 scheme of designations to indicate that the aircraft was an experimental prototype.

The **Lockheed C-36 Electra** from the Model-10A in 1937 was the forefather of the Lockheed family of military twin-finned Cargo Transports. It was very similar to Amelia Earhart's famous around-the-world low-wing light cabinplane with twin 400-hp engines and a top speed of 205 mph. Pilot and copilot sat in the forward cockpit and eight passengers were seated in the cabin. The Lockheed C-37 Electra was like the C-36 except

Lockheed C-36 Electra, first of the Lockheed twin-finned Transports. Similar to Amelia Earhart's civilian Electra in which she disappeared. (USAF Photo 13488)

for two 450-hp engines and a top speed of 220 mph. A total of thirty C-36 and C-37 Electras were delivered to the Air Corps. In January 1943 the "UC" for Utility Cargo Transport was introduced to classify all aircraft having a cargo capacity of less than 1,500 pounds and/or accommodations for less than ten passengers. Therefore, in January 1943 all C-36 and C-37 Electras became UC-36 and UC-37 Electras.

At ten o'clock on the morning of July 1, 1937, Amelia Earhart and Fred Noonan— reputed one of the best navigators in the country—took off from Lae, New Guinea, for Howland Island, a refueling stop 2,556 miles away. The next segment was to take them to Hawaii where they would begin the final leg home to California, but that dream was destined not to come true. Howland, a flyspeck in the vast Pacific, was a United States possession lying close to the Japanese mandated islands of Micronesia. After covering three-fourths of the 27,000 miles around the world, those two pioneer flyers and their airplane disappeared, never to be heard from again. The last radio contact was made at 8:44AM on July 2, 1937, a faint, garbled transmission. Exactly what happened on that flight remains shrouded in a cloak of intrigue. Five days later, on July 7, 1937, Japan invaded China as the first act of aggression prior to World War II.

The **Douglas DC-3** made its first flight on December 22, 1935, as an advanced version of the DC-2 with new engines, an expanded fuselage girth, different internal arrangements and availability of a large cargo door for loading heavy packages of freight.

The **Douglas C-38** in 1937 was called a DC-2½ because it had the wings of a DC-2 and the fuselage of a DC-3. A single example of the C-38 served as an experimental model for the later C-39.

The **Douglas C-39** delivered to the Air Corps in 1939 was also referred to as a DC-2½ because it had the DC-2 wings and the DC-3 fuselage with double-sized cargo door for loading large packages of freight. Two Wright R-1820 975-hp engines gave it a top speed of 210 mph and a range of 925 miles. Sixteen passengers, or a cargo load of 6,000 pounds, or a combination of the two could be carried in the fuselage cabin. Production totaled 35 C-39's for the Air Corps.

Immediately caught up in the war during Japanese attacks in the Philippines on December 8, 1941, C-39's heroically evacuated many surviving troops to Australia. In 1942 C-39's joined operation "Bolero," the hauling of personnel, supplies and equipment to the British Isles for a planned invasion of fortress Europe. In support of that "sky-train," the C-39 served as a shuttle between New England and Goose Bay, Labrador, the first leg of the treacherous North Atlantic route.

The **Lockheed C-40 Electra** in 1938 was developed from the civilian Model-12 light cabinplane. The C-40 was a low-wing monoplane with conventional 3-wheeled landing gear and the classic Lockheed twin-finned tail section. Twin Pratt & Whitney R-985 450-hp engines gave it a top speed of 220 mph. Pilot and copilot sat in the forward cockpit and five to seven passengers rode in the cabin. Production totaled 24 of the C-40's for the Air Corps. In January 1943 the C-40 was redesignated UC-40 to put it in the Utility Cargo Transport category.

In July 1941 the USAAF inventory listed only 216 Cargo Transports. In December 1941 after America entered World War II, the USAAF was granted authority to commandeer civilian aircraft for wartime duty. The mass influx of men and women into military service was causing logjams in transportation facilities all across America. Aircraft factories in the United States ceased their civilian work and volunteered an all-out effort toward converting and producing aircraft for transporting personnel, troops, equipment and supplies around the globe. The Faddis Act provided military officials with a legal instrument to stop any defense materials slated for overseas when they were deemed vital to this nation's survival.

The **Beechcraft C-43 Traveler** was the USAAF designation assigned to 158 civilian Model-17 light cabin biplanes commandeered for service in December 1941. Orders were submitted in 1942 for an additional 207 production model C-43's. Often referred to as the "Staggerwing" because of the negative tilt of the top wing backward from the lower wing, the C-43 had a standard tail section and conventional retractable landing gear with streamlined pants on the front legs. A single Pratt & Whitney R-985 450-hp engine gave it a top speed of 210 mph and a maximum range of 1,400 miles. Accommodations were provided for pilot and 3 or 4 passengers with 125 pounds of luggage. A total of 458 aircraft were delivered, including 365 USAAF C-43's, sixty-three Navy GB-1/2's and 30 for Lend-Lease to Britain. In January 1943 the C-43 became the UC-43 to put it in the new Utility category.

Beech Staggerwing Travelers served mostly within the continental limits of the United States, but after D-Day a few of them were sent to the European Theater of Operations (ETO) for use in administrative duties.

The **Beechcraft C-45 Expediter-Bugsmasher**, developed from the civilian Model-18 light cabinplane in 1940, became the most popular light Cargo Transport of WWII. Referred to affectionately by the men who flew it as the "Bugsmasher," the C-45 was a

Beechcraft C-45 Expediter, popular WWII and postwar light Cargo Transport nicknamed the "Bugsmasher." Pictured while parked at March AFB, California, on March 1, 1956. (USAF Photo C-155272-AC)

low-wing monoplane with conventional 3-wheeled landing gear and a twin-fin tail section. Pilot and copilot sat in a forward compartment, and the passenger cabin was designed for six people, but arrangements could be made to carry a few more. One baggage compartment was in the nose, and another hold plus lavatory was behind the main cabin. The F-2 Photo-Recon version was specially built for a 3-man crew with auto-pilot and 8 cameras for mapping of large areas during day or night. The AT-7 variant was a Navigator Trainer and the AT-11 was a Bombardier-Gunner Trainer. Navy versions were designated JRB-1/2. In January 1943 the C-45 became the UC-45 Utility Cargo Transport.

In 1948 the F-2 Photo-Recon planes became RC-45's and Utility Trainers became TC-45's. In 1949-50 approximately 900 WWII AT-7 and AT-11 Trainers were remanufactured as UC-45's. A total of 2,645 Expediters were delivered for service, and the last of them were retired in 1964 with 24 years of longevity.

The **Curtiss C-46** prototype was originally designated XC-55 at the time of its first flight on March 26, 1940, but by the time of first delivery in 1942 it had been redesignated as the C-46 Commando. The C-46 was a low-wing monoplane with retractable landing gear and a standard tail section. The 3- to 4-man crew included pilot and copilot in the cockpit up front and a roving crew chief plus loader or attendant when needed. Two Pratt & Whitney R-2800 2,000-hp engines drove either 3 or 4-bladed props to give it a top speed of 260 mph. Four windows along each side of the fuselage provided an outside view for 50 fully equipped combat troops. As a hospital ship it could accommodate 33 litter patients. A cargo load of 16,000 pounds could be carried and the ferry range stretched out to 2,990 miles with a full load of fuel. The normal cargo load for middle distances was 8,000 pounds.

The C-46 was produced in an array of minor and not-so-minor variations, and do-it-yourself field designers established for it the reputation of being the most-modified aircraft in World War II. Commandos were flown on all fronts by the USAAF, Navy and

Marines, but their claim to fame came while flying men, materiel and supplies from India across the Himalayan "Hump" into China. C-46's towed Assault Gliders on daring raids behind enemy lines. One version flown by pilots of the Chinese Air Force had gun ports in each of the 8 fuselage windows. Production totaled 3,180 C-46 Commandos for use in WWII.

After WWII many of the Commandos were sold as surplus to American airlines and to friendly nations abroad. Some USAF C-46's remained operational for use by Republic of Korea forces in the Korean war. When the Vietnam war intensified, C-46's were recommissioned for service and returned to the Air Force inventory for use by the 1st Air Commando Group. That action gave the Commando the distinction of having fought in three wars alongside the Douglas C-47 Gooney Bird, Douglas B-26 Invader and North American P-51 Mustang.

Curtiss C-46 Commando, largest twin-engined Cargo Transport of WWII. Achieved prominence flying the notorious China-Burma "Hump." (USAF Photo 21407)

The **Douglas C-47 Skytrain-Gooney Bird-Dakota** was developed from the Douglas DC-3 that made its first flight on December 22, 1935. First deliveries of the C-47 began to arrive at USAAF bases in January 1942. The DC-3 was named the Skytrain, the USAAF C-47 was nicknamed the Gooney Bird and the British named their version the Dakota. The C-47 was a low-wing monoplane having a standard tail section, conventional landing gear with retractable front wheels and seven windows along each side of the fuselage. Two Pratt & Whitney R-1830 1,200-hp engines drove 3-bladed props to produce a top speed of 220 mph and a maximum range of 1,500 miles. The 3-man crew consisted of pilot and copilot in the flight compartment near the nose and a trouble-shooting crew chief. Bench seats along each side of the fuselage could be stuffed with up to 32 passengers or 27 fully equipped combat troops. As a hospital ship it could accommodate 24 stretchers. The oversized cargo door allowed loading of large freight packages up to the aircraft's normal capacity of 6,000 pounds—but the Gooney Bird was known to haul twice that much cargo on short hops. The TC-47 was a Navigator Trainer.

Douglas C-47 Skytrain-Gooney Bird-Dakota. World renowned Cargo Transport that achieved undying fame for service in WWII, Korea, and Vietnam. (USAF Photo 32401)

The DC-3/C-47 series is the most famous Cargo Transport family ever built. While serving in over 40 countries around the globe, they have flown more miles, hauled more freight and carried more passengers than any other aircraft in history. During WWII the C-47 served in the roles of passenger airliner, troop carrier, cargo hauler, airborne ambulance, air-sea rescue, assault glider tug and VIP transport. Gooney birds flew the Himalayan "Hump" to supply the Flying Tigers at Kunming, participated in operation "Bolero," carrying men and materiel over the treacherous North Atlantic route to Britain, hauled liquor to the Officers in the Aleutian Islands, toted steel girders for a radio tower at Bluie West One, Narsasuak, ferried jackasses over the Pyramids of Egypt, furnished chow and boots to the Chindits in Burma, and once delivered a full load of Kotex to a battalion of troops at the front.

On November 8, 1942, C-47's out of England dropped paratroops during the invasion of Northwest Africa in the first American airborne assault of the war. On July 9, 1943, 370 Gooney Birds of the 51st and 52nd Troop Carrier Wings released gliders and dropped paratroops during the invasion of Sicily in the first use of Waco CG-4A Assault Gliders. On June 6, 1944, C-47's towed American assault troops in British-built Horsa Invasion Gliders during the D-Day landings in Normandy. On August 15, 1944, C-47's of the 12th Air Force dropped troops and supplies during the invasion of France from the south at Nice, Toulon and Marseilles on the Mediterranean coast. In September 1944, C-47's participated in dropping 35,000 troops behind enemy lines in Holland. During the Battle of the Bulge in Belgium during December 1944, C-47's of the 9th Troop Carrier Squadron dropped supplies and equipment to American troops surrounded at Bastogne.

On August 15, 1943, two C-47's were shot down while carrying men and equipment of the 35th Fighter Group to the forward base at Maralinan, New Guinea. On September 5, 1943, 79 Gooney Birds of the 54th Troop Carrier Wing out of Port Moresby, New Guinea, dropped 1,700 paratroops of the 503rd Regiment at Nadzab, New Guinea, in the first airborne assault of the Southwest Pacific offensive. On March 5, 1944, C-47's opened the Allied counter-offensive in North Burma by dropping Wingate's Raiders in gliders behind enemy lines at "Broadway," a clearing in the jungle 50 miles northeast of

Indaw. Nine days later General Wingate was killed in a plane crash near Broadway. On February 16, 1945, C-47's dropped 2,065 paratroopers at Corregidor in the Philippines.

General Eisenhower once commented, ''Four things won the Second World War—the Bazooka, the Jeep, the Atom-Bomb and the C-47 Gooney Bird.''

But the C-47's career did not end with World War II. In the Berlin Airlift of 1948-49, brought about by a Russian blockade of the capital city, 102 C-47's joined the fleet of airliners that flew 277,000 sorties and delivered two and a third million tons of supplies to defeat the Soviet tactic. When the Korean conflict broke out on June 25, 1950, Gooney Birds were the first to volunteer for front-line duty. On May 3, 1952, a C-47 fitted with skis made the first landing at the North Pole.

In 1965, when American jet Fighters were found to be too fast for some of the anti-guerilla jungle operations in Vietnam, Gooney Birds were fitted with three 7.62mm Miniguns poking out of the left rear windows, each having a firing rate of 6,000 rounds per minute. Officially designated the AC-47 Gunship, it carried 54,000 rounds of ammo to earn the affectionate nickname of ''Puff the Magic Dragon'' because of its firepower.

The C-47 Gooney Bird, C-46 Commando, B-26 Invader and P-51 Mustang were distinguished as serving side by side through three war periods—WWII, Korea and Vietnam.

It is estimated that 13,000 C-47 variants were produced for service, including 2,000 built in foreign countries under license.

The Douglas C-48, C-49, C-50, C-51 and C-52 in 1941 were DC-3 passenger airliners commandeered for wartime duty. They all had the same airframe as the C-47, but differed in internal arrangements, size of the door and type of engine. Accommodations included 18 to 24 passenger seats or 14 sleeping berths that could be used to carry litter patients. Conversions included 36 as C-48's, 138 C-49's, fourteen C-50's, one C-51 and five C-52's for a total of 194 aircraft drafted during the national emergency.

The **Douglas C-53 Skytrooper** in 1941 was another off-the-shelf variant of the DC-3 airliner with the same general structure as the C-47 Gooney Bird. Intended primarily for

Douglas C-53 Skytrooper, outstanding WWII long-range Troop Carrier. (USAF Photo 21905-AC)

use as a Troop Carrier, the C-53 had no large cargo door; however, about 4,000 pounds of small packages could be freighted. Up to 42 passengers could be crowded onto the benches along each wall, and 26 fully equipped paratroops could be accommodated. The C-53B had extra tanks that stretched its range to 3,000 miles. A total of 378 C-53 Skytroopers were delivered to the USAAF.

The Troop Carrier Command (TCC) was formed on June 1, 1942, under command of General F.S. Borum. On July 9, 1943, Skytroopers of the 51st Troop Carrier Wing out of North Africa towed Waco CG-4A Assault Gliders to invade Sicily as a prelude to the landings in Italy.

The Air Transport Command (ATC) was formed on July 1, 1942, with General Harold George commanding. C-53 Skytroopers immediately joined other Cargo Transports hauling personnel, equipment and supplies over the North Atlantic route from New England to Newfoundland to Greenland to Iceland to the British Isles in support of American buildups for the war in Europe. During WWII the ATC flew ten billion air miles and at peak strength employed 200,000 men and women in support of worldwide airlift activities.

The **Douglas C-54 Skymaster** was a development from the DC-4 civil airliner. When America entered WWII in December 1941, the USAAF commandeered 34 DC-4's already under construction and designated them C-54's. The first flight was made on March 26, 1942, and deliveries began to arrive at Air Transport Command bases in December 1942 of the first four-engine Cargo Transport in USAAF history. The C-54 was a low-wing monoplane with retractable tricycle landing gear and a standard tail section. Four Pratt & Whitney R-2000 1,350-hp engines drove 3-bladed props to produce a top speed of 285 mph. The crew of five included pilot, copilot, navigator, radio operator and crew chief. The original batch of C-54 Skymasters accommodated 26 passengers. The

Douglas C-54 Skymaster, the largest mass-produced 4-engine Cargo Transport of WWII. (USAF Photo 28821-AC)

C-54A had a large cargo door and carried 14,000 pounds of freight, including trucks and other heavy equipment. The C-54B was a hospitalship with 50 seats or 26 stretchers. Subsequent variants through the C-54M increased the payload to 32,000 pounds and extended the range out beyond 5,000 miles. Postwar variants that did not reach the production stage were the XC-114, XC-115 and YC-116, all built in 1946. The SC-54D was a Search and Rescue plane and the TC-54D was an electronics warfare trainer.

Flying the Atlantic, Pacific and Indian Oceans during WWII, C-54's made over 79,000 crossings while hauling men, materiel and supplies to every theater of operations, including the notorious Himalayan Hump between India and China. During operations across that "air bridge," a total of 600 planes and 1,000 crewmen were lost—only four C-54's were lost in combat. A single plush C-54 nicknamed "Sacred Cow" was specially built for President Roosevelt to use as his personal VIP Transport. The C-54 turned out to be the largest mass-produced Cargo Transport of WWII. The Lockheed C-69 Constellation in 1943 was larger, but only 22 of them were delivered in time for the war. The Douglas C-74 Globemaster would have been the largest, but it didn't get off the ground until after the war ended.

On June 25, 1948, Russian soldiers blocked all roadways, railways and waterways leading into West Berlin, and on the following day M/Gen. Curtis LeMay ordered all available Cargo Transports to begin operation of the "LeMay Coal and Feed Company," hauling supplies and equipment into Tempelhof airfield for distribution to West Berlin civilians. On July 29, 1948, M/Gen. Bill Tunner assumed command of the "Berlin Airlift" Task Force that eventually utilized some 300 C-54's, 102 C-47's, five C-82's, one C-74 and one YC-97. The single-day record of 12,940 tons by 1,398 sorties was set during the "Easter Parade" of April 1949. Premier Stalin halted the blockade on May 12, 1949, but the airlift continued until September 30, 1949, when Perry Immel piloted a C-54 on the last flight out of Rhein Main Air Base into Tempelhof.

On June 25, 1950, a C-54 became the first Air Force plane destroyed in the Korean war when it was strafed on the ground at Kimpo airport. On July 1, 1950, six Skymasters began a shuttle service from Itazuke Air Base, Japan, to Pusan, Korea, carrying Army Lt/Col. Charlie Smith and his 16,000-man task force that became the first American unit to fight on Korean soil. During the Pacific Airlift in support of the Korean conflict, C-54's operated back and forth between Tokyo and the States carrying personnel, cargo and war casualties.

Production totaled 953 C-54's for the USAAF and USAF.

The **Lockheed C-56 Lodestar** was the first in a series of Lockheed Model-18 plush executive cabinplanes commandeered by the USAAF in 1941 for use as Staff and VIP Transports. It was a mid-wing monoplane with hydraulic retractable landing gear and the classic Lockheed twin-finned tail section. Two Wright R-1820 1,200-hp engines gave it a top speed of 255 mph with a range of 1,700 miles. Pilot, copilot and crew chief sat in the pilot's cabin up front. Accommodations were provided for 17 to 22 seats, or seven tables were available for conferences in the main cabin. Lavatory facilities were provided in the rear. A total of 25 C-56's were drafted for active duty during WWII. Other members of the Lodestar family impressed for service were the C-57, C-59, C-60 and C-66.

The **Lockheed C-57 Lodestar** in 1941 was like the C-56 except for engine change and different internal arrangements. The C-57 had twin Pratt & Whitney R-1830 1,200-hp engines, a top speed of 270 mph and plush seats for 14 to 17 passengers. Twenty of the C-57's were commandeered for the war.

The **Douglas C-58 Bolo** in 1942 was a battlefield adaptation of the Douglas B-18A Bomber for use as a Cargo Transport. The C-58 was a low-wing monoplane with two Wright R-1820 1,000-hp engines, a top speed of 215 mph and a range of 1,200 miles. It was easily convertible from an 18-seat personnel transport to a cargo carrier with a 6,000-pound capacity—or the load could be mixed if necessary. The crew of three included pilot, copilot and crew chief. A total of only two conversions were made officially, but other Bolos were used as Cargo Transports without changing the designation.

The **Lockheed C-59 Lodestar** was built for export to Britain in 1942. It was exactly like the C-57 Lodestar except for the engines. A total of ten C-59's were Lend-Leased to England, but none were used by the USAAF.

The **Lockheed C-60 Lodestar** in 1942 was an improved version of its predecessor C-56, C-57 and C-59 Lodestars with the same airframe, retractable front wheels and two Wright R-1820 1,200-hp engines giving it a top speed of 270 mph. The C-60 was used as a 12-passenger Staff Transport. The C-60A was a plush VIP Transport with seven seats. The C-60B was a Troop Carrier with bench seats for 18 fully equipped paratroopers. The Lockheed B-34 and B-37 Ventura Bombers were also developments from the Model-18 Lodestar. A total of 161 C-60's were delivered to the USAAF during WWII.

The Lockheed-Vega C-60 Lodestar was the most-produced USAAF variant of the twin-finned Lodestar family during WWII. (USAF Photo 18571-AC)

The **Fairchild C-61 Forwarder** in 1941 was a direct conversion from their Model F-24 light cabinplane. It was a high-wing monoplane with standard tail assembly and 3-wheeled non-retractable undercarriage. A single Warner 165-hp engine gave it a top speed of 130 mph with a range of 560 miles. Four seats were provided in front-to-back pairs—the two front seats had dual controls and the two rear seats were for passengers. Two access doors were located one on each side of the enclosed cabin. In January 1943 the C-61 was redesignated UC-61 to classify it as a Utility Transport. The Navy version was designated GK-1 and the export version was named "Argus." Production totaled 981 C-61 variants for the war. The C-86 Forwarder in 1942 was a C-61 with an engine change.

The **Lockheed-Vega C-63 Hudson** in 1941 was a development from their Model-14 Super Electra airliner. The C-63 Transport and its sistership A-29 Attack Hudson were interchangeable for their double roles. Other variants were the A-28 Attack plane, AT-18 Advanced Trainer, C-111 Cargo Transport and the Navy PBO-1 Patrol Bomber. The C-63 was a mid-wing monoplane with retractable front wheels and the typical Lockheed twin-fin twin-rudder tail section. Two Wright R-1820 1,200-hp engines drove 3-bladed props to give it a top speed of 250 mph with a range of 2,000 miles. Seven windows along each side of the cabin gave an outside view for 17 passengers. As a cargo carrier it had a 4,000-pound freight capacity. A total of 364 Hudsons were delivered as C-63's, but some were converted back to A-29's for use on anti-submarine and coastal patrol duties.

The **Noorduyn C-64 Norseman** was built in Canada and delivered to the USAAF in 1942 as part of a Reverse Lend-Lease arrangement. The C-64 was a high-wing cabinplane with non-retractable landing gear and a standard tail section. A single R-1340 600-hp engine gave it a top speed of 160 mph. Eight seats were provided for crew and passengers. The Norseman was a rugged aircraft that could be fitted with floats or skis for operation on water or in snow-covered areas. The Norseman turned out to be the only Canadian light Cargo Transport to remain in production for the duration. On December 14, 1944, big-band leader Major Glenn Miller disappeared into the fog somewhere between England and France while flying as a passenger on a UC-64A. A total of 762 C-64's were procured by the USAAF for war service.

The **Lockheed C-66 Lodestar** in 1942 was the last of the Cargo Transports developed from the Model-18 civil liner. The C-66 was like the other Lodestars except for two Pratt & Whitney R-1830 1,200-hp engines and internal accommodations for eleven VIP passengers. All those different designations for the Lodestars were necessitated by the variety of engines, equipment and interior arrangements they had been customized with by their private and commercial owners prior to being drafted into service. Only one of the C-66 designation was refurbished for the USAAF.

The **Douglas C-67 Dragon** in 1942 waa s direct modification of the B-23 Bomber into a Cargo Transport. It was a low-wing monoplane with retractable front wheels and a standard tail section. Pilot, copilot and flight engineer made up the crew. Two Wright R-2600 1,600-hp engines gave it a top speed of 280 mph. The cabin could carry 21 passengers or 8,000 pounds of freight. About two dozen of the B-23's were converted to C-67's for hauling men and materiel in the war, of which twelve were used as glider tugs.

The **Douglas C-68** in 1942 was the military designation given to the Douglas DC-3A commercial airliner commandeered for service by the USAAF. It was similar to the C-47 Gooney Bird with two Wright R-1820 1,200-hp engines, a top speed of 230 mph and accommodations for 21 passengers or 10,000 pounds of cargo. Only two of the C-68's were drafted for the war.

The **Lockheed C-69 Constellation** was a wartime development from the Model-49 commercial airliner, which made its first flight on January 9, 1943, and was immediately drafted by the USAAF. The Constellation was distinguished as the largest and fastest Cargo Transport to serve during WWII; it was also the first of Lockheed's transports to have a triple-finned tail section. The C-69 was a low-wing monoplane with retractable tricycle landing gear, a pressurized cabin and a gross weight of 82,000 pounds. Crew consisted of pilot, copilot, navigator, flight engineer, radio operator and two flight attendants when required. Sixty-four passenger seats were provided and the cargo capacity was 32,500 pounds. Four Wright R-3350 2,200-hp engines drove 3-bladed props to produce a top speed of 330 mph. The C-69C had an airline interior and carried 43 passengers in plush seats for a range of 2,400 miles. Only 22 of the C-69's were produced for the USAAF.

Lockheed C-69 Constellation, largest and fastest limited-production 4-engine Cargo Transport in late days of WWII. (USAF Photo 29497)

The **Howard C-70 Nightingale** was commandeered in 1942 as a light personnel transport. It was a high-braced-wing cabinplane with conventional 3-wheeled landing gear and a standard tail section. The Pratt & Whitney R-985 450-hp engine gave it a top speed of 175 mph with a normal range of 700 miles. Pilot and copilot sat in the front two seats and two passengers with luggage could be accommodated in the cabin—without baggage, one more passenger could squeeze into the cabin. The Navy GH-1 was a transport like the C-70; the GH-2 Nightingale was an airborne ambulance with 3 seats and 2 litters; the NH-1 was an instrument trainer with four seats and 3 sets of controls. The USAAF received 20 C-70's, which were all redesignated UC-70's in January 1943.

The **Spartan C-71 Executive** was drafted for service in 1942 as a light VIP Transport and station-to-station shuttle taxi. It was a single-engine low-wing cabinplane with a standard tail section, retractable front wheels and a top speed of 210 mph. Pilot and copilot sat side by side with dual controls and three passengers occupied a bench seat in the rear. Sixteen C-71's were furnished to the USAAF. In January 1943 all of them were redesignated as UC-71 Utility Transports.

The **Waco C-72** light cabin biplane was commandeered in 1942 for use as a Staff Transport. The tail assembly was standard and the three-wheel landing gear had fixed legs with fancy pants. A single 400-hp engine drove a 2-bladed prop to give it a top speed of 185 mph. Pilot and copilot or student had dual control seats up front, and a side to side bench in the rear of the cabin provided three seats with dividing arm rests. The USAAF was supplied with 44 C-72's, which were redesignated UC-72 in January 1943.

The **Boeing C-73** was commandeered in 1942 from the Boeing Model-247 commercial airliner that made its first flight back on February 8, 1933. The Model-247 is often referred to as the forefather of modern commercial airliners, and it probably would have been even more famous had it not coincided with the arrival of the famous Douglas "DC" series. The C-73 was the first transport to have rubber deicer boots on the wing and tail. The C-73 was a low-wing monoplane with standard tail section and a conventional 3-wheeled undercarriage. Two Pratt & Whitney R-1340 600-hp engines drove 3-bladed props to produce a top speed of 200 mph. Pilot and copilot sat in the cockpit up front and ten forward-facing seats were located in the cabin, five on each side of the aisle. The C-73 was used for ferrying aircrews and for training combat cargo pilots. A total of 27 C-73's served with the USAAF. In 1943 an exported C-73 was field-modified in China to carry two machine guns in the nose and a swing gun in an upper turret for use by Chiang Kai-shek as his personal self-defending VIP Staff Transport.

The Model-247 was flown by Roscoe Turner and Clyde Pangborn in the 1934 Intercontinental race from England to Australia. That aircraft displayed a large "57" emblem on each side of the fuselage for advertisement of the Heinz 57-Varieties Food Company, sponsors of the plane. The Model-247 named "Adaptable Annie" is displayed in the Aviation Museum of the Smithsonian Institute as a fitting tribute to a great aircraft.

The **Douglas XC-74** prototype made its first flight on September 5, 1945—three days after the end of World War II. Therefore, the production contract was cut back to just 14 C-74 Globemaster-I Transports for delivery beginning in October 1945. Thus began a very inauspicious career for what would have been the largest long range Cargo Transport and Troop Carrier in the war. The C-74 had four Pratt & Whitney R-4360 3,500-hp engines, a top speed of 310 mph and a maximum range of 7,800 miles. Accommodations were furnished for 125 fully equipped combat troops or 115 stretcher patients. A wide variety of heavy freight packages could be loaded by an electric elevator in the middle of the fuselage into the main compartment that had a capacity of 60,000 pounds. The 6-man crew included pilot, copilot, navigator, flight engineer, radio operator and loadmaster, with a chef in the galley and a full relief crew on long trips. The C-74 Globemaster-I was the forerunner of the Douglas C-124 Globemaster-II. In 1948-49 a single C-74 took part in the Berlin airlift to break the Russian blockade. The last C-74 was delivered in April 1947.

Douglas C-74 Globemaster-I, huge 4-engine Cargo Transport, developed too late for service in WWII. (USAF Photo K-5375)

The **Boeing C-75 Stratoliner** in 1942 was an off-the-shelf conversion of the Model-307 commercial airliner that had made its first flight back in 1939. When the USAAF commandeered all the airliners, the civilian 307 crews asked, ''What are we going to do now?'' So the USAAF got permission to draft the crews with the airplanes. The Model-307 was distinguished as the first commercial airliner to have a pressurized cabin. It was an all-metal low-wing monoplane with retractable landing gear and a standard tail section. Four Wright R-1820 1,100-hp engines gave it a top speed of 250 mph. Pilot, copilot and crew chief made up the crew. As a personnel transport the C-75 could carry 33 passengers. As a sleeper it could accommodate 25 people in bunks and reclining seats. As a cargo carrier it could haul 6,600 pounds of freight in the compartment below the main cabin. A total of five C-75's, crews and all, were drafted by the USAAF.

Operating with the Air Transport Command across the hazardous North Atlantic route, Stratoliners were used to transport military VIP officials back and forth between America and the British Isles in support of operation ''Bolero,'' the buildup for a planned invasion of Hitler's fortress Europe.

The **Cessna C-78 Bobcat** in December 1941 was a commandeered version of their Model T-50 Transport cabinplane that had made its first flight in March 1939. The C-78 was a low-wing monoplane with a standard tail assembly and retractable landing gear. Twin Jacobs R-755 245-hp engines gave it a top speed of 195 mph with a range of 750 miles. The two front seats had dual controls and three passengers could ride in the rear of the cabin with room for 40 pounds of baggage per person. The AT-8 and AT-17 Bobcat Advanced Trainers were like the C-78 except for engines and internal fixtures. The JRC-1 was the Navy version. About twenty C-78's were impressed on the first request and 1,284 more were subsequently delivered to the USAAF on production contracts. In January 1943 the C-78 became the UC-78 to put it in the new Utility category.

The sistership Bobcat Advanced Trainers were used to train the WASPs—Women's Airforce Service Pilots—at the flying school in Houston, Texas. Formed in 1942 through the combined efforts of Hap Arnold and Jacqueline Cochrane, WASP was actually an offshoot of the Women's Air Ferry Service (WAFS) that had been organized through the efforts of Nancy Love. Volunteering to support the wartime emergency during those early months of WWII when there was a critical shortage of male combat pilots, the WASPs ferried trainers and tactical aircraft from the factories, flew as copilots ferrying heavy bombers, served on non-combat tow target duty, and some of them were checked out in jet aircraft. WASPs never became a part of the military service—they were hired as temporary Civil Service employees. Of approximately 25,000 female applicants, about 2,500 were accepted for training and only 1,074 graduated into flight duty assignments. Thirty-seven of those unsung heroines gave their life for their country, including Cornelia Fort, the first American aviatrix in history to be killed while serving on air-war duty.

The **Stinson-Vultee C-81 Reliant** was impounded from private individuals for use in WWII. It was a development from the Stinson Model-SR civilian cabinplane that dated back to 1933. The C-81 was a high-gull-wing monoplane with a standard tail section and conventional 3-fixed-wheel landing gear having streamlined housings on the front legs. A single Pratt & Whitney R-985 450-hp engine gave it a top speed of 175 mph with a range of 600 miles. The two front seats had dual controls and two or three passengers rode in the rear of the cabin that had access doors on both sides. A 200-pound load of baggage or freight could be carried. The Stinson AT-19 Reliant was an Advanced Trainer built for export to Britain. A total of 45 C-81's were commandeered by the USAAF. In January 1943 all C-81's were redesignated UC-81's to identify them as light Utility Transports.

The **Fairchild XC-82** prototype made its first flight on September 10, 1944, and delivery of the production model C-82 Packet began in September 1945—too late for WWII. The C-82 was a twin-boom high-wing monoplane with retractable tricycle landing

Fairchild C-82 Packet, post WWII twin-engine Cargo Transport. (USAF Photo 32847- AC)

gear and twin tailfins connected by a long tailplane. A central nacelle housed the crew, passengers and cargo, which was loaded through a large clam-shell door at the rear. Twin Pratt & Whitney R-2800 2,100-hp engines drove 3-bladed props to produce a top speed of 250 mph and a maximum range of 3,900 miles. As a Troop Carrier it could accommodate 42 fully equipped paratroops. As a hospital ship it could handle 34 stretcher patients. Heavy pieces of equipment such as tanks, howitzers and trucks up to a capacity of 11,500 pounds could be loaded through the large cargo doors. The normal crew of four included pilot, copilot, navigator and radio operator.

The proficiency of the C-82 as a glider tug was demonstrated in 1946 when Waco CG-4A and CG-15 Assault Gliders were towed simultaneously. C-82's out of Lawson Field at Fort Benning, Georgia, hauled post-WWII paratroop trainees with the 82nd Airborne Regiment. The Tactical Air Command and Military Air Transport Service used Packets for tactical and strategic airlifts. In 1947 a vastly improved C-82B became the prototype for the Fairchild C-119 Flying Boxcar. Five C-82's joined the armada of Cargo Transports to haul vehicles and heavy packages into Berlin during the Communist blockade of that city in 1948-49.

Production ended in September 1948 after 224 C-82 Packets had been delivered to the Air Force. The last one was retired in 1954.

The **Piper UC-83 Coupe** in 1943 was a development from the L-4 Grasshopper (Cub) light Liaison plane, which was itself a redesignation of the 0-59 Observation plane. The UC-83 was a high-wing monoplane with fixed wheels and a standard tail section. The Continental 75-hp engine gave it a top speed of 100 mph. Two side by side seats in the cabin had dual controls with access doors on each side and baggage space behind the cabin; when necessary a passenger could ride in the luggage area. Only one UC-83 was employed by the USAAF; however, many thousands of its variants were used by the War Training Service, Civil Air Patrol and Army ground forces during the war.

The **Douglas C-84** in 1942 was the designation given to Douglas DC-3B commercial airliners commandeered by the USAAF. The C-84 was similar to the C-47 Gooney Bird with two Wright R-1820 1,200-hp engines and a top speed of 230 mph. A slightly enlarged fuselage accommodated 28 passengers or 10,000 pounds of freight. A total of four C-84's were delivered to the USAAF.

The **Fairchild C-86 Forwarder** in 1942 was commandeered from the Fairchild Model F-24 light civilian cabinplane. It was exactly like the C-61 Forwarder except for a Ranger 6-cylinder engine. In January 1943 the C-86 became the UC-86 Utility Cargo Transport. It was a 4-seated high-wing monoplane with a top speed of 130 mph. Only a few C-86's were delivered to the USAAF.

The **Consolidated C-87 Liberator** in 1942 was a Cargo Transport version of the B-24 Heavy Bomber. The Army Ferrying Command was formed on May 21, 1941, and the following month they received their first long-range Transport, the B-24A, which was de-bombed to make room for freight, but was not assigned a distinctive designation. The Ferrying Command was the forerunner of the Air Transport Command that was organized on July 1, 1942.

The C-87 was a high-wing monoplane with a solid nose, retractable tricycle landing gear and a twin-finned tail section. Seven windows were cut into each side of the fuselage and a large cargo door replaced the porthole on the left rear of the fuselage. Four Pratt & Whitney R-1830 1,200-hp engines drove 3-bladed props to give it a top speed of 305 mph and a maximum range of 2,900 miles. The five-man crew included pilot, copilot, navigator, radio operator and crew chief. Up to 25 passengers or a 12,000-pound cargo load could be hauled. The C-87A in 1943 was an executive sleeper with ten berths. The C-87B was an armed version with two forward-firing 50-caliber machine guns. The C-87C would have had a single fin on the tail section, but it was still on the drawing board when WWII ended. The Navy version was the RY-3. About fifty C-87's were exported to England under the Lend-Lease Act. The C-109 was another Cargo Transport variant for the USAAF.

Those early transports of the Ferrying Command and Air Transport Command flew the perilous North Atlantic route from America to the British Isles ferrying men and materiel in support of the planned invasion of Europe. Armed C-87's flew the Himalayan "Hump" from India into China. C-87's hauled heavy loads all over the world, carrying supplies and equipment to men at the front.

Production ended in August 1945 after 291 C-87's had been delivered to the USAAF. Many more B-24 Bombers were converted to Cargo Transports like the C-87 without being given distinctive numbers. In September 1944 an entire wing of B-24's from the 8th Air Force hauled gas to Patton's tank corps in France.

The **Taylorcraft C-95 Grasshopper** in 1942 was a conversion from the L-2 light Liaison cabinplane, which was itself a redesignation of the 0-57 Observation plane. The C-95 was a high-wing monoplane with fixed wheels and a standard tail section. A single Lycoming 65-hp engine drove a 2-bladed prop to give it a top speed of 90 mph and a range of 230 miles. Pilot and copilot or trainee rode in side by side seats with a clear all-around view. In January 1943 the designation was changed to UC-95 when it was used as a communications and administrative aircraft. Only seven C-95's were procured by the USAAF, and in late 1943 all of them were turned over to the Army where they were designated L-2 for use as enemy artillery spotters.

The **Fairchild C-96** in 1942 was commandeered by the USAAF from the 1928 vintage Model FC-2 civilian cabinplane, which was similar to the Air Corps C-8 Photography Transport of 1930. The C-96 was a high-wing monoplane with three fixed wheels and a standard tail section. The wings folded backward for parking in hangars with narrow doors. A 410-hp engine gave it a top speed of 140 mph. As a Photo-Recon plane its 3-man crew included pilot, copilot and photographer. As a light personnel transport it carried a pilot, copilot and five passengers. Three of them were supplied to the USAAF, and in January 1943 they were redesignated UC-96.

Drafted for use in the Photo-Recon role, the C-96 had the distinction of being the oldest Photography plane in service. Shortly thereafter they were converted for use as light personnel transports and station taxis until retired near war's end.

The **Boeing XC-97** and **YC-97** prototypes were delivered to the Air Force for evaluation in 1948 as blendings of the wings, tail, engines and landing gear from the B-29

The Boeing KC-97 Stratofreighter was the first USAF Cargo Tanker and the last Boeing piston-engine military aircraft. (USAF Photo 48426-AC)

Bomber with the main components of the B-50 Bomber. During trials in May 1949 a single YC-97 workhorse hauled 500 tons of cargo in support of the Berlin Airlift to help break the Soviet blockade. Deliveries of the production model C-97A Stratofreighter began to arrive at Air Force bases in October 1949. The C-97 was a mid-wing monoplane with retractable tricycle landing gear, a fat-bellied fuselage and standard tail section. Four Pratt & Whitney R-4360 3,500-hp engines drove 4-bladed propellers to give it a top speed of 375 mph with a maximum range of 4,300 miles. The crew of four included pilot, copilot, navigator-radioman and flight engineer. As a personnel transport the C-97A could carry 134 fully equipped combat troops. As an airborne ambulance it could accommodate 83 litter patients with attendants and medical supplies. As a cargo hauler a capacity load of 53,000 pounds of freight could be carried. The cargo door was large enough for tanks and trucks to be driven into the compartment. Three VC-97's were delivered to the Strategic Air Command for use as Airborne Command Posts. A total of 76 C-97's were built for the Air Force.

The Boeing KC-97 Tanker was delivered in July 1951 as the first multi-purpose aircraft that could be rapidly converted for the roles of troop carrier, cargo freighter, airborne hospital or inflight refueler with a transfer rate of 600 gallons per minute. The flying boom attached underneath the tail had movable fins and was steered into a receptacle by the boom operator. During the Korean conflict KC-97A's were flown by SAC and MATS in support of the war effort. KC-97C hospital ships evacuated casualties from Japan to California. In August 1953 KC-97 Tankers participated in operation "Longstride," refueling F-84G Thunderjets deploying from Turner AFB, Albany, Georgia, to England and French Morocco. In January 1957 KC-97's refueled three B-52 Stratofortresses during their record breaking non-stop flight around the world.

Production ended on July 18, 1956, the same date that the first all-jet KC-135 Stratotanker rolled off the production line. The KC-97G turned out to be the last piston-engined aircraft to be constructed by Boeing. A total of 888 C-97's and KC-97's were built for the Air Force. Upon being retired from the USAF, KC-97's were transferred to the Air National Guard—in 1965 the Hayes Aircraft Company installed J47 turbojets in the ANG Tankers to increase their effectiveness when refueling jet Fighters.

The **Consolidated-Vultee XC-99** first flew on November 23, 1947, as a modified version of the B-36 Bomber. Only one XC-99 was built. It was delivered to Kelly Field, San Antonio, Texas, in May 1949 and put to work for the Air Materiel Command until retired in 1957. The XC-99 had a double-deck fuselage with accommodations for 400 combat troops, 300 stretchers or 101,000 pounds of cargo. Six Pratt & Whitney R-4360 3,000-hp engines gave it a top speed of 300 mph on a range in excess of 8,000 miles. Had the XC-99 reached production it would have been the largest prop-driven Cargo Transport in Air Force history, but that honor is held by the Douglas C-133 Cargomaster.

The **Northrop C-100 Gamma** in 1942 was a civilian Model-2D long range speedster commandeered by the USAAF. The C-100 was a low-wing monoplane with a standard tail section and fixed wheels having streamlined spats. Pilot and one passenger sat tandem style in the enclosed cockpit at the rear of the fuselage. The single Wright R-1820 710-hp engine drove a 3-bladed prop to give it a top speed of 230 mph and a range of 2,800 miles with a tailwind. Distinguishing features of the Gamma were the park-bench ailerons mounted on supports at the trailing edge of each wingend. Only one C-100 was drafted for the war—it was redesignated UC-100 on January 1, 1943, to fit in the new Utility category.

The original single-seat Northrop Gamma in 1933 was flown by Howard Hawks to establish many cross-country speed and endurance records; it was named "Sky Chief" under sponsorship by the Texaco Oil Company. Lincoln Ellsworth named his two-seat Gamma "Polar Star" after his successful Antarctic exploration in 1935. That aircraft is exhibited in the Smithsonian Institute. Howard Hughes and Jackie Cochrane also flew Northrop Gammas.

The **Lockheed C-101 Vega** in 1942 was a civilian Model-5 light cabinplane commandeered for military service. It was a high-wing monoplane having a standard tail section and three fixed wheels with streamlined housings. Pilot and copilot sat up front and five passenger seats were provided at the rear of the cabin. The single Pratt & Whitney 550-hp engine was equipped with a 2-bladed propeller. Back in 1931 the Air Corps procured two other Vegas under the designations of YC-12 and YC-17. Only one example of the C-101 was impressed for service, and it was redesignated as the UC-101 in January 1943.

On April 15, 1928, Ben Eielson and Hubert Wilkins, after 20 hours flying time in a Lockheed Vega, made the first eastward crossing of the Arctic wastelands from Point Barrow, Alaska, to Spitsbergen, a Norwegian island in the Arctic Ocean. On July 15, 1933, Wiley Post departed Floyd Bennett Field, New York City, in a Lockheed Vega named "Winnie Mae" to begin his famous solo flight around the world. After traveling 15,596 miles through Germany, Siberia and Alaska in 7 days 18 hours and 49 minutes he landed back at Floyd Bennett Field to receive the full treatment in a ticker tape parade. In his younger days Wiley Post worked in the oil fields of Oklahoma where he lost an eye in a drilling accident. His airplane was named after Winnie Mae Hall, the daughter of his boss. On August 15, 1935, Wiley Post and Will Rogers, the famous cowboy comedian, were killed when their pontooned Lockheed hybrid monoplane crashed near Point Barrow, Alaska.

The **Boeing XC-105 Cargo Transport** in 1943 was a direct redesignation of the XB-15 Bomber prototype after trials in the Bomber role were completed and the plane was in good shape for flying. The XC-105 was a huge four-engine mid-wing monoplane with a top speed of 195 mph. It could accommodate 64 passengers or 21,000 pounds of freight. Only one XB-15 experimental Bomber was built; therefore only one XC-105 was modified for the USAAF. It was retired shortly after the end of World War II.

The **Boeing YC-108** in 1941 was a modification of the B-17 Heavy Bomber. The YC-108 was a mid-wing monoplane with conventional 3-wheeled retractable landing gear and a standard tail section. The crew included pilot, copilot, navigator, radio operator and crew chief. Four Wright R-1820 1,200-hp engines gave it a top speed of 315 mph and a normal range of 2,500 miles. It had a large cargo door for loading 10,000 pounds of freight. Windows along each side of the fuselage provided an outside view for up to 38 passengers.

One YC-108 retained the four guns in the nose and tail for use as General MacArthur's VIP Transport and personal plane. Another YC-108 was rigged with extra tanks in 1943 and tested for hauling fuel across the Himalayan "Hump" to B-29 Bombers flying missions out of Chengtu, China. Only four YC-108 Flying Fortresses were converted for the USAAF.

The **Consolidated C-109** in 1943 was a conversion from the B-24 Heavy Bomber and was similar to the C-87 Cargo Transport. The C-109 was a high-wing monoplane with retractable tricycle landing gear and a twin-finned tail section. Four Pratt & Whitney R-1830 1,200-hp engines gave it a top speed of 305 mph and a range of 2,900 miles. The 5-man crew consisted of pilot, copilot, navigator, radio operator and flight engineer. The C-109 was specially rigged with extra tanks in the nose, bomb-bay and fuselage holding 3,000 gallons of fuel.

The C-109 was selected in preference to the YC-108 for use as a "flying filling station" hauling fuel across the "Hump" to B-29 Superfortress Bombers operating out of Chengtu, China. While the tanks were being emptied at the destination, they were simultaneously filled with an inert gas to prevent spontaneous explosions. Approximately 107 of the C-109 conversions were made by the USAAF.

The **Lockheed C-111 Super Electra** in 1944 was commandeered by the USAAF from the Model-14 civil airliner. Very similar to the C-63 Hudson, the C-111 was a mid-wing monoplane with retractable front wheels and the classic twin-finned tail section. Two Wright R-1820 1,200-hp engines gave it a top speed of 250 mph and a maximum range of 2,000 miles. The 3-man crew included pilot, copilot and crew chief. Seats were available for 17 passengers, or the interior could be rearranged to hold 2,500 pounds of cargo. A total of only three C-111's were drafted for service when the USAAF needed all the transports it could muster in support of the Allied invasion of Europe, while simultaneously chasing the Japanese from island to island across the Pacific. On July 10-14, 1938, Air Corps Lt. Hiram Thurlow flew as copilot with Howard Hughes in a Lockheed Super Electra, covering 14,824 miles while circling the globe in 91 hours 17 minutes.

The **Douglas YC-116 Skymaster-II Transport** in 1946 was a heavy C-54 with engine change, thermal deicers and a longer range. The YC-116 was a low-wing monoplane with standard tail section and retractable tricycle landing gear. The crew of five included pilot, copilot, navigator, radio operator and flight engineer. Four Allison V-1710 1,620-hp engines gave it a top speed of 260 mph and a range of 4,000 miles. Fifty passengers or 26 stretchers with medical attendants or 14,000 pounds of cargo could be accommodated. The YC-116 arrived too late for WWII at a time when the USAAF was overstocked with Cargo Transports; therefore, only two were delivered for service trials and the contract was cancelled.

The **Douglas C-117 Skytrain-II** in 1945 was the last military variant of the famous DC-3 family of Cargo Transports. The C-117 was a low-wing monoplane with retractable front wheels and a standard tail section. Two 1,200-hp engines gave it a top speed of 230 mph with a range of 1,600 miles. Pilot, copilot and flight engineer made up the crew. Arrangements were available for 18 passengers or 6,000 pounds of cargo.

The C-117A and C-117B had plush interiors with reclining seats for use as Staff Transports, and the C-117C was a VIP carrier. In 1953 the C-117 was redesignated VC-117 to put it in the VIP Cargo Transport category, and a few of them were still in service during the Vietnam war. Eighteen C-117's were built at the factory and eleven more were converted from VC-47's for a total of 29 Skytrain-II transports delivered to the USAAF.

The **Douglas C-118 Liftmaster** first flew on February 15, 1946, as the XC-112, which was developed from the C-54 by way of the DC-6 commercial airliner. Deliveries of the C-118A began to arrive at Air Transport Command bases in late 1946. The C-118 was a low-wing monoplane with retractable tricycle landing gear and a standard tail assembly. The crew included pilot, copilot, navigator, radio operator and flight engineer. Four Pratt

Douglas C-118 Liftmaster, four-engined MATS Cargo Transport of the Korean war era. (USAF Photo 154466-AC)

& Whitney R-2800 2,500-hp engines gave it a top speed of 370 mph and a maximum range of 4,910 miles. As a personnel transport the C-118 could carry 76 passengers, or as an airborne ambulance it could accommodate 60 stretchers. The cargo version could haul 27,000 pounds of freight, which was loaded through large doors fore and aft of the wings.

Reporting for duty with air terminals when the Military Air Transport Service (MATS) was formed on March 1, 1948, Liftmasters began providing comfortable air travel to enlisted men and their dependents for the first time in history—but the men travelling alone still had to ride "last class" in troop ships. C-118's were used by major Air Commands as Staff Transports and VIP carriers. A VC-118 in 1947 was named "Independence" for its role as a Presidential Transport by President Harry Truman, a lifetime resident of Independence, Missouri, a suburb of Kansas City. That example had a plush interior with 25 reclining chairs and 12 sleeper bunks. The Military Airlift Command (MAC) used C-118's for aeromedical evacuation during the Vietnam war. Production totaled 101 C-118's for the Air Force. In 1965 all but eleven of the survivors were retired from active duty and issued to other military agencies after a career spanning almost twenty years.

The **Fairchild C-119 Tactical Transport** made its first flight in November 1947 as an enlarged version of the C-82 Packet. Deliveries of the production model C-119B Flying Boxcar began to arrive at Air Force bases in December 1949 as the biggest and most powerful Cargo Transport of its time. The C-119 was a twin-boom high-wing monoplane with retractable tricycle landing gear and twin tailfins connected by a long tailplane. A large central nacelle housed the crew, passengers and cargo. Each of the large clamshell cargo doors at the rear of the nacelle had an embedded standard door, allowing two rows of paratroops to jump in unison. An electrically operated monorail provided for rapid aerial delivery of twenty 500-pound paracans through a hatch in the bottom of the fuselage. Wheeled equipment was loaded via ramps through the hinged rear cargo doors. Two

Fairchild C-119 Flying Boxcar, noted Korean war twin-boom Cargo Transport nicknamed the "Dollar Nineteen." (USAF Photo 39357-AC)

Pratt & Whitney R-4360 3,500-hp engines drove 4-bladed reversible-pitch props to give it a top speed of 280 mph and a range of 2,000 miles. Nicknamed the "Dollar Nineteen," its 5-man crew included pilot, copilot, navigator, radio operator and crew chief. Accommodations were provided for 62 passengers or 40 fully equipped paratroopers. The normal cargo load was 10,000 pounds, but up to 20,000 pounds could be hauled for a short distance. The entire door section could be removed if the situation demanded that everything be shoved out at once. The YC-119H Skyvan in May 1952 increased the span by 38'8" and had a cargo capacity of 27,200 pounds. The C-119J was used by TAC for paratroop drops and aerial resupply. The XC-120 Packplane was an experiment with detachable belly pods and the XC-128 was a proposed variant that never got off the drawing board. The R4Q was the Marine version.

The Flying Boxcar got its baptism under fire in Korea where 4,000 troops of the 187th Regimental Combat Team plus all their equipment and supplies were dropped in the first airborne attack of the war at Sukchon-Sunchow on a daring mission to rescue a northbound trainload of American prisoners of war. When the 1st Marine Division was cut off by the bugle-blowing Chinese Communists at Chosin reservoir, supplies from the C-119's kept them alive and fighting for ten days. Then, when the Marines broke out of the trap, a 32-ton bridge was air-dropped for helping them across an impassable gorge to the fleet at Hungam. That was the first time in history that an entire bridge had been dropped by air.

On August 19, 1960, a C-119 of the 6593rd Air Recovery Group accomplished the first mid-air recovery of an Orbital Vehicle when Discoverer-XIV was snatched out of the sky at 8,600 feet as it floated down by parachute.

The AC-119 Gunship joined the Vietnam war in 1968 armed with four 7.62mm Miniguns and two 20mm cannon. Known by its code name of "Spooky," the Gunship achieved fame by furnishing firepower on countless occasions in support of "grunts" down on the ground. The AC-119K had two turbojet boosters installed under the wings to help get the heavy loads of ammunition off the ground when operating out of short landing strips.

Production ended in 1955 after 1,112 C-119's had been delivered to the Air Force. The last five were retired to Arizona and Arkansas in 1975 after 28 years of service spanning the Korean and Vietnam wars.

The **Lockheed C-121 Super Constellation** in 1949 was a development from the Model-79 commercial airliner, which was itself an advanced version of the WWII C-69 Constellation. The first deliveries had individual airplane names for use as personal VIP Transports. The "Columbine" was for General Eisenhower, General MacArthur was given the "Bataan," and the "Dewdrop" was used by General Vandenberg. They were also referred to as PC-121 for Personal Cargo Transport and VC-121 for VIP Cargo Transport.

The C-121 was a low-wing monoplane with retractable tricycle landing gear and a triple-finned tail section. Four Wright R-3350 3,500-hp engines drove 3-bladed props to give it a top speed of 380 mph, and wingtip tanks extended the range to 2,100 miles. Sixteen windows down each side of the fuselage provided an outside view for up to 106 passengers. As a troop carrier it could accommodate 72 fully equipped combat troops. As an airborne ambulance it could handle 75 ambulatory patients or 47 stretchers. As a cargo transport it could carry a 40,000-pound payload. The crew normally consisted of pilot,

Lockheed C-121 Super Constellation, triple-finned Cargo Transport and electronics surveillance aircraft that spanned the Korean and Vietnamese wars. (USAF Photo 35150-AC)

copilot, navigator, radio operator and flight engineer, supplemented by flight attendants and medical specialists when necessary.

The C-121C in 1951 was developed from the Model-1049 to stretch the long-range capability on trans-oceanic flights for MATS to 3,500 miles. Its last crossing of the Pacific was made on October 25, 1963. The RC-121C and EC-121C joined the Air Defense Command's 552nd Airborne Early Warning and Control (AEWC) Wing at McClellan AFB, California, starting in October 1953. They introduced the large dorsal (back) and ventral (belly) radomes and added about 12,000 pounds of sophisticated electronics surveillance equipment. The VC-121E had a plush interior for use as a VIP Transport by President Dwight Eisenhower.

First deliveries of the RC-121D were made to ADC's 551st AEWC Wing at Otis AFB, Massachusetts, beginning in May 1954. During the Vietnam war the RC-121D was modified into the EC-121H by adding another ton of modernized equipment including an airborne computer and a 250-mile radar system capable of pinpointing an enemy target up to an altitude of 60,000 feet. As many as 27 specialists made up the crew. The EC-121H was used in Vietnam along "MacNamara's Wall" to monitor transmission devices placed on the ground for detecting enemy movements and conversations at night. As an Air Combat Controller the EC-121H detected distant targets on its wide-range radar set and steered American jet Fighters toward the blips. When the Fighters picked up the blips on their short-range radar, they relayed the coded phrase, "I've got Judy," and closed in for the kill. Normal missions lasted 12-15 hours, but some jobs kept the crews out as long as 20 hours.

A total of 160 examples of the C-121 Super Constellation were delivered to the Air Force, including all variants.

The **Fairchild C-123B Provider** originated in 1948 with the Chase YC-122, a powered version of their CG-18 non-powered Cargo Glider for transporting airborne assault troops into combat. The Chase XC-123 Avitruc adapted from their G-20 Glider in 1949, was the second step of development—it had a 67-troop capacity. The Chase XC-123A was equipped with four J47 turbojet engines and took to the air on April 21, 1951, as the first Air Force all-jet Cargo Transport. Deliveries of the C-123A production model began to arrive in 1952. Orders for the Chase C-123B were cancelled in 1953 after Kaiser-Frazer acquired a majority interest in the Chase Company. Then, Fairchild submitted a bid and won the contract to resume production of the C-123B, which first flew on September 1, 1954, and deliveries began to arrive at the 309th Troop Carrier Group in July 1955. One of the prototype XC-123B's was fitted with two J44 turbojets during development and made its initial flight on February 7, 1955, as Fairchild's first all-jet Cargo Transport. The jet-equipped C-123B was rigged with skis for use on rescue missions and logistics support along the Distant Early Warning (DEW) line stretching from Greenland through Canada to Alaska along the 70th parallel. The Stroukoff YC-123E was an amphibian for operating on all types of surfaces. The Stroukoff YC-134 Pantobase assault transport in 1958 was an experimental amphibian developed from the YC-123E by Stroukoff, the original designer of the Chase C-123 who formed his own aircraft company. The C-123K in May 1966 had two J85-GE auxiliary turbojet engines for quick takeoffs from short landing strips.

The C-123B was a high-wing monoplane with retractable tricycle landing gear and a standard tail section. The crew of two to four men consisted of pilot, copilot, flight engineer and loadmaster. Two Pratt & Whitney R-2800 2,500-hp engines gave it a top speed of 245 mph and a range of 1,470 miles. A payload of 30,000 pounds could be carried, including heavy artillery pieces and trucks. As a troop carrier or airborne ambulance it could accommodate 60 fully equipped combat troops or 50 litters plus six sitting patients and six medical attendants. The C-123 had no engine nacelles or internal fuel tanks; the engines were bolted to the wings and fuel was carried in tanks underneath. Field maintenance was easily performed, eliminating many trips back to the depot.

Providers entered the Vietnam war in 1962 and carried out a wide variety of special assignments, including airborne drops of troops, ammunition, food and livestock. Missions included chemical spraying, mercy trips, rescue, air evacuation, and delivery of rubberized nylon bladders filled with fuel.

On May 12, 1968, Lt. Colonel Joe Jackson, the Aircraft Commander of a C-123 Provider on a war mission, became the 5th Air Force crewman to earn the Congressional Medal of Honor in Vietman. That also distinguished him as only the 51st aircrew member in history to receive the Nation's highest award. Flying with the 834th Air Division out of Danang, Joe Jackson and his crew—Jesse Campbell, Ed Trejo and Manson Grubbs—departed on a sortie to help evacuate 1,000 "grunts" from Kham Duc, a Special Forces camp near the Laotian border 44 miles southwest of Danang that was being overrun by the Viet Cong. Just when the Command Post thought everyone had been rescued and was about to order the camp destroyed, it was discovered that three Combat Control Team members—John Gallagher, Mort Freedman and Jim Lundie—had been left behind after their Jeep was blown up. Responding instantly in the highest tradition of Air Force professionalism from his altitude of 9,000 feet directly above the camp, Colonel Jackson nose-dived his aircraft into an intense barrage of anti-aircraft fire, landed on the pock-marked strip with rockets bursting all around, plucked the three men out of a ditch under heavy fire from automatic weapons, took off through a deadly wall of cross-fire and pointed the ship's nose toward safety at Danang. On January 16, 1969, President Johnson read the citation, ''for conspicuous gallantry, his profound concern for his fellow man, and his intrepidity at the risk of his own life above and beyond the call of duty,'' and hung the medal around his neck. Hell! He should have been presented with the Pentagon! On March 17, 1941, Joe Jackson and Andy Waters held up their right hands together in Atlanta, Georgia, and were sworn in as 21-dollar-a-month Privates in the Army Air Corps.

Production ended in September 1969 after nine Chase YC-122's, ten Chase C-123's, three Stroukoff C-123E's and 300 Fairchild C-123B's had been built for the Air Force.

Fairchild C-123 Provider (Avitruc) Assault Transport, celebrated for its all-purpose versatility during the war in Vietnam. Shown here is the first example, making its initial flight on September 1, 1954. (USAF Photo 152104-AC)

The **Douglas C-124 Globemaster-II** was an extensively modified version of the C-74 Globemaster-I, using the old wings, tail unit and power plant of its predecessor while expanding the fuselage in length and girth. The YC-124 prototype made its first flight on November 27, 1949, and deliveries of the production model C-124A began to arrive at Air Force Bases in May 1950. They immediately found themselves caught up in the Korean conflict. The C-124C added an APS-42 radar set in a nose-mounted radome. A single example of the YC-124B was developed for the Air Research and Development Command (ARDC); it was fitted with T34-P 5,500-hp engines and became the YKC-124B experimental Tanker. A later improved version of that aircraft became the prototype XC-133 Cargomaster.

The C-124 was a low-wing monoplane having a standard tail section and nose-loading doors with a walk-up ramp. Four Pratt & Whitney R-4360 3,800-hp engines gave it a top speed of 304 mph. A cargo load of 26,000 pounds could be carried on a 4,000-mile trip, or 50,000 pounds on a 2,300-mile range; the maximum payload was 74,000 pounds. As a troop carrier the interior could be double-decked to carry 220 fully equipped combat troops. As an airborne ambulance the C-124 could handle 136 stretcher patients and 52

Douglas C-124 Globemaster-II, Korean war Cargo Transport and airborne ambulance. Caught while making its maiden flight out of the Long Beach plant on November 27, 1949. (USAF Photo 36892-AC)

ambulatories with 15 medical attendants. The 8-man crew included pilot, copilot, navigator, radio operator, radarman, flight engineer, loadmaster and a flight attendant. Backup crewmen could be accommodated on long flights.

Final delivery of 448 C-124's was completed in May 1955, and they continued in service until retired to the Air Force Reserve in the middle of 1961.

The **Northrop XC-125** prototype made its first flight on August 1, 1949, and deliveries of the production model C-125A began to arrive in early 1950. The C-125A was intended as a STOL powered replacement for powerless Assault Gliders. The C-125B was an Arctic Rescue version with special preheating equipment for efficient warm-up of the engines while operating in cold climates. Raiders were assigned to Sheppard AFB, Wichita Falls, Texas, where they served as mechanical trainers until 1955.

The C-125 was a high-wing monoplane with a tailplane mounted halfway up the fin. The landing gear was of the fixed 3-wheel type that looked like a reverse tricycle arrangement. The crew of 2-4 included pilot, copilot, navigator and radio operator. Three Wright R-1820 1,200-hp engines mounted two in the wings and one in the fuselage nose supplied the power for a top speed of 207 mph. As a general purpose light transport the C-125 could carry 32 fully equipped combat troops or haul an 11,000-pound load of freight.

Production totaled 23 C-125 Raiders delivered to the Air Force, none of which were used in combat.

The **Cessna LC-126 Light Cargo Transport and Trainer** was an off-the-shelf purchase of the Model-195 civilian cabinplane adapted for Air Force use. The LC-126A could be rigged with wheels, skis or floats for use as an Arctic Rescue plane. The LC-126B was routed to the Air National Guard for use as a utility trainer. The LC-126C was the most-produced variant for use as an instrument trainer. First deliveries to the Air Force began in 1949.

The LC-126 was a single engine high-wing monoplane with conventional 3-wheeled landing gear and a standard tail assembly. The Jacobs R-755 300-hp engine drove a 2-bladed prop to give it a top speed of 175 mph. Pilot and copilot or trainee had side-by-side seats with dual controls. Three passengers sat on a bench seat in the cabin and their bags were loaded in a luggage compartment to the rear. The bench could be removed to accommodate up to 600 pounds of small freight packages. Production totaled 83 of the LC-126 variants delivered to the USAF.

The **Lockheed C-130** made its first flight on August 23, 1954, and delivery of the production model C-130A Hercules to TAC's 463rd Troop Carrier Wing began in December 1956. Designed to give the Air Force a Cargo Transport that could haul heavy loads and large equipment while operating in and out of short, rough runways, the C-130 proved effective in deploying entire military organizations to and from overseas bases. The RC-130 is a Photo-Recon variant; the DC-130 is used to launch and control drone targets for air defense weapons practise; the GC-130 is a drone director for ARDC; the JC-130 was used for satellite recovery of SAMOS capsules; and the C-130D was fitted with skis and ATO rockets for use in Antarctica. The most versatile variant is the HC-130, which has been used for search and rescue, weather recon, recovering Discoverer satellite capsules and recovery of NASA manned space flight modules. One example was fitted with two Allison turbojets for Short Take-Off and Landing (STOL) experiments. The KC-130 Tanker variant was used by the Marines.

The C-130 is a high-wing monoplane with retractable tricycle landing gear, a standard tail section and an unusually long nose. The 5-man crew includes pilot, copilot, navigator-radio operator, flight engineer and radar operator. Four Allison T56-A 4,050-hp engines give it a top speed of 385 mph. A cargo load of 25,000 pounds can be hauled 4,300 miles or a load of 35,000 pounds can be carried on a range of 3,500 miles. The maximum payload is 40,000 pounds and the unloaded ferry range stretches out to about 4,900 miles. The Hercules carries 92 passengers or 64 fully equipped troopers or 74 litters

Lockheed C-130 Hercules, world-famous Cargo Transport that saw service as an all-purpose STOL aircraft in Vietnam. (USAF Photo 154561-AC)

with medical attendants. Also nicknamed "Herk" and "Herky Bird," the C-130 has been used around the globe by USAF, Navy, Marines and Coast Guard units. It has been chosen by the air arms of many friendly nations to fulfill their needs for a genuine all-around workhorse.

On December 16, 1963 a 13-man team of Army and Air Force parachutists jumped from a C-130 at 43,000 feet and delayed the chute openings until below 2,500 feet to claim the world record for mass free falls. In 1964 a C-130 made 21 takeoffs from and landings onto the deck of the aircraft carrier *Forrestal* without JATO assistance or the use of arresting equipment. In 1967 the Hercules was converted into the most powerful Gunship of the Vietnam war. Designated the AC-130 Gunship-Hercules, it was armed with four 7.62mm miniguns and four 20mm Vulcan multi-barreled cannon.

On July 3, 1976, at a secret base outside of Tel Aviv, squads of Israeli commandos boarded three C-130 Transports that carried them to Entebbe airport in Idi Amin's Uganda on a raid to rescue Air France Flight 139, its crew of 12 and 102 of the original 244 passengers who boarded the plane in Athens, for a journey to Paris, via Tel Aviv, and were skyjacked along the way by Palestinian guerrillas.

On April 24, 1980, six Air Force C-130's took off from Egypt and participated in "Project Delta," the aborted attempt to rescue 50 American hostages being held captive by Iranian militants at the U.S. Embassy in Tehran. A Navy RH-53 Helicopter collided with one of the C-130's in the "Desert-One" staging area, resulting in 8 deaths and 5 injuries to members of Army Colonel Charlie Beckwith's 90-man "Blue Light" rescue team, referred to affectionately as "Chargin' Charlie's Angels."

During 1980, Air National Guard prop-jet C-130's were equipped with new modular airborne forest fire fighting systems developed by FMC Corporation in San Jose, California. Each "Flying Fire Truck" carries 30,000 pounds of liquid fire suppressants and can spray 300,000 square feet in ten seconds, leaving the area fireproofed and fertilized for regrowth.

A total of 928 C-130's have been built, including 404 for the USAF.

The **Convair C-131 Samaritan** joined the Air Force on April 1, 1954, as a military version of the Model CV-240 civilian airliner. The C-131 was also a progressive development from the Consolidated-Vultee T-29 Flying Classroom. The cargo door on the C-131 was on the left front side and the door on the T-29 sistership was on the right. Joining MATS as the first air evacuation hospitalship to have a pressurized cabin, the C-131A carried iron lungs, chest respirators and a variety of modern special medical equipment. The C-131B was patterned after the Model CV-340 with 48 seats and sophisticated electronics devices for use as testbeds and for retrieving missile nose cones. The YC-131C was used as a testbed for Allison T56-A turboprop engines. The VC-131D was derived from the Model CV-440, having improved sound proofing and a plush interior with 44 reclining chairs for use as a flying laboratory and VIP Staff Transport. The TC-131E was an Electronics Counter Measures (ECM) trainer used by the Strategic Air Command. The RC-131F was a Photo-Recon variant, and the C-131G was used by the Airways and Air Communications Service (AACS) in checking flight aids.

The C-131 was a low-wing monoplane with tricycle landing gear and a standard tail section. The crew consisted of pilot and copilot with a nurse and two medical technicians. Two Pratt & Whitney R-2800 2,500-hp engines gave it a top speed of 295 mph with a

The Convair C-131 Samaritan aero-medical Cargo Transport, developed from the commercial Convair-Liner, was the first USAF hospitalship to have a pressurized passenger cabin. Seen here during its first flight on March 26, 1954. (USAF Photo 49773-AC)

range of 1,500 miles. Up to 48 seats were provided for crew and passengers. The airborne ambulance carried 27 stretchers or 37 ambulatory patients seated facing to the rear.

A major air disaster occurred on December 17, 1960, when 53 personnel were killed in the crash of an overloaded C-131 Samaritan at Munich, Germany.

A total of 122 C-131's of all variants were delivered to the Air Force.

The **Douglas C-133 Cargomaster** was an extensively modified version of the Douglas YC-124B Globemaster-II for use as a Strategic Transport. The C-133A made its first flight on April 24, 1956, and deliveries to the 1607th Air Transport Wing, MATS Terminal, Dover AFB, Delaware, began to arrive in August 1957. The C-133B was the largest production model prop-driven Cargo Transport in Air Force history, with a total weight in excess of 300,000 pounds and a maximum payload of over 100,000 pounds. Cargomasters equipped with an all-weather radar system strengthened the MATS worldwide capability, flying routes to Japan, Africa, Europe and the Middle East. All the Inter-Continental Ballistics Missiles (ICBM) in the Air Force aresenal—Thor, Titan, Atlas, Jupiter—could be loaded through the mammoth rear cargo doors. On December 16, 1958, a C-133 broke the international payload-to-height record by lifting a 117,900-pound payload to an altitude of 10,000 feet.

The C-133 was a high-wing monoplane with a standard tail section and paired wheel sets supported by a nose gear. The 5-man crew included pilot, copilot, navigator, systems engineer and loadmaster. On long flights a full backup crew slept at the rear of the upper deck. Four Pratt & Whitney T34-P 7,500-hp engines gave it a top speed of 360 mph. Designed primarily as a long range freight carrier, the C-133 had a range of 2,250 miles with a 90,000-pound load of cargo, or 50,000 pounds could be hauled on a range of 4,030 miles. The ferry range stretched out to 4,360 miles. As a troop transport it could carry over 200 fully equipped combat troops, or accommodations could be provided for a combination of stretchers, ambulatory patients and medical attendants. Production was completed in April 1961 after 49 C-133's had been delivered to the USAF.

Douglas C-133 Cargomaster, largest prop-driven Cargo Transport in USAF history. (USAF Photo 156690-AC)

The **Boeing KC-135 Stratotanker** was developed from the Model-707 civilian airliner that was designed as a private company-funded project to become the first jet transport completed and flown in the United States. The Model-707 made its first flight on July 15, 1954; the KC-135 made its initial flight on August 31, 1956, and deliveries began to arrive at the 93rd Air Refueling Squadron (SAC), Castle AFB, California, in June 1957 as the first jet-powered Cargo Tanker in Air Force history. Originally, the "flying boom" technique used a rigid 47-foot boom, with all the refueling equipment on the lower deck—the upper deck was for passengers and cargo. Later versions used the "probe and drogue" method whereby the Fighters rigged with probes hooked up with drogues trailing from the boom. The KC-135A was a Cargo Tanker version, the KC-135B was a SAC Airborne Command Post (ACP), and the JKC-135 and NKC-135 were used in test programs.

The KC-135 is a low-swept-wing monoplane with retractable landing gear and a standard tail section. The 4-man crew includes pilot, copilot, navigator and boom operator. Four Pratt & Whitney J57-P 13,750-pounds thrust turbojets or TF39-P 18,000-pounds thrust turbofans give it a top speed of 585 mph. The transfer fuel load of 31,200 gallons can be delivered to the receiving aircraft at a rate of 1,000 gallons per minute. As a troop carrier the KC-135 can accommodate 80 fully equipped combat troops in the upper deck. As a cargo hauler it can carry 50,000 pounds of freight. With a full 120,000-pound transfer fuel load, the range is 1,150 miles, and the ferry range stretches out to 9,200 miles.

In 1975 Boeing began installing new wing skins on KC-135A's in a program designed to extend the aircraft's lifespan by about 25 years. In June 1977 work was started on the development of NASA-styled winglets to be fitted on outer wing panels of all KC-135's in service. The first flight with winglets was made on July 24, 1979, and it is predicted that future refinements will produce an eight percent saving of fuel.

The **Boeing C-135 Stratolifter** was developed from the KC-135, differing notably by deletion of the refueling equipment and modification of the interior. The C-135 made its first flight on May 19, 1961, and initial deliveries to the MATS Terminal at McGuire

AFB, New Jersey, began in June 1961 of the first strategic jet Cargo Transport in USAF history. The C-135 could carry 89,000 pounds of cargo or 126 fully equipped combat troops or 44 litters and 54 ambulatory patients with medical attendants. Included in the combat cargo loads were the capability of carrying 376 boxes of ammunition or 1,090 cases of "C" rations. The C-135 was the Cargo Transport version, the RC-135 was a Photo-Recon plane, the VC-135 was a VIP Staff Transport, and the WC-135 was a Weather Recce variant. The EC-135 was designed as an Electronics Airborne Command Post for TAC, a "Looking Glass" ACP for SAC, and a "Night Watch" ACP for Headquarters Command. Normal crew consisted of pilot, copilot, navigator, systems operator and flight engineer.

On April 17-18, 1962, Dave Crow flew his C-135B to a new payload-to-height record by lifting 66,139 pounds to a height of 47,170 feet, and Major Hamann's crew established the payload-speed record by carrying 30,000kg around a 2,000km course at a speed of 616 mph.

Stratolifters were struck by two major air disasters when 75 people were killed in a crash at Clark Air Base in the Philippines on May 11, 1964, and 85 people died in a crackup at the Los Angeles airport, California, on June 25, 1965.

Production ended in 1965 after 820 of all the KC-135 and C-135 variants had been delivered to the Air Force.

Boeing KC-135 Stratotanker, first USAF all-jet Cargo Tanker, is photographed here during a flight with the 93rd Air Refueling Squadron out of Castle AFB, California, June 1957. (USAF Photo 157131-AC)

The **Boeing VC-137 VIP Cargo Transport** was developed from the Model-707 air-liner simultaneously with development of the KC-135 Stratotanker. The VC-137A made its first flight on April 7, 1959, and delivery of three aircraft began shortly thereafter to the 1254th Air Transport unit (MATS), Andrews AFB, Maryland. The VC-137B was a redesignation of the VC-137A for flying Special Air Missions (SAM) by MATS aircrews. The VC-137C served as Air Force One, flying the President of the United States, cabinet members and foreign heads of state on diplomatic missions around the globe. The VC-137 set many point-to-point speed records flying from the East Coast across the Atlantic Ocean.

The VC-137 was a low-swept-wing monoplane with retractable gear and a standard tail assembly. Four Pratt & Whitney JT3D 18,000-pounds thrust turbofan engines gave it a top speed of 625 mph. Up to 49 VIP passengers or a payload of 53,300 pounds could be carried on a maximum range of from 3,000 to 6,000 miles. The 4-man flight crew included pilot, copilot, navigator and engineer. An 8-seat forward compartment was a communications center with galley and toilets. The central section was an airborne Headquarters with conference tables, swivel chairs, projection screen and catnap bunks. The rear section had 14 double-reclining chairs, tables, galley and toilets.

Within a matter of minutes after President John F. Kennedy was assassinated by Lee Harvey Oswald on November 22, 1963, Lyndon B. Johnson and Jacqueline Kennedy boarded Air Force One at Dallas, Texas, for the flight back to Washington. Two hours after the fatal bullets had struck, Vice President Johnson was sworn in as President with Jackie at his side—her skirt still bloodstained. That was the only time in history a President of the United States took the oath of office aboard an airplane. Only four VC-137's were delivered in the initial 1959 batch. In 1970 Boeing was awarded a contract to design an Airborne Warning And Control System (AWACS) aircraft. Two VC-137D prototypes were developed, the first flight being made on February 9, 1972. Those two airplanes were modified and redesignated E-3A Electronics AWACS aircraft, and first flew in February 1975.

At 2:55PM EST on Sunday, January 25. 1981, a VC-137—appropriately nicknamed "Freedom One" for the occasion—landed at Stewart Airport near the West Point Academy in New York, thus completing the final leg of a long journey that returned the 52 freed American "Hostages" to home soil for the first time since their imprisonment when Iranian terrorists took over the U.S. Embassy in Tehran on November 4, 1979. Their ordeal in captivity had lasted 444 days.

The **Lockheed C-140 Jetstar** compact jet Transport and Utility aircraft made its first flight in August 1961, and deliveries of the C-140B began to arrive at the 1254th Special Air Missions unit, Andrews AFB, Maryland in December 1961. The C-140A was delivered to the Air Force Communications Service (AFCS) in July 1962.

The C-140 was a low-swept-wing monoplane having a plus-type tail section and retractable tricycle landing gear with paired wheels on all three housing units. Four Pratt & Whitney J60-P 3,000-pounds thrust engines were mounted in lateral pairs on both sides of the fuselage just forward of the tail assembly. The fail-safe fuselage had five windows and five seats along each side of the cabin. The top speed was 570 mph and a maximum cargo load of 3,000 pounds could be hauled for an extended range of 2,000 miles. Accommodations in the Transport version used by SAM for MATS provided for a 2-3 man crew and ten passengers. The AFCS version could carry 8 passengers or a 5-man crew for inspecting Military Navigation Aids worldwide.

Production totaled 16 C-140's for the USAF. The T-40 was a proposed Trainer variant, ten of which were allocated in the Fiscal 1962 budget, but the T-40 never reached the production stage.

The **Lockheed C-141A Starlifter** made its first flight on December 17, 1963, and became operational with the Military Airlift Command (MAC) in April 1965. The world's most powerful Cargo Transport ushered in a new epoch of strategic airlift, giving the

Lockheed C-141 Starlifter Strategic Airlift Cargo Transport snapped during its initial flight out of Dobbins AFB, Georgia on December 17, 1963. The late president John F. Kennedy pressed a button at the White House on August 22, 1963, electronically opening factory doors at the Marietta plant for the C-141A roll-out. (USAF Photo 168991)

USAF a capability of rapidly deploying entire military organizations to any place on earth. In addition to troops, equipment, supplies and weapons necessary to a fighting unit, the C-141A can airlift all missiles employed by the armed services. The Starlifter can be slowed down to 130 mph allowing air-dropping of paratroops and aerial delivery of very heavy loads of field equipment and ammunition. The C-141 is the first jet Transport from which Army paratroops have jumped, and it also set the record for air drops when 70,195 pounds of cargo were parachuted to earth in a field exercise.

The C-141A is a high-swept-wing monoplane having a T-type tail section and hydraulically retractable tricycle landing gear with a 2-wheel nose unit and 4-wheel bogie main units. The 4-man crew includes pilot, copilot, navigator and systems engineer, with accommodations for a full backup crew when needed. Four Pratt & Whitney TF33-P 21,000-pounds thrust turbofan engines mounted in underwing pods give it a top speed of 570 mph. Arrangements can be varied to accommodate 154 passengers, or 123 fully equipped combat troops, or 80 litters and 16 ambulatory patients with 8 medical attendants. The range varies from 3,600 miles with a full load of 70,850 pounds to 5,500 miles with 20,000 pounds to an unloaded ferry range of 7,000 miles. With slight modification the payload can be upped to 86,000 pounds when carrying a Minuteman Inter-Continental Ballistic Missile (ICBM). Teamed with the Lockheed C-5A Galaxy, Starlifters annually participate in operation "Reforger," an exercise in overseas deployment of units to Europe, and a practical demonstration of the art of strategic airlift.

Production totaled 284 C-141A Starlifters for the Air Force.

On January 8, 1977, a stretched YC-141B prototype rolled out of the factory at Marietta, Georgia, for trials. The YC-141B was converted from a C-141A by inserting two fuselage extension plugs that increased the length by 23 feet and expanded the cube cargo capacity by thirty percent to handle a payload of 89,150 pounds. Inflight refueling capability was added and fuel consumption was improved. The first modified production model C-141B was delivered to the USAF on December 4, 1979. Immediate plans called for delivery of 80 C-141B's in 1980, and it is proposed that eventually all 271 active duty

C-141A's will be converted to C-141B's. The fleet's effectiveness will be increased by about 20 percent with little added expense in costs of operation.

The **Ryan-Hiller-LTV XC-142** prototype Tiltwing VTOL/STOL Cargo Transport made its first flight on September 29, 1964, and gave its first all-around capability demonstration in February 1965. Designed as the largest tri-service Vertical/Short Take-Off and Landing (VSTOL) aircraft, the XC-142 was intended for rapid transport of combat troops from ocean vessels or unimproved landing strips into and out of rugged terrain. The engines were permanently fixed in wings that tilted upwards through 100 degrees; at maximum tilt the engines exhausted straight downward. The XC-142 could fly conventionally with normally extended wings, or it could take off almost vertically when wings were at full tilt. It could fly backwards or hover in one spot and rotate through 360 degrees. While in the hover attitude it could point the nose straight up, open the rear cargo door and perform like a dump truck to unload freight packages right into the arms of troops waiting on the ground.

The XC-142 was a high-wing monoplane with a tail section that had the tailplane mounted one-third the way up the fin and a tail rotor for hover stability. The landing gear was of the retractable tricycle type with six wheels. Four General Electric T64-GE 2,850-hp turboprop engines gave it a top forward speed of 430 mph. Pilot and copilot sat in the flight deck with dual controls. The main cabin could accommodate 32 fully equipped assault troops or 24 litters with four medics. As a cargo carrier it could haul 8,000 pounds of freight. The combat radius with a full load was 470 miles; the ferry range stretched out to 3,800 miles.

Only five XC-142's were delivered for tri-service testing—no production contracts were made. The XC-142 turned out to be the highest designation assigned in the "C" category prior to rolling the numbers back to zero in the tri-service scheme of 1962.

CARGO TRANSPORTS— 1962 TO THE PRESENT

The **Lockheed C-5A Galaxy** originated in 1963 as the XC-4 when the Air Force ordered a Strategic Transport with a gross weight about 700,000 pounds. The prototype XC-5 made its first flight on June 30, 1968, and delivery of the production model C-5A to the Military Airlift Command began in December 1969. The C-5A was another giant step forward in the USAF science of Strategic Airlift, already proven to represent the most awesome air armada in the history of military Cargo Transports. That Goliath of an aircraft is large beyond comprehension—when viewed flying directly overhead, it gives one the feeling of being beneath an aluminum overcast. A bit of trivia: the Galaxy is almost twice as long as the first Wright Brothers' flight at Kitty Hawk. The Galaxy got off to a good start by immediately justifying all the time and money spent on its upbringing—two M-48 tanks were hauled in one demonstration, and three CH-47 Chinook helicopters in another; dry runs proved it can be stuffed with ten Pershing missiles or up to 110 ordinary vehicles.

The C-5A is a high-wing monoplane with a T-type tail section and retractable under-carriage having 28 wheels. The 5-man crew is composed of pilot, copilot, navigator, systems engineer and a loadmaster. A full backup crew is carried on long missions. Four

Lockheed C-5A Galaxy, the world's largest jet-propelled Cargo Transport snapped over Charleston, South Carolina, March 13, 1972. (USAF Photo 182554)

General Electric TF39-GE turbofan engines generate 41,000 pounds of thrust each to give it a top speed of 570 mph. As a cargo carrier it can haul up to 265,000 pounds of freight, and as a personnel transport arrangements can be made to accommodate 345 troops and their gear. The range varies from 3,000 miles with a full load to 6,500 miles with a 100,000-pound payload.

Once a year the C-5A joins the Lockheed C-141 Starlifter to participate in operation "Reforger," a joint Army and Air Force exercise in Strategic Airlift of dual-based forces from the United States to Europe. Those planes carry far more than men and equipment when they make the journey across the Atlantic; they bring to Europeans a symbolic reminder that our pledge to help maintain freedom is as strong today as it was when the North Atlantic Treaty Organization (NATO) was formed in 1949. During Reforger of 1974, over 12,000 personnel and 2,500,000 pounds of equipment and supplies were transported for deployment.

Tragedy always stalks a great aircraft. On April 4, 1975, a C-5A Galaxy carrying a load of war orphans crashed after takeoff near Saigon (now Ho Chi Minh City) in South Vietnam, killing 172 of the passengers, to rank 8th on the list of the world's major air disasters.

Production ended in May 1973 after 81 C-5A's had been delivered to the USAF. In 1979 the Air Force still had 77 C-5A's operational, and Lockheed maintains a standby readiness to resume production in event of a national emergency.

The **Beechcraft VC-6 King Air** was developed from the Model-90 civilian business cabinplane that made its first flight on January 20, 1964. The VC-6A was delivered to the 1254th Special Air Missions (SAM) unit at Andrews AFB, Maryland, for use as a VIP Cargo Transport. The VC-6B delivered to the 1254th in 1974 had a wider span, but otherwise the two variants were similar. The left side door has a built-in airstairs, and an emergency exit is provided on the right fuselage wall. The Army U-21 King Air Utility aircraft was identical to the VC-6A.

The VC-6 is a low-wing monoplane with retractable tricycle landing gear and a standard tail section having a swept-back tailplane. Two Pratt & Whitney PT-6A 550-hp turboprop engines drive 3-bladed propellers to produce a top speed of 280 mph. Pilot and copilot sit side by side with dual controls in the forward cockpit. Passenger arrangements can vary from four plush reclining chairs to eight comfortable seats, with a refreshment bar, baggage racks and toilet facilities. The ferry range stretches out to 1,575 miles with a minimum load.

Only two VC-6 transports were delivered to the Air Force; one was a VC-6A and the other a VC-6B.

The **DeHavilland C-7 Caribou** was a derivative of the Canadian DHC-4 civilian Short Take-Off and Landing (STOL) tactical transport. The DHC-4 made its first flight on July 30, 1958, and was delivered to the Army for trials as the YAC-1 Caribou in October 1959. The Army designation was changed from YAC-1 to CV-2B under the tri-service scheme of 1962. On January 1, 1967, when the USAF assumed responsibility for handling all Army tactical and strategic deployments, all surviving CV-2B's were transferred to the Air Force where the designation was again changed, this time to the C-7A Caribou. In Vietnam the C-7A was used to transport troops and cargo from major supply bases to forward fire camps.

The C-7A is a high-wing monoplane with retractable tricycle landing gear and a plus-type tail assembly. Two Pratt & Whitney R-2000 engines with 1,450-hp each give it a top speed of 215 mph. Pilot and copilot sit up front, and arrangements can be made to accommodate 32 passengers or 26 fully equipped combat troops or 22 stretchers with 4 ambulatory patients and 4 medical attendants. A maximum cargo load of 6,000 pounds can be hauled, including a couple of Jeeps. Combat radius with a full load is 240 miles, and the ferry range when empty is 1,300 miles.

A total of 134 C-7A's were redesignated from Army CV-2B's in 1967. No subsequent production orders were submitted.

The **DeHavilland C-8 Buffalo** was originally developed from the Canadian DHC-5 for the Army as the CV-7 powered assault and utility transport that was delivered for trials in April 1965. In 1967 all CV-7's were transferred to the USAF and redesignated C-8A. The Bell-DeHavilland XC-8A Buffalo was a DHC-5 borrowed from DeHavilland by Bell for STOL experiments. The XC-8A made its first flight on March 31, 1975, and reported for development tests to the 4950th Test Wing at Wright-Patterson AFB, Dayton, Ohio, in April 1975. Designed as an Air Cushion Landing System (ACLS) Cargo Transport, the XC-8A can take off from swamps, ice covered ponds, ships, water, soft soil, snow fields or rough landing strips. Lifting power is supplied by downward blowing fans mounted on the sides of the fuselage to literally create an ''air cushion'' that flings the aircraft upward almost vertically. Another version of the C-8A used by NASA was extensively modified and equipped with two Rolls Royce turbofans generating 9,000 pounds of thrust each.

The Bell XC-8A is a high-wing monoplane with long-legged 6-wheel retractable tricycle landing gear and a T-type tail section. Two General Electric T64-GE 2,850-hp engines give it a top speed of 290 mph. Pilot, copilot and crew chief make up the crew. The Buffalo has an 18,000-pound cargo capacity and can accommodate up to 41 passengers or 35 combat troops or 24 litters with medics. The combat range is 750 miles when fully loaded.

Production totaled four C-8A's transferred from the Army, one Bell XC-8A ACLS for the Air Force, and two jet-powered C-8A's for the National Aeronautics and Space Administration (NASA).

The **McDonnell-Douglas C-9 Nightingale Aeromedical Transport** was developed from the DC-9 civil airliner that made its first flight on February 25, 1965. The production model C-9A rolled out of the factory in June 1968 and delivery began on August 10, 1968, to the 375th Aeromedical Wing (MAC) at Scott AFB, Illinois. Rushed immediately into the Vietnam war for transporting casualties to the West Coast, the C-9A has an intensive care compartment with separate atmosphere and ventilation controls for special handling of the seriously wounded. The C-9B for the tri-service is a Navy transport. The VC-9C was delivered in 1975 to the Special Air Missions unit at Andrews AFB, Maryland, for use as a VIP Transport.

The C-9A is a low-wing monoplane with retractable 6-wheel tricycle landing gear, a long nose, a T-type tail assembly and twin fanjets mounted in pods on either side of the fuselage rear just forward of the tail section. Two Pratt & Whitney JT8D-9 turbofan engines produce 14,500 pounds of thrust each to give it a top speed of 570 mph. Pilot, copilot, engineer, two nurses and three aeromedical technicians make up the crew. Accommodations are provided for 45 sitting patients or 40 litter patients or combinations of the two. The payload capacity is 26,000 pounds and the maximum range is 1,570 miles.

Production totaled 38 C-9's, including 24 for the Air Force and 14 for the Navy.

The **McDonnell-Douglas KC-10 Advanced Tanker-Cargo Aircraft (ATCA)** is an improved development from the DC-10 civilian jet airliner. An order was submitted in 1978 for 20 production model KC-10A Extenders to be delivered beginning in the Fall of 1980 for use in Cargo Transport or inflight refueling Tanker roles, or in a combination of the two when deploying entire mobilization units to overseas staging areas. Future plans call for the KC-10 to replace the fleet of Boeing KC-135 Stratotankers; each KC-10 will be able to do the work of about two KC-135's. The basic internal fuel system and the tanks carrying the transfer fuel are inter-connected, allowing use of the entire fuel load for long range missions. Five passenger doors and a large cargo door allow efficient loading of personnel and equipment into the main compartment. Transition from the KC-135 to the KC-10 will be a long process because recent modifications have extended the KC-135 lifespan into the next century.

The KC-10A is a three-engine low-wing monoplane with retractable tricycle landing gear having ten wheels. The tail section is a standard configuration with one engine mounted on top of the fuselage at the base of the tail fin; the other two engines are housed in underwing pylons. The crew includes pilot, copilot, systems engineer, navigator and boom operator, plus flight attendants, medical technicians and scientific specialists when needed. Three General Electric CF6-50 turbofan engines generate 51,000 pounds of thrust each to produce a top speed of 610 mph. Internal arrangements can be varied for accommodating 250 to 380 passengers. The capacity payload of 170,000 pounds can be hauled for a distance of 3,800 miles and the ferry range extends to beyond 7,000 miles.

The worst American air disaster in history occurred on May 25, 1979, when a civilian DC-10 jetliner crashed on takeoff from O'Hare International Airport at Chicago, Illinois, killing all 271 people aboard and two on the ground. Production of the KC-10 is scheduled

McDonnell-Douglas KC-10 Extender; Advanced Cargo Tanker Aircraft (ATCA) designed as replacement for KC-135 sometime in the future. (USAF Photo 80-306)

to continue throughout the 1980's, as funds are allocated on an annual basis until all the KC-135's have been phased out. As of July 1, 1979, the KC-10 was still in the development stage.

The **Beechcraft C-12 Super King Air** was a direct conversion from the Model-200 light cabinplane that made its first flight on October 27, 1972. Delivery of the C-12A began in 1975 to Military Assistance Groups (MAG) on Air Attache duty.

The C-12 is a low-wing monoplane with a T-type tail section and retractable tricycle landing gear having five wheels. Pilot and copilot sit side by side with dual controls in the forward cockpit separated from the flight deck by a sliding door. Two Pratt & Whitney PT6A 850-hp engines drive 3-bladed props to give it a top speed of 330 mph. The normal passenger load is eight, but internal arrangements can be altered to accommodate 13 people. A baggage area in the rear holds 410 pounds of luggage. The cabin seats are removable for conversion to a cargo carrier hauling 1,950 pounds of freight. Wingtip fuel tanks give it a maximum range of over 2,100 miles.

By July 1979 a total of 154 C-12's had been built, including 30 C-12A's for the USAF, 80 Army variants and 44 UC-12B's for the Navy.

The **Boeing YC-14 Advanced Medium STOL Transport (AMST)** was designed in competition with the McDonnell-Douglas YC-15 as a potential replacement for the fleet of Lockheed C-130 Hercules Cargo Transports. The YC-14 made its first flight on August 9, 1976, and evaluation tests began immediately. The engines are mounted on top of the wings near the fuselage so that they exhaust over the wing's upper surface. The engines jut forward noticeably from the wings to offer significant noise reduction.

The YC-14 is a high-wing monoplane with a T-type tail section and retractable 10-wheeled tricycle landing gear. Pilot and copilot make up the crew. Two General Electric CF6 turbofan engines produce 51,000 pounds of thrust each to give it a top speed of 505 mph. Up to 150 combat troops can be carried. A capacity cargo load of 81,000 pounds can be hauled under normal power operations, or 27,000 pounds can be loaded for STOL operations. The range with a full load of fuel is 3,190 miles.

By July 1979 only two prototype YC-14's had been delivered to the Air Force for trials.

The **McDonnell-Douglas YC-15 Advanced Medium STOL Transport (AMST)** was developed in competition with the Boeing YC-14 AMST aircraft. The YC-15 made its first flight on August 26, 1975, and operational trials began shortly thereafter. The YC-15 represents quite a different aerodynamic approach from that of the YC-14; however, both aircraft have demonstrated the capability of taking off within the 2,000-foot specification requirement.

The YC-15 is a high-wing monoplane with T-type tail assembly and retractable 10-wheeled tricycle landing gear. Pilot and copilot ride in the forward cockpit. Four Pratt & Whitney JT8D turbofan engines produce 16,000 pounds of thrust each to give it a top speed of 500 mph. Up to 150 fully equipped combat troops can be accommodated in the cabin. A capacity cargo load of 62,000 pounds can be carried on a combat radius of 460 miles, and the ferry range extends to 2,990 miles. One YC-15 was redesigned by adding 22'3" to the span and substituting a CFM56 22,000-pounds thrust turbofan at one of the outboard positions.

By July 1979 only two YC-15 prototypes had been delivered to the Air Force for service trials.

The Fiscal 1981 Defense Budget included funds for planning future development of a C-X Cargo Transport capable of hauling heavy equipment, such as the Army's 60-ton XM-1 Tank. Designed to reinforce the Rapid Deployment Force (RDF) in support of NATO and other overseas movements, specifications call for a Short Take-off and Landing (STOL) capability, a compartment about half the size of the Lockheed C-5 Galaxy, a payload capacity of about 135,000 pounds, top speed near 500 mph, an unrefueled range of around 2,750 miles and a cruising altitude above 26,000 feet.

ELECTRONICS AIRCRAFT FOR THE USAF— 1962 TO THE PRESENT

The **Boeing E-3 Airborne Warning And Control System (AWACS)** aircraft was originally designated EC-137D, which was itself a development from the Boeing-707 jetliner. The EC-137D prototype flew for the first time on February 9, 1972, and the redesignated E-3 prototype took to the air in February 1975. Delivery of the production model E-3A Sentry began on March 24, 1977, to the 552nd Airborne Warning and Control Wing, Tinker AFB, Oklahoma. Distinguishing features are an elliptical rotodome thirty feet in diameter mounted on six-foot tall struts rooted into the top rear of the fuselage, new engine pylon fairings, multiple antennas, specially located windows, doors and hatches, completely new electrical wiring and provisions for receiving inflight refueling. Internal arrangements provide separate compartments for crew and surveillance technicians. The sophisticated radar, communications, sensing, display and navigational devices afford high and low level surveillance of all types of airborne vehicles. The range of accurate detection is 230 miles for low altitude targets and 300 miles for higher flying aircraft.

The E-3A is a low-swept-wing monoplane with retractable ten-wheeled landing gear and a standard tail section. The four-man flight crew includes pilot, copilot, engineer and

Boeing E-3 Sentry, Airborne Warning And Control System (AWACS), with a radar rotodome atop the fuselage. Photo taken during a flight over Edwards AFB, California in 1976. (USAF Photo KE-63050)

navigator. The systems crew is composed of 13 specialists to monitor the computerized electronics devices. Four Pratt & Whitney TF33-PW turbofan engines generate 21,000 pounds of thrust each to produce a top speed of 625 mph. The maximum unrefueled range extends out to 7,475 miles and when carrying a capacity payload of 89,000 pounds the range is 4,300 miles.

Designed primarily for use by the Tactical Air Command (TAC) and Air Defense Command (ADC), the E-3A is used in a dual role. The TAC system is a Command and Control Center for quick reaction tactical deployment, and the ADC version is a survivable Early Warning Air Command Post for tracking alien aircraft and directing North American Air Defense (NORAD) forces over the United States and Canada. During surveillance the rotodome is hydraulically driven at an angular velocity of six revolutions per minute while the pulse doppler technology emits beams at a high-frequency rate. Computers operate in a real-time immediate-display environment to detect enemy vehicles while simultaneously throwing out an anti-detection screen that protects against being picked up by enemy devices.

The hierarchy of USAF aircraft developed from the Boeing-707 series resembles a who's-who in aeronautical technology represented by the KC-135 Stratotanker in 1957, the VC-137 VIP Transport in 1959, the C-135 Stratolifter in 1961, the EC-137D AWACS in 1972, and the E-3A AWACS in 1975. As of July 1, 1979, a total of 16 E-3A's had been built for the Air Force and money had been allocated for delivery of six more.

The **Boeing E-4 Advanced Airborne Command Post (AACP)** was developed from the Boeing-747 civilian jetliner that made its first flight on February 9, 1969. Designed to replace EC-135 Stratolifters in service with the National Military Command, the first production model E-4A was delivered to Andrews AFB, Maryland, in December 1974. The E-4B was delivered to Offutt AFB, Nebraska, in 1977 after responsibility for operating AACP aircraft was transferred to the Strategic Air Command. The E-4B has an advanced command and control system plus room for a larger wartime staff. The aircraft has three decks: an upper deck for crew rest and comfort, lower deck front and rear for data processing and communications equipment, and the main deck for command and staff work areas, with eating and sleeping facilities. One purpose and mission of an AACP aircraft is to provide the President of the United States, his staff and military leaders with airborne capabilities for running the government and conducting a war in the event of a surprise nuclear attack against America—thus its nickname, "Doomsday," coined by the news media. The E-4 AACP is also referred to as the "Flying Whitehouse."

The Boeing-747 Jumbo Jet was designed as a versatile cargo and personnel carrier with optimum characteristics for both roles. The finished product was the most advanced airliner of its day, seating up to 490 passengers and carrying a capacity payload of 225,000 pounds.

The E-4 AACP is a low-swept-wing monoplane with retractable 18-wheel landing gear and a standard single-finned tail section. Four Pratt & Whitney JT-91 engines develop 41,000 pounds of thrust each to produce a maximum speed of 630 mph. The range with a 186,000-pound payload is 6,000 miles, and the unloaded ferry range extends to 7,600 miles. The normal flight crew of pilot, copilot, engineer and navigator is supplemented by a varied number of electronics, data processing and communications specialists.

Boeing E-4A Advanced Airborne Command Post (AACP) pictured in flight over Edwards AFB, California, July 18, 1974. (USAF Photo KE-50891)

On February 11, 1977, Commander-in-Chief Jimmy Carter got the thrill of his airborne life when he flew home to Plains, Georgia, in the E-4B AACP. Along the way, all that costly machinery was explained to him for the first time since he had taken the oath of office. As of July 1, 1979, four E-4's had been delivered to the USAF and two more were budgeted for.

BIRTH OF THE JETS

When warclouds darkened over the European continent in the late 1930's, aeronautical engineers began to focus their attention on a new type of aircraft propulsion. Airplane designers realized that piston engines had just about reached their peak power proficiency level. It was anticipated that the loss of propeller efficiency as blade tips approached sonic speed would prohibit any appreciable increase in airplane speeds above the existing record of 469 mph held by a German Messerschmitt ME-109.

The jet engine was invented in England by Frank Whittle, who demonstrated his gas turbine for the first time on April 12, 1937. Meanwhile, a German designer named Hans von Ohain completed research on his jet engine and was hired by Ernst Heinkel to produce a turbojet suitable for flight. Heinkel and Ohain produced the Heinkel HE-178, which took to the air on August 27, 1939, as the first jet-powered aircraft in history. At the request of Adolf Hitler, Heinkel also developed the first experimental military combat jet Fighter, the HE-280, which made its initial flight for the Luftwaffe on April 2, 1941.

Back in Britain, the Gloster E28/39 experimental aircraft, powered with a Whittle W-1 turbojet engine, made its first flight on May 15, 1941, as the Allies' initial entry into the age of the jets. That prototype was the predecessor of Britain's first Military jet aircraft, the Meteor MK-I Fighter.

As a result of negotiations on the part of Gen. Hap Arnold, in June 1941 an example of the Whittle W-IX turbojet engine was transported to the United States aboard a Boeing B-17 Flying Fortress under an international cloak of top secret intrigue. Using the Whittle W-IX engine as a model, on September 4, 1941, General Electric began design studies of America's first jet engine, the GE-I-A. On September 5, 1941, Bell Aircraft at Buffalo, New York, began development of three prototype airframes. Bell was already working on an XP-59 prop-driven, twin-boom pusher type Pursuit that was about to be abandoned. The designation XP-59A was assigned to the jet project as a shroud, pretending that it designated a part of the prop-driven XP-59 development. On October 2, 1942, with Bob Stanley at the controls and Col. Larry Craigie supervising, the XP-59A, fitted with the GE-I-A turbojet engines, took off from a dry lake bed in the desert at Muroc Field (later Edwards AFB), California, to become the USAAF's first airborne jet aircraft.

Adolf Hitler, confident in the early stages of WWII that the Allies would capitulate quickly, ignored the advice of airplane designers to proceed rapidly with development of jet-powered aircraft for the war. Had he accepted their suggestion sooner, the outcome of WWII might have been reversed.

The house of Messerschmitt joined the competition with their liquid rocket powered ME-163 Komet Fighter, which went zooming through a formation of B-17 Bombers over Berlin one day in March 1944, causing the startled American aircrewmen to look at each

Bell P-59A Airacomet, America's first jet-propelled aircraft. Photographed on the ground while its performance details were classified TOP SECRET. (USAF Photo 28715)

other in amazement. They would soon become accustomed to seeing an occasional jet Interceptor in the skies over Germany when the Messerschmitt ME-262 became operational.

A political struggle developed in Germany between Heinkel and Messerschmitt, each seeking favor in the eyes of "der Fuehrer." Messerschmitt was given the nod to proceed with development of a jet combat plane for use in the war. The choice indicated that Hitler was losing his touch as the world's foremost strategist because Heinkel was well ahead of Messerschmitt in technical knowledge of jet aircraft.

The first successful German jet combat plane was the Messerschmitt ME-262 prototype that made its first flight in July 1942. Reichsmarshall Goering steadfastly argued in favor of developing the ME-262 as a Fighter, and Hitler was equally insistent that it be developed as a Bomber. That high-level disagreement in policy was one more delay that granted a little time to the Allies. The ME-262A-1 Schwalbe (Swallow) was a Fighter, and the ME-262A-2 Sturmvogel (Stormbird) was the Bomber version.

While those differences were slowing development in Germany, the British were progressing steadily with their Gloster MK-I Fighter, sixteen of which reported to RAF Squadron-616 in July 1944 as the world's first combat-ready jet aircraft. Rushed into action as a counter-weapon to German Flying Bombs that were terrorizing the London area, the Meteor MK-I got its first air kill of the war by knocking down a V-1 Buzz Bomb on August 4, 1944. Although Germany had made the first successful jet aircraft flight, Britain had made up the lost ground.

In November 1943 Hitler decided to favor the Bomber version of the ME-262, which made its combat debut along the Western Front around the first of August 1944, two weeks after the English Meteor joined the battle. Approximately 1,433 of the ME-262's were built, but only about a hundred or so participated in a brief but frightful appearance at the front, being used mostly against elements of Patton's tank corps during the closing battles of WWII. The 2nd Staffel KG-51, Kommando Edelweis, was one of the first Bomber outfits to be equipped with the ME-262; that unit was knocked out of action in April of 1945 and transferred to Salzburg-Maxglam where it surrendered on May 3, 1945. The Meteor and the Swallow never engaged each other in head-to-head dogfights.

Back in America an order was placed with Bell for thirteen YP-59A jet Fighters, the first being flown in October 1943 with two GE-I-16 turbojets. One of those YP-59A's was traded to England for one of their Meteor MK-I Fighters in a friendly exchange of scientific know-how. The P-59A Airacomet production version began to arrive in August 1944 at the 412th Fighter Group, March Field, California, America's first and only all-jet aircraft unit at that time. Two turbojet engines were required on all the world's early jet planes because no one had invented a single engine powerful enough to supply the needed thrust. The P-59 was never used in combat, but was valuable as a trainer in those pioneer days.

Plans for development of America's XP-80 jet-powered Pursuit were turned over by Wright Field to the Lockheed Company in July 1943. The design team was led by Clarence Johnson, destined to achieve fame for his work on the P-80, the F-104 Starfighter, the U-2 Gray Ghost and the SR-71 (YF-12) Blackbird. The prototype XP-80 got off the ground for the first time on January 8, 1944, and an order was submitted for 5,000 production P-80 jets for use against Japan in the Pacific war. When the Atom Bomb brought an abrupt cessation of hostilities in August 1945, the anticipated probability of a long war with Japan ceased to exist. The contract for 5,000 aircraft was immediately reduced to 917 postwar P-80A Shooting Star Fighters; the first deliveries began to arrive at the 412th Fighter Group on December 3, 1945, three months after the "Rising Sun" had set.

The Age of the Jets was officially underway and America became the world leader in fierce competition that saw Charles Yeager break through the so-called "sound barrier" in a Special Research skyrocket Bell XS-1 on October 14, 1947. During hypersonic experiments in June 1962 and August 1963 the North American X-15 Special Research aircraft flew at a speed of 4,159 mph and reached an altitude of 354,200 feet—Joe Walker was at the controls on both of those historic flights. Honorary "Astronauts Wings" were awarded to Joe Walker, Bob White and R.A. Rushworth for having flown the X-15 above 50 miles (264,000 feet). Such trials with jet aircraft supplied the technical knowledge that allowed America to put a "man on the moon with the man in the moon."

JET FIGHTERS — WW II TO 1962

Development of American jet airplanes began with Fighter type aircraft in World War II when orders were submitted for the Bell P-59 Airacomet, Lockheed P-80 Shooting Star, Republic P-84 Thunderjet and North American P-86 Sabre Jet, all classified as Pursuits under the designation scheme of that time. On June 11, 1948, when the new USAF scheme of aircraft designations changed all "P" Pursuits to "F" Fighters, survivors in those four designs were redesignated F-59, F-80, F-84 and F-86. The Northrop P-89 Scorpion was ordered as a Pursuit, but the designation had already been changed to F-89 by the time of its first flight. All higher numbered Fighters were ordered with "F" designations. Propeller-driven veterans of WWII that were carried forward into the United States Air Force as Fighters were the F-24 Dauntless, F-38 Lightning, F-40 Warhawk, F-47 Thunderbolt, F-51 Mustang, RF-61 Black Widow, QF-63 Kingcobra and F-82 Twin Mustang. They are discussed in the section dealing with the Pursuits.

F-59 (P-59) AIRACOMET

The Bell XP-59A prototype was ordered on September 4, 1941, and made its first flight at Muroc Field, California, on October 2, 1942, with Bob Stanley at the controls on the first Jet flight in American history. The YP-59A limited-production model, equipped with 1,600 pounds thrust turbojet engines, first flew in August 1943, and deliveries for service trials began in January 1944. The production model P-59A Airacomet was delivered to the 412th Fighter Group, 4th Air Force, March Field, California, starting in August 1944. The P-59A was the first operational jet aircraft delivered to the USAAF, and the 412th Fighter Group, America's first all-jet unit, was stocked with the best people available to conduct the tests. The P-59B followed shortly thereafter, but it had only minor modifications. .

The P-59 was a twin-jet mid-wing monoplane with a slender fuselage, standard tail section and retractable tricycle landing gear. The pilot sat in an enclosed cockpit above the wing's forward edge. Two General Electric J31-GE-5 turbojet engines were installed at the juncture of the fuselage and the underside of the wings. Each engine generated 2,000 pounds of forward thrust to produce a top speed of 410 mph. Nose armament consisted of two 37mm cannon or one cannon and three 50-caliber machine guns. Underwing racks could be loaded with 1,000 pounds of assorted weapons to be carried on a combat range of 525 miles.

Operational tests revealed that the Airacomets were a little too unstable for use in the Fighter role. However, their pioneer existence was subsequently justified, since they provided a wealth of technical knowledge through service in the experimental roles of familiarization trainer, drone director, scientific research and engine development.

A total of 66 P-59 Airacomets were delivered to the USAAF. Production ended in 1945 with cancellation of outstanding contracts in favor of the preferred Lockheed P-80 Shooting Star, soon to start rolling out of the factory. In June 1948 the surviving test aircraft were redesignated F-59 Airacomet in the new Air Force designation scheme.

F-80 (P-80) SHOOTING STAR

The Lockheed XP-80 prototype made its first flight on January 8, 1944, at Muroc Field, California, with Milo Burcham at the controls. The YP-80A was delivered for service trials in October 1944 and the production model P-80A Shooting Star began to arrive at the 412th Fighter Group, March Field, California, in December 1945—too late for World War II. The P-80B was similar to the P-80A except for a more efficient power plant. The Shooting Star came as either a Fighter-Interceptor or a Fighter-Bomber. The name "Shooting Star" is the English translation from the Indian word "Tecumseh." Chief Tecumseh was the great Shawnee leader who was killed by General William Henry Harrison's troops during the War of 1812.

The P-80 was a single-jet, low-straight-wing monoplane with conventional tail section, retractable tricycle landing gear and the cockpit located just ahead of the wing's leading edge. The J33-A turbojet engine generated a thrust of 4,600 pounds to produce a top speed

Lockheed F-80 Shooting Star, first production jet Fighter in Air Force history. Four aviation cadets out of Williams AFB are flying in close formation over the Arizona desert. (USAF Photo 43679-AC)

of 605·mph. Armament consisted of six 50-caliber nose guns. A bomb load of 4,000 pounds could be hauled on a combat range of 1,200 miles. Alternate weapons available were ten air-to-surface missiles, an assortment of rockets, or six napalm tanks. Fuel tanks holding 165 gallons each were hung downward at the wingtips—later models increased the tiptank capacity to 260 gallons each. The FP-80 (F-14) was the first Jet Photo-Recon plane and the TP-80C in March 1948 was the first officially designated Jet Pilot Trainer.

Two YP-80A Shooting Stars arrived in Italy just before Benito Mussolini was executed and Adolf Hitler committed suicide in late April 1945. They were not used in combat, but were sent as a display of strength if needed in a prolonged war.

On August 6, 1945, Richard Bong, America's all-time leading Fighter Ace and winner of the Congressional Medal of Honor in WWII, was killed when the engine of his YP-80A failed during a test hop for Lockheed out of Burbank, California. On June 19, 1947, Al Boyd set a new world speed record of 624 mph in a P-80A. That was merely a prelude to many speed records destined to come in the Age of the Jets. In June 1948 all surviving P-80 Pursuits were redesignated F-80 Fighters in the new Air Force scheme of airplane designations. At long last America joined the rest of the world in officially calling them ''Fighters.''

The Lockheed F-80C was delivered in October 1948 as the first Shooting Star to come out of the factory with the ''F'' designation. It was like the predecessor P-80's except for a few minor improvements. The TF-80C two-seated Trainer became the first T-33 Silver Star. The DF-80 was a Drone Director for the QF-80 Drone. Other experiments included a rocket gun in the nose, ramjets at the wingtips and landing skis for Arctic operations. On July 20, 1948, sixteen F-80's completed the first mass-jet trans-Atlantic crossing on a deployment for overseas duty at Furstenfeldbruck.

Available in quantity when South Korea was invaded by the North Korean Communists on June 25, 1950, Shooting Stars arrived at the front from Japan within 48 hours after the outbreak of hostilities. On June 27, 1950, Bob Wayne of the 35th Fighter Squadron, 8th Fighter Group, scored the USAF's first Jet victory in history when he shot down two prop-driven IL-10 Fighters. F-80's provided air cover for the landing of 70,000 counter-invading Americans at Inchon on September 15, 1950. Generally outclassed by the Russian MIG jet Fighters in dogfights, Shooting Stars proved to be potent weapons in the ground attack role. When Mao Tse Tung's bugle-blowing Communist troops came swarming out of the hills on November 2, 1950, isolating United Nations ground forces into pockets, F-80 Fighter-Bombers flew around the clock to help prevent wholesale slaughter. Lt. Russ Brown, flying his F-80C on November 8, 1950, achieved the distinction of scoring America's first all-jet air victory in history when he attacked and destroyed a Russian MIG-15 jet Fighter.

Louis Sebille became the first Fighter pilot to be awarded the Congressional Medal of Honor in Korea for repeatedly attacking Communist forces in his damaged plane until it crashed into the ground near Hamchang on August 4, 1951. Charles Loring was awarded the CMH for heroic actions in his F-80 Fighter-Bomber on November 22, 1952, plunging to his death while on a dive-bombing run at an enemy gun emplacement near Sniper Ridge, North Korea.

Production totaled 1,731 Shooting Stars, including 933 as P-80's and 798 as F-80's for the USAAF and USAF, respectively.

F-84 THUNDERJET, F-84F THUNDERSTREAK, RF-84F THUNDERFLASH

The Republic XP-84 prototype first flew on February 28, 1946, and the YP-84A limited-production model was delivered for tests in October 1946. The first production model P-84B Thunderjet began to arrive in June 1947 at the 14th Fighter Group, Dow Field, Bangor, Maine. The P-84 (F-84) family evolved through three distinctly different aircraft which should have carried separate designations, but because of the ways in which funds were channeled in those days, they were only given different names.

The P-84 was a single-jet, mid-straight-wing monoplane with standard tail section and retractable tricycle landing gear. The pilot sat in a bubble cockpit centered above the wing's forward edge. The J35-GE turbojet engine with an intake in the nose generated 5,600 pounds of thrust to produce a top speed of 620 mph. Four 50-caliber machine guns on the fuselage and two wing guns carried a total of 1,800 rounds of ammo. A bomb load of 4,000 pounds was carried on underwing racks—weapons could be varied among air-to-air missiles, 32 rockets, and 500-pound or 1,000-pound bombs. Two 230-gallon underwing tanks plus two wingtip tanks stretched the ferry range out to 2,000 miles.

On September 20, 1946, the XP-84 set an American speed record of 611 mph. Thunderjets pioneered inflight air refueling by both the probe-and-drogue method and the flying boom technique. In June 1948 all P-84 Pursuits were reclassified as F-84 Fighters. The F-84E was the first to use the probe-and-drogue method of inflight refueling for Fighters. On September 22, 1950, Dave Schilling, a Fighter Ace in WWII, made the first non-stop Jet westward crossing of the Atlantic Ocean in an F-84E, being refueled three times enroute from England to America. The F-84G was the first single-seat Fighter capable of carrying a 1,200-pound tactical nuclear weapon. During operation "Fox Peter One" in July 1952, F-84G's pioneered non-stop overseas deployments when the 31st Fighter-Es-

Republic F-84 Thunderjet, first Fighter-Bomber to carry a nuclear weapon. Seen here on a non-stop flight out of the 49th Fighter-Bomber Wing, Japan, to Bangkok, Thailand, illustrating the global mobility of USAF Fighters on the occasion of King Adundet Phumiphon's birthday, December 7, 1953. (USAF Photo 84812-AC)

cort Wing flew from Turner, AFB, Albany, Georgia, to bases in the Far East, being refueled by KB-29 Tankers along the way.

On December 7, 1950, F-84D Thunderjets got their baptism of fire in Korea with the 27th Fighter-Escort Wing, flying cover for B-29 Bombers on raids against enemy targets. In air-to-air combat, the Thunderjets were inferior to Russian MIG-15 Fighters. F-84's were subsequently used effectively as Fighter-Bombers by units of the Tactical Air Command. On August 29, 1952, Art Gebaur, flying an F-84 with the 49th Fighter-Bomber Wing, was killed while on a low-level bombing mission against enemy machine gun nests in North Korea.

Early attempts to negotiate the end of the Korean war in 1953 were stalemated by the stubborn attitude of North Korean arbitrators. In efforts to force resumption of the talks, the first two mass Jet Fighter bombings in military history were unleashed when 59 Thunderjets from the 136th Fighter-Bomber Wing attacked the Toksan Dams on May 13, and another wave of 90 aircraft bombed the Chasan Dams on May 16. Those two raids contributed to achieving the desired objective: negotiations were immediately stepped up, resulting in the signing of a cease-fire agreement on July 27, 1953, ending the Korean Conflict with a standoff at the 38th parallel.

Production of the Thunderjet ended in July 1953 after 2,427 had been built for the USAF and 2,030 had been delivered to NATO forces, adding up to a final total of 4,457 constructed.

The republic F-84F first flew as the XF-96A prototype on June 3, 1950. The XF-96A was resignated YF-84F, and took to the air on February 14, 1951, as the first swept-wing Fighter-Bomber in Air Force history. The production model F-84F Thunderstreak was delivered to SAC as a Fighter-Escort and to TAC as a Fighter-Bomber for trials in November 1952, but did not become fully operational until the spring of 1954, too late for the Korean war.

The F-84F was a single-engine, mid-swept-wing monoplane with a swept-back tailplane mounted about one-fourth the way up the fin. The pilot's cockpit blended into the fuselage top above the wing's leading edge. The J65-W turbojet engine with an intake in the nose had a 7,200-pound thrust to produce a top speed of 695 mph. Weapons packages consisted of four nose guns and two wing guns, plus a variety of 8 missiles, 24 rockets and four bombs, adding up to a 6,000-pound load with a combat radius of 1,000 miles.

Production of the Thunderstreak totaled 2,710, including 1,410 for the USAF and 1,300 for NATO Allies.

The Republic RF-84F Thunderflash made its first flight on February 10, 1952. Deliveries began to arrive at SAC's 91st Strat-Recon Squadron and Tactical Air Command units in March 1954 as armed Photo-Recon combat planes, but too late for Korea.

The RF-84F was similar to the F-84F in appearance, differing mainly by having a solid nose and two engine intakes at the juncture of the fuselage with the wing's blunted forward edge. The J65-W turbojet engine with 7,800 pounds thrust produced a top speed of 680 mph. Six cameras were installed in the nose and four guns were housed in the outer walls of the intakes. Two underwing drop tanks extended the range to 2,200 miles.

During development in May 1953, YRF-84F's were used as satellites to Consolidated-Vultee GRB-36 "motherships" in experiments with the idea of extending the Photo-Recon range capability; that was before air refueling techniques had been refined for the Fighters. The YRF-84F was carried by the GRB-36 on a retractable belly hook to within striking distance of the target. Then the YRF-84F was launched to fly the final leg under its own power. Meantime, the mothership circled over the planned rendezvous point until the satellite returned for recovery and a free ride home. Simultaneous advancements in air refueling capabilities rendered the mothership concept obsolete before the trials were completed.

Coming along between wars, the Thunderflash was used sparingly by the Air Force. Many were turned over to the Air National Guard and others were exported to Allied nations under the Military Assistance Program. One of the friendly units to fly the RF-84F Thunderflash was the 3rd Air Brigade of the Italian Air Force.

Production ended in January 1958 after 715 RF-84F's had been built, including 335 for the USAF and 380 for NATO countries.

In 1953 the famous USAF "Thunderbirds" aerobatics team began their career, flying the straight-winged F-84G out of Luke AFB, Arizona, and in 1955 they changed to the swept-wing F-84F. In 1956 they moved to Nellis AFB, Nevada, and switched to the F-100C Super Sabre. In 1964 they adopted the F-105 Thunderchief for a brief period before switching back to the F-100D. In 1969 the Thunderbirds changed to the F-4E Phantom-II. In 1974 they switched to the T-38 Talon, which was Northrop's two-seated Trainer variant of the F-5B Freedom Fighter. On January 18, 1982, four USAF Thunderbird pilots—Norm Lowry, Willie Mays, Pete Peterson and Mark Melancon—were killed when their four T-38 Talons crashed into the desert near Las Vegas, Nevada, while flying in formation on a routine training mission.

F-86 (P-86) SABRE JET

The North American XP-86 prototype made its first flight on October 1, 1947, and the production model P-86A Sabre Jet was delivered to the 1st and 4th Fighter Groups of the Tactical Air Command in May 1948. On June 11, 1948, P-86A Pursuits became F-86A

North American F-86 Sabre Jet, renowned jet dogfighter of the Korean war. Photographed at Elmendorf AFB, Alaska on May 18, 1955. (USAF Photo 171283)

Fighters in the new scheme of designations. The Sabre Jet was distinguished as being the first successful swept-wing jet military aircraft in USAF history. Deliveries of the F-86A to Air Defense Command Fighter-Interceptor Wings began during 1949, and eventually twenty ADC wings were equipped with Sabre Jets. The F-86C with four 20mm nose cannon originally carried the designation of YF-93, and the F-86D Sabre Dawg was originally designated YF-95. The TF-86F was a two-seated Operational Trainer and the RF-86 was a Photo-Recon version with two K-22 and one K-17 camera in the fuselage.

The F-86 was a single-engine, low-swept-wing monoplane with a swept-back tailplane and retractable tricycle landing gear. The pilot sat in a bubble type cockpit above the forward edge of the wings. The J47-GE engine generated 5,970 pounds of thrust to produce a top speed of 700 mph. Six 50-caliber machine guns carrying 1,600 rounds of ammunition, were fitted on the sides of the nose. A bomb load of 2,000 pounds could be carried on underwing racks for a combat radius of 640 miles. Up to 24 Mighty Mouse rockets could be carried on underwing pylons and in a retractable fuselage tray. Two underwing drop tanks extended the normal range to 1,250 miles.

Performance records established by the F-86 are too numerous to recount. On May 18, 1948, an experimental Sabre Jet became the first jet Fighter to break through the sound barrier in a shallow dive. On September 15, 1948, Dick Johnson set the world speed record of 671 mph, and on August 17, 1951, Fred Ascani set a closed course record of 636 mph. In November 1952 Slade Nash broke the speed record with a mark of 699 mph, and in July 1953 Bill Barnes upped it to 716 mph.

The F-86 Sabre Jet flew to fame during the Korean war as the best jet dogfighter in modern times. Ordered into combat with the 4th Fighter-Interceptor Wing on November 8, 1950, F-86A's got into their first scrap over Sinuiji on December 17, 1950, shooting down four Russian MIG-15 Fighters in the first swept-wing jet air battle in history. The 51st Fighter-Interceptor Wing entered combat in September 1951. The psychological effect on the morale of the American fighting men down on the ground was analogous to a Cavalry troop coming over the hill to rescue a wagon train in days of the Old West.

Joseph McConnell, flying with the 16th Fighter Squadron, 51st Fighter-Interceptor Wing, was America's Jet Ace of Aces in Korea. His fifth air victory came on February 16, 1953, and on May 18, 1953, he destroyed his 14th, 15th and 16th MIG-15 Fighters. After surviving 106 combat missions in the war, Joe McConnell was killed while test-flying a F-86H on August 26, 1954.

On May 20, 1951, James Jabara became the first Jet Ace in history when he shot down his fifth and sixth MIG-15 in his Sabre Jet with the 334th Fighter Squadron. He later returned to Korea for a second tour and ran up his total to 15 air victories, second highest in the war. Ranked third in Korea was Manuel Fernandez, with 14 victories and one assist.

George Davis was awarded the Congressional Medal of Honor for fearlessly attacking twelve MIG-15 Fighters in his F-86 near the Yalu river on February 10, 1952, losing his life in the dogfight that followed.

Frank Gabreski of World War II fame as America's third leading all-time Fighter Ace, reported for combat in Korea to become a two-war Ace. Flying Sabre Jets with the 4th and 51st Fighter Wings he shot down seven MIG-15 Fighters. Glenn Eagleston, a P-51 Mustang Ace with the 9th Air Force in Europe during WWII, scored two kills in his Sabre Jet while Commanding the 4th Fighter-Interceptor Wing. During the Korean war Sabre Jets flew 87,200 sorties and shot down 814 Russian Fighters. Final statistics for the war claimed 984 enemy aircraft destroyed, including 823 MIG's and 161 prop-driven Fighters. America lost 58 Sabres and 36 prop-driven Fighters in dogfights, 671 aircraft to ground fire, and 206 planes to other causes, adding up to a total of 971 USAF planes destroyed in Korea.

Jackie Cochrane became the first supersonic woman in history on May 18, 1953, when she broke the sound barrier in an Air Force F-86 and flew on to set a new women's speed record of 652 mph.

Production ended in December 1956 after 5,800 F-86 variants had been built, including 5,375 for the Air Force and 425 furnished to Allies under the Military Assistance Program (MAP).

F-89 SCORPION

The Northrop XP-89 prototype was ordered in December 1946 and the designation was changed at the factory to XF-89 by the time of its first flight on August 16, 1948. The production model F-89A Scorpion began to arrive at Air Defense Command bases in July 1950 as the first Fighter-Interceptor to be fitted with "decelerons," which were ailerons that acted as air brakes.

The F-89A was a twin-jet mid-straight-wing monoplane with retractable tricycle landing gear, a tail section that resembled a plus-sign, and permanent wingtip tanks. Pilot and systems operator sat front to rear in tandem cockpits just above the engines, which were mounted side by side in a fuselage housing beneath the cockpit floor. The two J35-A turbojets generated 5,200 pounds of thrust each to produce a top speed of 635 mph. Six 20mm cannon, carrying 1,200 rounds of ammunition, were mounted in the nose. A weapons load of 3,200 pounds could be carried underneath, including bombs and up to 16 rockets on a normal combat range of 1,100 miles.

The F-89B had an autopilot to go with the radar equipment and was distinguished as the first all-weather multi-seat jet Fighter-Interceptor in Air Force history. The F-89C had an

Northrop F-89 Scorpion, the USAF's first all-weather multi-seated jet Fighter-Interceptor and the last jet Fighter to have a wingspan greater than its length. Snapped over the Alaskan range near Anchorage on July 9, 1954. (USAF Photo 151990-AC)

alcohol deicer tank on one of the bomb racks. The F-89C was the most-produced of all the sub-types, having two underwing fuel tanks and 52 Folding Fin Air Rockets (FFAR) in wingtip housings where the tip-tanks were on earlier models. The F-89H was the first to carry Hughes GAR-1 Falcon air-to-air guided missiles, three at each wingtip position. The F-89J had an advanced fire control system and was fitted with Douglas Genie air-to-air nuclear-tipped unguided missiles.

The job of an Interceptor is straightforward: when early warning installations indicate that enemy aircraft are approaching, the crew is scrambled to a combat-ready plane, makes a rapid takeoff, a quick climb, a swift pass, and fires the weapons and destroys the enemy before he reaches the bomb release point. A few split seconds can determine the success of the mission—and possibly decide our nation's fate. During exercises in Canada, F-89's acted as Interceptors against the speedy Boeing B-47 Stratojet Bomber. The Scorpion was a little faster than the B-47, but only fast enough to make one pass. If it missed on the first swipe, all the crew could do was helplessly wave goodbye to the Stratojet. Those performance facts helped Pentagon officials convince Congress to allocate funds for development of a much faster Fighter-Interceptor.

Phase-out began in 1957 with transfer of F-89's to the Air Training Command, Air National Guard and Air Force Reserve units. They were replaced with Convair F-102 Delta Daggers. The F-89 turned out to be the last USAF jet Fighter to have a wingspan (59'8") that exceeded its length (53'10"). Production ended in 1958 after 1,052 Scorpions had been delivered to the Air Force.

F-94 STARFIRE

The Lockheed YF-94 prototype made its first flight on July 1, 1949, as an advanced development from the TF-80C Shooting Star pilot trainer. First deliveries of the production model F-94A Starfire began to arrive at the 319th Fighter Squadron, 52nd Fighter Group, Air Defense Command, in June 1950. Designed as an all-weather Interceptor and

Lockheed F-94 Starfire, the first production model all-weather Fighter-Interceptor-Nightfighter equipped with engine afterburner. (USAF Photo 151990-AC)

Nightfighter, the F-94 was distinguished as the first production Fighter with after-burner capability.

The F-94A was a single-engine, low-straight-wing monoplane with standard tail section, retractable tricycle undercarriage and permanently fitted downward hanging 165-gallon tiptanks. Pilot and systems operator sat in tandem under a long canopy forward of the wings. Engine intakes were located on the fuselage sides below the cockpit. The J48-P engine delivered 8,750 pounds of thrust to produce a top speed of 585 mph. A radar set weighing 950 pounds was housed in the nose. Armament consisted of four 50-calibre machine guns under the nose, and the combat range reached out to 1,200 miles.

The F-94B had 230-gallon tanks on the wingends and blind-flying equipment for Operational Training. The F-94C (originally designated YF-97) was quite a different airplane, with a thinner wing, larger nose, swept-back tailplane, and an advanced fire control system. Armament was changed to 24 Mighty Mouse rockets in a ring around the nose, plus 24 more available in wingtip pods, twelve on each side. Each Mighty Mouse rocket was four feet long and weighed 18 pounds—that added up to 864 pounds of lethal firepower. The YF-94D was a one-seater with racks underneath for close support ground attacks, but the production contract was cancelled when the Korean war came to an end.

The Russian Blockade of Berlin during 1948-49, coupled with the detected explosion of a Soviet nuclear weapon in September 1949, spurred America to concentrate on developing Interceptor aircraft. The USAF rush-ordered F-94's from Lockheed because they could be built quickly and economically by using F-80 and T-33 parts already available. Starfires never saw combat in Korea because they were designed only for home defense in the Interceptor role.

Production totaled 854 F-94's, ending in 1953. After retirement from front-line service, many Starfires were assigned to Air National Guard units.

F-100 SUPER SABRE

The North American YF-100A prototype made its first flight on May 25, 1953, and the production model F-100A Super Sabre was delivered to the 479th Fighter Wing (TAC) at George AFB, California, in September 1954. The Super Sabre was distinguished as the first combat plane in Air Force history that could cruise at supersonic speeds in level flight. It was the first of what came to be known as the "Century" series of Jet Fighters.

The F-100A was a single-jet, low-swept-wing monoplane with retractable tricycle landing gear, single tailfin and a two-piece tailplane mounted at the bottom of the fuselage. The pilot sat in an enclosed cockpit on top of the fuselage ahead of the wing's forward edge. The J57-P turbojet engine generated 14,250 pounds of thrust to produce a top speed of 860 mph. Four 20mm cannon were fitted in the fuselage nose and a bomb load of 7,500 pounds could be carried on six underwing hardpoints. The weapons load could be varied among an assortment of hard and soft bombs, missiles, rockets and napalm for a combat range of 1,200 miles. "Hard" bombs were so called because they could penetrate the ground before detonating, being highly effective against an enemy burrowed down deep in his bunkers. "Soft" bombs were rigged with a chute to slow their descent, allowing a timing device to detonate them as frag-bombs just before hitting the ground.

On October 29, 1953, Frank Everest flew a YF-100A to set a new world speed record of 755 mph. The F-100B was an entirely different aircraft designed as a Tactical Bomber. It was redesignated YF-107, but never reached the production stage, losing out to the Republic F-105 Thunderchief in operational trials. The F-100C was flown by the Thunderbirds, famous USAF aerobatics team. On August 20, 1955, Harold Hanes, flying a F-100C, set the first official world supersonic speed record of 822 mph, exceeding Mach-1 as he broke through the sound barrier. The term "Mach" is used in aeronautical jargon to mean "the speed of sound, regardless of altitude." Because of the changes in density with height, the actual speed of sound in miles per hour varies at different altitudes. The F-100F (TF-100C) was an Operational Trainer.

On May 16, 1962, twelve F-100C Super Sabres of the 510th Tactical Fighter Squadron were ordered to Thailand by President Kennedy as the first combat Fighters to serve in the Southeast Asia war. Moving on over to South Vietnam, flying out of Danang, Phan Rang, Tuy Hoa and Cam Ranh Bay, Super Sabres were put on 24-hour around the clock alert for use in the Fighter-Bomber role, or anything else needed. When a field Commander called in to the 7th Air Force Tactical Air Control Center (TACC) for air support, the order to "scramble" went out, and within minutes F-100's were dropping their ordnance exactly where the Forward Air Control (FAC) pilots called for it. Rules handed down from higher Headquarters in that war firmly stated, "If you've got no mark, you've got no target," meaning that any target attacked must first be identified, approved, and marked in some manner by the FAC boys flying in liaison planes. When production ended in October 1959 a total of 2,394 Super Sabres had been delivered to the Air Force. The last F-100 was retired after a training mission on November 10, 1979, with the 181st TFG (ANG) at Hulman Field, Indiana.

North American F-100 Super Sabre, the first USAF Fighter to cruise at supersonic speed. Achieved fame as a Fighter-Bomber during the Vietnam war. (USAF Photo 155153-AC)

F-101 VOODOO

The McDonnell F-101 had no prototype as such; it was a development from the XF-88 Penetration Fighter that made its first flight in October 1948. Ordered in 1951 by the

Strategic Air Command as a long-range Escort Fighter to fly cover for B-36 Global Bombers, the production model F-101A Voodoo was delivered for trials on September 29, 1954, as the heaviest single-seat Fighter up to that time. Because of a change in strategic plans that called for an all-jet Bomber force and the gradual phasing out of the B-36, SAC cancelled the contract. But the Voodoo was an excellent aircraft; therefore, the contract was switched over to the Tactical Air Command, and initial examples of the F-101A were delivered to the 81st Tactical Fighter Wing. The F-101B was a two-seated Operational Trainer. The F-101C carried a thermo-nuclear weapon for use on Tactical Strike missions. On December 12, 1957, a F-101C flown by Adrian Drew set a world speed record of 1,208 mph. As an early warning Interceptor the Voodoo was the best aircraft of its time.

The F-101 was a twin-jet, low-swept-wing monoplane having retractable tricycle landing gear and T-shaped tail section with a swept-back tailplane at the top of the fin. The pilot's cockpit was located atop the nose section. Two J57-P turbojet engines generated 10,000 pounds of thrust each to produce a top speed of 1,050 mph. The engines were mounted on the lower fuselage sides, with intakes at the blunted forward edge of the wings. Armament consisted of four nose-mounted 20mm cannon. Weapons stores were carried at attachment points underneath and could be varied among Hughes Falcons, Douglas Genies or air-to-ground rockets. The normal combat range of 1,700 miles could be extended to 2,800 miles with a full load of fuel.

The YRF-101A Photo-Recon Voodoo made its first flight on May 10, 1956, and deliveries to Tactical Recon units began in 1957. Although designed strictly as an unarmed surveillance aircraft, the RF-101A retained all the good features of the Fighter-Interceptor version. Six cameras in the long nose consisting of a framing camera, a strip camera and Hycon trimetrogon cameras, provided day and night photography capability. The 363rd Tactical Reconnaissance Wing at Shaw AFB, South Carolina, was equipped with the RF-101C.

In 1962 Voodoos snapped the pictures over Cuba showing the Russian Surface to Air Missiles (SAM) and long range rockets. Those photographs created an international incident and caused President John Kennedy to take drastic precautionary measures during the ''Missiles of October'' crisis. Firmness on the President's part caused the Cubans and Russians to back down and withdraw the weapons.

Flying out of Thailand during the war in Vietnam, RF-101 Voodoos gathered vital information during surveillance flights over North Vietnam and along the border.

Production ended in March 1961 after 807 Voodoos had been delivered, including 511 F-101's and 296 RF-101's. A few were supplied to the Nationalist Chinese at Formosa and some were transferred to the Air National Guard.

F-102 DELTA DAGGER

The Convair YF-102 prototype took to the air on October 24, 1953, and the YF-102A delivered for service trials promptly broke through the sound barrier during its first test flight on December 20, 1954. The production model F-102A Delta Dagger was delivered in June 1955 to the 327th Fighter-Interceptor Squadron. The F-102 was distinguished as the first production Delta-wing Fighter and the first Interceptor to be armed with rockets and missiles only—there were no guns. The term ''Delta'' was used to describe the wings because of their resemblance to the fourth letter (delta) of the Greek alphabet. The

Convair F-102 Delta Dagger, first production delta-winged Fighter and the first Interceptor to have a complete manned weapons system. Photo taken on a flight with the 31st Fighter-Interceptor Squadron, a veteran combat unit guarding the Alaskan skies out of Elmendorf AFB. (USAF Photo 158415-AC)

TF-102A delivered in November 1955 was a side-by-side, two-seated Operational Trainer.

The F-102A was a single-jet, low-delta-wing monoplane with retractable tricycle landing gear and a single-finned tail section having a tiny tailplane near the top of the fin. The pilot sat in a cockpit forward of the wing's leading edge. The J57-P turbojet engine generated 17,200 pounds of thrust to produce a top speed of 825 mph. The engine was mounted inside the fuselage with side intakes near the cockpit. Six Hughes Falcon missiles were carried in a belly bay, and 24 folding fin rockets could be fitted in the bay doors. Normal combat range was 1,450 miles.

The Delta Dagger was the first Interceptor to be delivered as a complete weapons system—the weapons, the electronics equipment and the plane itself functioned as a unit. The F-102 could be flown remotely from ground control facilities through its Remote Control Flight System (RCFS). All the pilot had to do was take off and land the plane; the technical experts down on the ground took care of the rest. During emergencies and under certain operational circumstances the pilot had overriding capabilities. The secret to the weapons system was its ability to sense the infra-red glow from the enemy's jet engine exhaust and lock in for the kill. During a brief stint in Vietnam, F-102's demonstrated great potential for fighting in a war of advanced technologies.

Production ended in April 1958 after 1,100 Delta Daggers had been built, including 989 F-102A's and 111 TF-102A's.

F-104 STARFIGHTER

The Lockheed XF-104 prototype first flew on February 7, 1954, at Edwards AFB, California, and during service trials of the YF-104A on April 27, 1955, Joe Ozier exceeded a speed of Mach-2. Deliveries of the production model F-104A Starfighter Interceptor began to arrive at the 83rd Fighter-Interceptor Squadron (ADC), Hamilton AFB, California, in January 1958. Referred to as a "missile with a man in it," the F-104A had the narrowest wingspan of any jet Fighter in Air Force history. The wings were seven and a half feet long from fuselage to wingtip and only about a sixteenth of an inch thick at the leading edge. The F-104B was a two-seated Operational Trainer for ADC. The F-104C Fighter-Bomber was first delivered to the 831st Air Division (TAC), George AFB, California, in October 1958. The F-104D was a two-seated Operational Trainer for TAC. The NF-104A was a rocket-boosted high altitude trainer, and the F-104N was used for astronaut proficiency check-rides. The F-104G and TF-104G were built for export and were also built under license overseas.

The F-104 was a single-jet, mid-straight-wing monoplane with T-type tail assembly, retractable tricycle landing gear and a long sharp-pointed nose. The pilot's cockpit was located in the nose forward of the front wheel. The J79-GE turbojet engine generated 15,800 pounds of thrust to produce a top speed of 1,450 mph. The engine was mounted inside the fuselage with an intake on each side in front of the wings. A six-barrel 20mm rotary cannon was installed in the nose and four Sidewinder air-to-air missiles or Bullpup air-to-ground missiles could be fitted under the fuselage and at the wingtips. Weapons load could be varied among 2,000 pounds of bombs under the fuselage and 2,000 pounds under the wings, including 38 rockets or fire-bombs, for a combat radius of 750 miles. Two 195-gallon underwing fuel tanks and two tiptanks stretched the ferry range to 2,200 miles.

The F-104A was distinguished as being the first Air Force Fighter to hold both the speed and altitude records simultaneously. On May 7, 1958, Howard Johnson set a new altitude mark of 91,249 feet and nine days later Walt Irwin flew at a speed of 1,404 mph. Iven Kincheloe, Korean Jet Ace and test pilot with the Bell X-2 Transonic Special Research aircraft, was killed on July 26, 1958, in the crash of a F-104 out of Edwards AFB, California. On December 14, 1959, Joe Jordan climbed a F-104C to 98,124 feet in 15.1 minutes to set a new climb-to-altitude record, and in November 1963 Major R.W. Smith pushed his NF-104A up to 118,860 feet. On May 11, 1964, Jacqueline Cochrane became the fastest woman flyer in history when she piloted a F-104G at 1,429 mph. The Starfire was so fast that rumors suggested it even outran its own bullets. That sent the designers back to the drawing boards!

Starfighters were deployed in the Cuban "Missiles of October" political crisis in 1961-62. In 1965 a Squadron of F-104C's from the 479th Tactical Fighter Wing was deployed to Southeast Asia for a brief tour in combat over Vietnam. They were used to counteract the effect of Russian MIG-21 Fighters sent to the combat zone in 1966.

A total of 705 F-104's were produced, including 303 for the USAF and 402 for Allies under the Military Assistance Program. Many of the Starfighters were transferred to the Air National Guard. An additional 1,600 F-104's were manufactured under license by friendly nations.

F-105 THUNDERCHIEF

The Republic YF-105A prototype took to the air on October 22, 1955, and promptly broke through the sound barrier during that initial flight. The YF-105A was delivered for service trials in May 1957 and the production model F-105B Thunderchief Tactical Fighter began to arrive at the 4th Tactical Fighter Wing, Seymour Johnson AFB, Goldsboro, North Carolina, in May 1958. Chosen in preference to the North American YF-107 Fighter-Bomber, the Thunderchief was the last and finest of Republic's "Thunder" line of

Republic F-105 Thunderchief, first Fighter-Bomber to exceed a speed of Mach-2 and a workhorse during the Vietnam war. Pictured during a flight with the 335th Tactical Fighter Squadron near Eglin AFB, Florida on November 21, 1958. (USAF Photo 160815-AC)

Military Fighters. On December 11, 1959, Joe Moore set a 100km course mark of 1,216 mph. The RF-105B was a Photo-Recon version and the F-105D Fighter-Bomber featured the newly designed Thunderstick all-weather computerized radar navigation and weapons delivery system. On August 10, 1961, a Thunderchief dropped 14,000 pounds of ordnance on a practice target to establish a record for the heaviest load ever carried aloft by a single-engine aircraft. The F-105F was a two-seated Operational Trainer and Tactical Fighter combined.

The F-105 was a single-jet, mid-swept-wing monoplane with retractable tricycle landing gear and a tail section that featured a swept-back fin and swept-back tailplanes near the bottom of the fuselage. The cockpit was located forward of the wings above the nose wheel. The J75-P turbojet engine generated 26,500 pounds of thrust to produce a top speed of 1,400 mph. The engine was installed in the fuselage rear with scoop-type intakes at the points of the wings. A 20mm Vulcan nose cannon carried 1,030 rounds of ammo and had a firing rate of 6,000 rounds per minute. The weapons load of 12,000 pounds was distributed 8,000 pounds internally and 4,000 pounds underneath. Attachments were provided for Sidewinder air-to-air and Bullpup air-to-ground missiles, nuclear weapons, rocket packs, firebombs, mines, scatter bombs and toxic bombs. Normal combat range was 1,350 miles, and underwing tanks stretched the ferry range out to 2,100 miles.

Reporting for war duty in Vietnam during 1965, F-105's parlayed great speed, maneuverability, long range, terrific punching power and an ability to perform blindfolded into the most-feared Fighter-Bomber of the war. Its ability to survive strikes against heavily defended ground targets around Hanoi and Haiphong established for the Thunderchief a reputation for ruggedness that the folks at Republic boastfully compared to that of their tough P-47 Thunderbolt of WWII fame. Specializing in low altitude bombings of pin-pointed targets, F-105's flew more than half of the strike missions over North Vietnam. Combat losses were less than one percent per sortie, but even that low attrition rate added up to quite a few losses after many busy months of dodging Russian SAM projectiles. The crews in Vietnam referred to their F-105's affectionately as "Thud" and "Superhog."

Production ended in 1964 after 833 Thunderchiefs had been delivered to the Air Force. On June 27, 1980, the last F-105 was retired from the 562nd Tactical Fighter Squadron, 35th Tactical Fighter Wing, George AFB, California, and transferred to the 116th Tactical Fighter Wing, Air National Guard, Dobbins AFB, Georgia. From its first flight until its last gasp, the Thunderchief's career spanned 24 years and 8 months.

F-106 DELTA DART

The Convair-General Dynamics F-106 prototype made its first flight at Edwards AFB, California, the day after Christmas in 1956 as an advanced development from the F-102B Delta Dagger. Deliveries of the production model F-106A Delta Dart Interceptor began to arrive at Air Defense Command bases in July 1959 as replacements for the F-102 being phased out. The F-106B had two tandem seats and full combat capability for use as an Operational Trainer.

The F-106A was a single-jet, low-delta-wing monoplane with retractable tricycle landing gear, single fin at the tail and no tailplane. The engine was mounted inside the fuselage rear with side intakes at the wing's leading edge. The pilot's cockpit was located on top of the nose section. A J75-P turbojet engine generated 16,100 pounds of thrust to produce a top speed of 1,530 mph. Four Falcon missiles were fitted in a retractable bay and two Genie nuclear-tipped unguided rockets were hung on underwing racks. A late version experimented with an internally-mounted Gatling cannon. Combat radius with full weapons load was 800 miles, and the ferry range reached out to 2,400 miles.

The Hughes MA-1 Guidance and Fire Control system worked in conjunction with the Semi-Automatic Ground Environment (SAGE) facility. The pilot took the plane off the ground; then SAGE took control of the aircraft all the way to target and back; then the pilot took over and brought it in for a landing. During the mission the pilot monitored the system, but only had overriding capability during certain emergencies. Advances included an infra-red device for tracking and honing-in on the enemy, a supersonic ejection seat and vertical display instrument panels.

Charged with defending the country against sneak attacks by nuclear-armed aircraft, the F-106 had to be quick, agile, fast and long of range. On December 15, 1959, Joe Rogers set a new speed record of 1,526 mph in the skies above Edwards AFB, California. In 1960 Frank Forsythe made a completely automatic flight that covered 2,400 miles without air refueling, sitting back to enjoy the ride as SAGE did all the work.

Production ended in July 1961 after 320 F-106's had been delivered to the Air Force, including 257 one-seat Interceptors and 63 two-seated Trainers.

F-111 TACTICAL FIGHTER

The General Dynamics F-111A prototype made its first flight as the TFX Tactical Fighter on December 21, 1964. Delivery of the production model F-111A Tactical Fighter to Tactical Air Command units began in April 1965. The RF-111A is a Photo-Recon version. The Navy F-111B is capable of aircraft carrier operations and stowage. The FB-111A Strategic Bomber was delivered to the Strategic Air Command in 1969 with a wider span for use on long-range Strike missions.

General Dynamics F-111 Tactical Fighter, Vietnam war veteran and first Fighter to have variable-sweep wings, shown here in 1967 with weapons mounted on wing pylons. When wings are in swept-back position, pylons swivel to keep the weapons pointed straight ahead. (USAF Photo 178125)

The F-111A is a twin-engine, high-variable-sweep-wing monoplane with retractable tricycle landing gear having double tires on the nose wheel. The two TF30-P turbofan engines generate 19,000 pounds of thrust each to produce a top speed of 1,665 mph. The engine housings blend into the fuselage sides with intakes beneath the wing's forward edge. The tail section has a single fin in the middle and tailplanes on the outside of each engine housing. Pilot and systems operator sit side by side in the cockpit located above the nose wheel. Armament consists of a 20mm cannon in the internal weapons bay and two air-to-air missiles underneath. The 27,000 pounds of weapons are carried on 36 attachments in the weapons bay and underwing. Maximum ferry range stretches out to 3,800 miles with a full load of fuel.

Submitted jointly by the Air Force and Navy, specifications for the F-111 called for an aircraft with variable-geometry wings that could fly at double-sonic speeds, range out to 4,000 miles, have a ground-support loiter time of six hours, and land at a speed of 120 mph with a capability of operating in and out of short, rough landing strips—quite an order! But the F-111 filled it.

The job of the F-111A Tactical Fighter calls for versatility in the roles of Fighter, Interceptor, Fighter-Bomber, Attack, Strike and long-range Reconnaissance. A computerized system controls the wings, allowing them to be fully extended during takeoff, then gradually adjusted for optimum efficiency throughout the entire operational range, then extended again for landing. The F-111 can fly from its strategically located bases to any designated target on the globe, delivering a load of either thermo-nuclear or conventional ordnance. Slight troubles were experienced with the split-second timing in varying the wing-sweep, but the engineers straightened that problem out.

The F-111A made its combat debut in Vietnam on March 17, 1968, when six of them from the 474th Tactical Fighter Wing out of Nellis AFB, California, were deployed to Takhli, Thailand, for participation in operation "Combat Lancer." Flying single-plane sorties against targets over North Vietnam, three of the six aircraft were lost during the first month of combat. Later in the war 48 F-111's of Colonel Bill Nelson's 474th TFW returned to Vietnam for a second tour. During five months of combat those Tactical Fighters flew 4,012 sorties while proving to be even more effective than the designers had hoped; six aircraft were lost to all causes.

A total of 491 F-111's were built, including 455 for TAC and 36 for the Navy. An additional 76 FB-111A's were delivered to SAC units as relief for the aging B-52 Stratofortress.

JET FIGHTERS—
1962 TO THE PRESENT

F-4 PHANTOM II

The McDonnell F-4A made its initial flight on May 27, 1958, as a Navy Interceptor. In January 1962 the USAF procured two F-4's from the Navy under the designation F-110 as a continuation of the Century series of jet Fighters. Then, under the new tri-service scheme, the production model F-4C Phantom II retained the wing folding mechanism and carrier arrester gear for delivery to the Air Defense Command on May 27, 1963. The F-4C was the first Air Force version because the F-4A and F-4B were for the Navy and Marines. The RF-4C was the Photo-Recon variant with panoramic cameras, infra-red tracking device, inflight film processing and a new nose that stretched the length by 2 feet 9 inches. The F-4D and F-4E were like the F-4C except for engine change and a few modications. The F-4G was used in the USAF Wild Weasel program, an exercise in anti-radar defense systems.

The F-4C is a twin-jet low-swept-wing monoplane having retractable tricycle landing gear and a standard tail section with downward angled tailplanes. Each J79-GE turbojet

McDonnell F-4C Phantom laden with napalm bombs prior to takeoff on a test flight from Nellis AFB, Nevada, March 1962. (USAF Photo KE-22308)

engine generates 10,900 pounds of thrust to produce a top speed of 1,585 mph. The engines are installed on the fuselage sides with intakes forward of the wings. Exhaust tubes are located under the fuselage between the tail and the wing's trailing edge. Aircraft Commander (AC-pilot) and Weapons System Operator (WSO) sit tandem style in the cockpit located above the engine intakes. Armament consists of a Vulcan multi-barrel 20mm cannon plus eight Sparrow or Sidewinder air-to-air missiles under the fuselage and on underwing pylons. The 11,000-pound ordnance load can be varied among bombs, mines, smoke bombs or napalm canisters carried on racks under the wings for a combat radius of 785 miles. One 600-gallon fuel tank under the fuselage and two 370-gallon underwing tanks stretch the ferry range to 2,300 miles.

F-4C Phantoms made their combat appearance in Vietnam on June 5, 1965, flying Escort cover for F-105 Thunderchief Fighter-Bombers on Strike missions over North Vietnam. The early Phantoms were equipped only with missiles, and strict regulations required that any target be visually identified before being attacked. At double-sonic speeds, by the time some of the enemy aircraft were identified they were already inside minimum missile range, but still within gun range—and they didn't have any guns! That's when the 20mm cannon got installed in a pod under the fuselage.

On April 26, 1966, an F-4C downed a MIG-21 in a dogfight to score the first all-supersonic air victory in history. In a two-seated Fighter like the Phantom, when an enemy aircraft is destroyed, both the pilot and the systems operator are credited with the air victory. Aerial combat was not a major feature of the war in Vietnam; therefore, total statistics of air victories were not very large.

Charles de Bellevue, flying as the systems operator in F-4C Phantoms with the 55th Tactical Fighter Squadron out of Udorn, Thailand, was acclaimed as the Air Force Ace of Aces in the Vietnam war with six air victories against Russian MIG jet Fighters. Jeff Feinstein was the 2nd ranking WSO with five victories to his credit. Dick Ritchie was the only F-4 pilot to become an Ace, flying as the Aircraft Commander during five air victories.

Production ended with delivery of the last Phantoms in May 1966. Over 2,000 F-4's were built for the USAF and friendly foreign nations, plus others built overseas under license.

F-5 FREEDOM FIGHTER

The Northrop XF-5 prototype first flew as the Model N-156C on July 30, 1959, in a parallel development with the T-38 Talon jet Pilot Trainer. Deliveries of the F-5A Freedom Fighter began in April 1964 to the Tactical Fighter Training Wing at Williams AFB, Phoenix, Arizona, for training foreign students in operation and maintenance of the aircraft. One squadron of the Tactical Air Command was equipped with the single-seated F-5A combat Fighter in 1964. The F-5B in February 1964 traded the cannon for a second tandem seat, but retained the bomb racks for use as a Fighter-Bomber. The RF-5A was the Photo-Recon version with an entirely new nose section. The F-5 is a twin-jet, low-wing monoplane with retractable tricycle landing gear and a standard tail section. The cockpit is located in the nose section just behind the front wheel. Two J85-GE turbojet engines generate 4,080 pounds of thrust each to produce a top speed of 1,000 mph. The engines are mounted on the fuselage sides with intakes near the cockpit and exhaust outlets in the

Northrop F-5A Freedom Fighter armed with a mixed bomb load during performance evaluation near Edwards AFB, California on May 21, 1964. (USAF Photo KE-22607)

tail. Armament consists of two 20mm cannon located in the nose and two Sidewinder missiles on the wingtips. Underfuselage and underwing racks accommodate 6,200 pounds of ordnance that can be varied among a mixture of rockets, missiles, bombs and extra gun packs. The combat radius is 200 miles with a full load of weapons, but can be stretched to 570 miles with a lighter load. The ferry range with three 150-gallon drop tanks is about 1,760 miles.

Outstanding all-around flying qualities throughout its entire flight range enable the Freedom Fighter to lift more payload per pound of airframe weight than any other similar aircraft. As an Interceptor, the F-5A can go from brakes-off on the runway to 40,000 feet in less than four minutes. Deliveries to Iran, South Korea and Greece began in February 1965, followed by export models under the Military Assistance Program to Spain, Norway, Nationalist China, Israel, South Vietnam, Thailand, Canada, the Netherlands and Ethiopia.

The F-5A entered combat in Vietnam on October 23, 1966, for use in close support of ground units and proved highly effective on COunter-INsurgency (COIN) missions.

The F-5E Tiger-II made its first flight on August 11, 1972, and was delivered to the 425th Tactical Fighter Squadron in April 1973. Equipped with laser-guided "smart" bombs, the Tiger-II was selected above its competition as a replacement in the international market for Fighter aircraft. The F-5E was used initially by the Air Training Command and Tactical Air Command to train foreign students in operation of the aircraft. The F-5E is nicknamed "Aggressor" when flown as an enemy aircraft against operational Fighter-Trainers at USAF combat training schools. The F-5F is a two-seated version of the F-5E.

Ethiopian F-5's inflicted heavy damage on the independence-seeking Eritrean Liberation Forces (ELF) during a revolutionary uprising in February 1975. In retaliation, members of the ELF kidnapped four Americans to hold as hostages while bargaining against the flow of American arms and ammo into Ethiopia. Those captives were later released, undernourished, but happy to be alive.

On February 15, 1978, as part of an unveiled $4.8 billion jet Fighter package for the Middle East, America proposed its first sales in history to Egypt—Mr. Sadat was to receive fifty of the F-5E Tiger-II's at a total price of 400 million dollars.

YF-12 BLACKBIRD

The Lockheed YF-12 was developed in a Top Secret project as the Lockheed Model A-11 (a company label). After Gary Powers was released from a Russian prison where he was sentenced for the highly publicized "U-2 Incident," he was given the honor of making the first official flight in the A-11 on April 26, 1962. President Johnson revealed the A-11's existence during a news conference on February 29, 1964, and it was officially redesignated as the YF-12A Interceptor on April 27, 1964.

Unofficially nicknamed the "Blackbird", the YF-12A pioneered the use of the wonder metal titanium. The fuselage sides were flattened forward of the wings to provide additional lifting surface. A large belly fin at the tail extended downward for stability while in flight and folded to the side out of the way when the landing gear was lowered. A Hughes fire-control radar set was installed in the nose and infra-red sensors were mounted on the fuselage sides.

The YF-12A was a twin-jet delta-wing monoplane with retractable 8-wheeled tricycle landing gear. Pilot and Air Interception Officer (AIO) sat in tandem cockpits with individual canopies. Two J58-P turbojet engines generated 30,000 pounds of thrust each to produce a top speed of 2,100 mph. The engines, each with a large tailfin at the rear, were mounted in the delta-wing about halfway between the fuselage and wing tips. Armament consisted of 18 Hughes nuclear-tipped air-to-air missiles carried in two retractable weapons bays under the fuselage on a range of 3,000 miles. The service ceiling extended to above 100,000 feet, and the loaded weight topped out at around 150,000 pounds.

On May 1, 1965, with Bob Stephens and Dan Andre as the crew, the Blackbird took off from Edwards AFB, California, and roared to an official speed record of 2,062 mph, and they climbed it to above 80,000 feet. Announcement of that feat shocked the aviation world into disbelief and immediately provoked international political discussions concerning its full capabilities and planned purpose. Advances displayed by that futuristic aircraft won for its designer, Clarence Johnson, the coveted Robert J. Collier award for outstanding achievements in aviation.

Only two of the YF-12A Blackbirds were built, and flight evaluation for the Interceptor role ended in 1966. In 1968 the two YF-12A's were turned over to the USAF/NASA Advanced Supersonic Technology program for use in hypersonic research alongside the North American X-15. In 1976 NASA put them to work conducting tests for the Space Shuttle program. The YF-12B was a replacement for one of the YF-12A's, and the YF-12C became the prototype for the SR-71 Strategic Reconnaissance plane. That adds up to a total of four aircraft built.

F-15 EAGLE

The McDonnell-Douglas XF-15 prototype made its initial flight on July 27, 1972. The first production model was the TF-15A Eagle, a two-seated Trainer delivered on November 14, 1974. The 555th Tactical Fighter Squadron of WWII, Korea and Vietnam fame was the first outfit to fly the F-15A single-seated Fighter version. Eagles became operational with the 58th TFTW at Luke AFB, Arizona, and the 1st TFW at Langley AFB, Virginia, during 1976.

The F-15A is a twin-jet, high-swept-wing monoplane with retractable tricycle landing gear and a twin-finned tail section. The pilot sits in a cockpit forward of the nose wheel.

McDonnel-Douglas F-15A Eagle, modern Air Superiority Fighter that, reputedly, can accelerate while flying straight up! Shown here on January 30, 1976, in a 45-degree climb wearing its "Ghost Grey" color scheme. (USAF Photo KE-59993)

The F100-PW turbofan engines generate 25,000 pounds of thrust each to produce a top speed of 920 mph. The engines are housed in the fuselage with intakes on either side of the cockpit and straight-through exhaust tubes to the rear of the tail section. Armament consists of a 20mm 6-barrel cannon carrying 940 rounds of ammo. Air-to-air missiles include 4 Sidewinders underwing and 4 Sparrows under the fuselage. Attachment points underneath can accommodate 16,000 pounds of ordnance varied among a wide assortment of modern weaponry. The ferry range stretches out to 3,450 miles.

Designed as an Air Superiority Fighter for use in a variety of roles, the F-15 came on the scene as the first aircraft in the Air Force arsenal that could accelerate while flying straight up! Role capabilities include interceptor, escort, fighter sweep and strike missions. The Eagle has a thrust-to-weight ratio of one-to-one, and the secret to its long-range ability lies in two Fuel And Sensor Tactical (FAST) packs attached to the engine intakes. The Head Up Display (HUD) armament system has a snap-shoot tracer line that informs the pilot whether he is going to hit his target, even before he pulls the trigger! Optional equipment includes cameras and sensors for Photo-Recon duty, Wild Weasels for surface-to-air missile evasion, laser-light television system for nocturnal missions and infrared tracking system for locating and destroying enemy aircraft.

During operation "Streak Eagle" in January 1975, an exercise in dexterity, Dave Peterson, Roger Smith and Willard MacFarlane flew their F-15's to eight new world records in climb-to-height performance, reaching as high as 30,000 feet in 3 minutes 28 seconds.

In early 1976 Israel proposed to purchase 25 Eagles for $600 million in order to counter the Russian MIG-23 Fighters being sent to Syria in increasing numbers. By 1977 it was estimated that the Syrian Air Force numbered about 500 Russian-built combat planes. On February 15, 1978, America proposed the sale of 60 F-15's to Saudi Arabia at the going rate of $2.5 billion. In that same package was an offer of fifteen more Eagles for Israel— to "promote the peace-negotiating process."

By July 1, 1979, a total of 444 F-15's had been funded for the Air Force. Future plans call for an inventory of 749 Eagles of several versions.

F-16 COMBAT FIGHTER (FIGHTING FALCON)

The General Dynamics YF-16 prototype made its initial flight on February 2, 1974. In April 1975 the YF-16 was selected in preference to its competitor, the Northrop YF-17. Deliveries of the F-16A single-seat Fighter and F-16B two-seated Trainer were made in 1976 to USAF units for evaluation under operational conditions before deciding whether or not to continue the program.

The F-16A is a single-jet, mid-swept-wing monoplane with standard tail section and retractable tricycle landing gear. The cockpit is located in the nose ahead of the wing's leading edge. The F100-PW turbofan engine generates 25,000 pounds of thrust to produce a top speed of 1,500 mph. The engine is housed in the fuselage with a fish-mouth air scoop beneath the cockpit and exhaust tube at the rear. Armament consists of a 20mm cannon on the lower left side of the fuselage carrying 500 rounds of ammunition. Six Sidewinder air-to-air missiles can be fitted two at the wingtips and four underneath. A capacity bomb load of 15,200 pounds can be carried at 18 attachment points for a combat radius of 600 miles. The weapons load is limited to 10,500 pounds with a full load of fuel. The ferry range stretches to 2,300 miles with extra tanks and no weapons.

In an age of super planes, the F-16 came on the scene as the world's most advanced lightweight Fighter—it has a maximum takeoff weight of 33,000 pounds. And its maneuverability, speed, range, weapons systems and Top Secret technology apparently justify the claim. The HUD (Head Up Display) gunsight system is computerized to provide the pilot with a simulated trace of the path his weapon will take when the trigger is pulled. The simulated path is then superimposed on the actual target so that he can tell at a glance if he is going to get a hit, before he squeezes the trigger. Technological advances include a Pave Penny laser tracking device, electronic bomb ejection racks and missile launcher, Electronics Counter-Measures (ECM) jammer, chaff dispenser, radar guidance system and an Automatic Tracking Laser Illumination System (ATLIS) pod installed under the air intake.

In 1975 the NATO countries of Norway, Denmark, Belgium and the Netherlands ordered 306 F-16's to replace their Lockheed F-104 Starfighters. In 1976 the Shah of Iran signed an agreement to purchase 160 F-16's as replacements for F-4 Phantoms and F-5 Freedom Fighters. America okayed the deal because Iran presented a strategic location for keeping an eye on the Soviet Union and monitoring their attempts to exercise political hegemony over the Persian Gulf nations. However, that philosophy received a major setback when the Shah was overthrown in February 1979. As part of the arms package proposed to Congress on February 15, 1978, Israel was to receive 75 F-16's along with fifteen F-15 Eagles at a total cost of $1.9 billion.

General Dynamics F-16 Air Combat Fighter research aircraft with an AIM-9E Sidewinder on each wingtip, caught over Edwards AFB in California, March 30, 1976. (USAF Photo KE-63051)

In May 1977 Congress approved a compromise program for the USAF, authorizing an order to be placed for delivery through 1979 of 72 single-seat F-16A Fighters to the 388th Tactical Fighter Wing, Hill AFB, Salt Lake City, Utah, and 30 two-seated F-16B Trainers for a Tactical Fighter Training Squadron, also at Hill. As of January 1979 a total of 15 examples had been built, including two prototypes, 6 trial F-16A's, two trial F-16B's and 5 production model F-16A's. Plans called for 1,184 one-seaters and 204 two-seaters for a total of 1,388 to bolster the Air Force arsenal.

On August 5, 1980, Pratt & Whitney was awarded a $1.2 billion contract to begin building 574 new engines for F-15 and F-16 Fighters, some of which are slated for export to Belgium, Denmark, Israel, the Netherlands, Norway and Saudi Arabia. The Martin Marietta Corporation at Orlando, Florida, was awarded a $94.1 million contract on September 18, 1980, for project LANTIRN (Low Altitude Navigation Targeting Infra-Red at Night), a night-fire and bad-weather navigation pod designed for F-16 Fighters and A-10 Attack planes.

GLIDERS — 1941 TO 1962

Two new aircraft designations were introduced in 1941, the "CG" for Cargo Glider and "TG" for Trainer Glider categories. The "AG" for Assault Glider was added in 1942 and "PG" for Powered Glider was introduced in 1943. When the USAF was formed in 1948 all surviving gliders were redesignated into the "G" for Glider category, but all of them were subsequently declared obsolete by 1955.

The successful use of Gliders by Germany during the Blitzkrieg of Crete in the Mediterranean Sea on May 20, 1941, convinced the Allies that Assault Gliders could be used as effective military weapons. What the Allies didn't know at the time—but were to learn the hard way—was that effective use of gliders during an invasion requires a number of ideal conditions in order to avoid disaster.

The **Waco CG-4A Cargo Glider** made its first flight in 1942 and subsequently became the most popular troop carrying assault glider of WWII. Built of wood with steel tubing and a fabric skin, two or three CG-4A's were normally towed behind each C-46, C-47 or C-53 Cargo Transport and cast free near the target area. Then the glider pilots took control and thus began a mad scramble to find a spot large enough to land. With a sky full of gliders heading down, no time to spare, and being shot at all the while, competition for

Waco CG-4 Cargo Glider, outstanding WWII powerless Assault Glider nicknamed "Hadrian" by the British. Caught in a glide on August 1, 1944. (USAF Photo 42972-AC)

available landing space was sometimes fierce. Upon landing in enemy territory the pilot and copilot became instant infantrymen alongside the other assault troops on board who exited through an opening formed by the upward-hinged cockpit section.

The CG-4A was a powerless high-wing monoplane with a standard tail section and conventional 3-wheeled landing gear. It had a maximum towing speed of 120 mph and stalled out at 44 mph. Fifteen troops were carried, including the pilot and copilot. The capacity payload of 3,800 pounds could include a Jeep, a quarter-ton truck or a 75mm howitzer and its gun crew. The never-exceed takeoff weight was 7,500 pounds, but they fudged on that a few times.

CG-4A's got their baptism under fire during the invasion of Sicily on July 9, 1943, being towed by C-47 and C-53 Transports. Over a third of the gliders on that raid went down in the Mediterranean Sea and the rest were scattered over southern Sicily's rocky terrain. On March 5, 1944, CG-4A's towed by C-47 Gooney Birds landed Wingate's Raiders at "Broadway," a jungle clearing in Burma 150 miles inside the Japanese lines. During operation Market Garden, a diversionary action at Arnhem, Holland, on September 17, 1944, every Allied CG-4A that participated was either destroyed or lost to the enemy. CG-4A's led the D-Day landings in Normandy, came to the aid of troops stranded at Bastogne during the Battle of the Bulge, took part in the invasion of France from the south at Marseilles, and launched the final crushing drive that began with the crossing of the Rhine on the way to Berlin. Glider pilots referred to themselves as "conceived in error, trained in a scandal, forced into combat unprepared, and delivered to the wrong place at the wrong time under adverse circumstances."

Production totaled 12,400 CG-4A's for the USAAF and about 940 for Britain, where they were nicknamed "Hadrian."

The **Waco CG-13** in 1943 was the largest of the WWII Gliders with a span of over 85 feet. It carried 30 fully equipped combat troops at a towing speed of 205 mph. As a cargo plane it could haul up to 10,200 pounds of freight. Only 136 CG-13's were delivered to the USAAF.

The **Waco CG-15 "Hadrian"** in 1945 was the second most important Assault Glider of WWII. It was an improved version of the CG-4A, differing mainly in the 21-foot reduction in span while retaining the capability of carrying 15 or 16 assault troops. The gross weight of 8,000 pounds included a top payload of 4,300 pounds that could be towed at a speed of 180 mph. Production totaled 430 of the CG-15's before contracts were cancelled at war's end in August of 1945. All told, more than 16,000 Gliders were built for the USAAF and about 7,000 glider pilots were trained for WWII.

Shortly after WWII one CG-15A was converted into the XPG-3 Powered Assault Glider by installing two Jacobs R-755 engines giving it a powered flight endurance of three hours. It was designed to have a double capability of either being towed or flying under its own power, or a combination of the two. Postwar cutbacks precluded further development of the PG-3, therefore only one aircraft was delivered for trials.

The **Christopher AG-1** and **Timm AG-2 Assault Gliders** were cancelled while still on the drawing board; therefore, the USAAF had no successful gliders in the "AG" category.

The **Airspeed MK-I Horsa** was a British Invasion Glider that made its first flight in 1941. It was named after the Danish Jutland invader "Horsa," who mounted an invasion of the British Isles in 449AD. The MK-I was procured by the USAAF under a Reverse Lend-Lease arrangement for use by American assault troops stationed in Great Britain. The MK-II Horsa was the version used by the British. It was a foot longer than the MK-I, had a hinged nose, and was towed by ropes attached to the nose wheel.

The USAAF MK-I was a powerless high-wing monoplane with the tailplane mounted about one-third the way up the fin. The tricycle landing gear had jettisonable wheels for landing on rough terrain. The tail section was detachable by detonating with a hammer the four explosive bolts located one at each corner of the juncture with the fuselage. Pilot and copilot sat in a clear-view nose cone with full instrumentation. The front exit was a large cargo door on the left side of the fuselage just behind the cockpit. Towing ropes were attached at main gear hookup points. Twenty-five to thirty fully equipped combat troops were accommodated, or up to 7,120 pounds of cargo could be hauled. The loaded gross weight was 15,500 pounds with a towing speed of 150 mph and a gliding speed of 100 mph.

On November 19, 1942, two British MK-II Horsas carrying 15 combat toughened Commandos each got their baptism of fire during the daring but unsuccessful raid against a heavily defended Nazi heavy water plant in southern Norway where German scientists were conducting experiments for development of an atomic bomb. Both gliders were destroyed and all of the Commandos were either killed or captured. American MK-I Horsas worked alongside Waco CG-4A Assault Gliders during the attack on Sicily, the raid at Arnhem, the D-Day invasion of Normandy and the victorious crossing of the Rhine.

Approximately 2,600 MK-I Horsas were procured from Britain by the USAAF during World War II.

Trainer-Gliders were delivered beginning with the Frankfort TG-1 sailplane in 1941 and ending with the highest numbered Aeronca XTG-33 in 1944. Civilian gliders and sailplanes ordered off-the-shelf into production for military service included the Schweizer TG-2 and TG-3, Laister-Kauffman TG-4, and Pratt TG-32, all unpowered two-seaters with dual controls. Most produced of the training gliders were the Aeronca TG-5 (L-3), Taylorcraft TG-6 (L-2) and Piper TG-8 (L-4), all unpowered variants of their light Liaison planes. Conversion from a Liaison plane to a Trainer-Glider involved removing the engine, substituting a long nose with a third seat for the instructor, enlarging the tailfin, and shortening the legs supporting the three wheels. Two trainees occupied the tandem seats in the cabin—all three seats had full controls and separate instrument panels. The Flying Training Command and Troop Carrier Command employed training gliders to instruct glider pilots in the art of flying Assault-Invasion gliders into combat. Deliveries totaled about 1,250 training gliders for use by the Air Corps and USAAF.

HELICOPTERS—
1935 TO THE PRESENT

G—GYROPLANE: 1935 TO 1941

The first Army Air Service experiments with Helicopters were the DeBothezat Autogyro of 1921 and the Berliner Gyroplane of 1923, both tested on a limited basis for possible adoption as Observation planes; neither was ordered for production. In 1935 the "G" for Gyroplane category was added to the Air Corps scheme of aircraft designations.

The Kellett YG-1 Gyroplane in 1935 was the first successful Air Corps helicopter and the only autogyro to carry a "G" designation. It was a development from the civilian Model KD-1 Wingless Autogyro with a single Jacobs 225-hp engine and a top speed of 125 mph. It looked like a cross between a winged aircraft and a gyroplane, having no tail rotor and two tandem seats on top of the fuselage. A total of nine YG-1's were delivered for service trials.

R—ROTATING WING: 1941 TO 1948

In 1941 the "R" for Rotating Wing category replaced the "G" for Gyroplane designation. The Kellett XR-2 and XR-3 prototypes in 1941 were direct redesignations of two

Kellett YG-1 Gyroplane, the Air Corps' first successful helicopter and the only autogyro to carry a "G" designation. (USAF Photo 27676-AC)

surviving YG-1 Gyroplanes with an overhead rotor for lift and a tractor-type propeller for forward acceleration. In 1942 seven XR-2's were delivered as YO-60 Observation aircraft, thus becoming the first helicopters to carry an "O" designation.

The Vought-Sikorsky R-4 Rotating Wing in 1942 was the first full-production helicopter in USAAF history. It had a 180-hp engine with a top speed of 75 mph and a range of 130 miles. Pilot and copilot sat side by side with dual controls. Both the main and tail rotors had three blades to provide forward and backward flight. The undercarriage was of the 3-wheel divided type. During WWII the R-4 was used as a coastal patrol chopper. In 1943 it was put through rigorous trials under cold conditions in Alaska and hot humid weather in Burma, passing all the tests with flying colors. A total of 132 R-4's were delivered to the USAAF, and in 1948 all survivors were redesignated H-4 Helicopters in the new Air Force classification scheme.

The Vought-Sikorsky R-5 Rotating Wing in 1944 was the first USAAF helicopter to go into use with the Air Rescue Service (ARS). It had a 450-hp engine with a top speed of 105 mph and a 360-mile range. The two-man crew sat in tandem seats and an exterior litter rack on each side of the fuselage provided air evacuation for two patients. The main rotor overhead supplied both vertical lift and forward flight. The small tail rotor provided hover stability. A tricycle landing gear up front allowed for soft landings on any surface.

The Air Rescue Service was organized during WWII to immediately begin plucking pilots out of the water and evacuating wounded troops from battlefields around the world. Those daring exploits achieved for the ARS the distinction of being one of the most decorated combat units of the 20th century. The ARS-men lived by a simple motto: "If we lose 'em, we want to get 'em back." Appreciated by pilots of every breed, when an ARS crew walked into the club they weren't allowed to buy a drink. In 1948 all of the R-5's became H-5 Helicopters, with production continuing. Counting all versions of the R-5 and H-5, a total of 131 were delivered to the USAAF and USAF.

The Sikorsky R-6 Rotating Wing in 1944 was an improved version of the R-4 with a new fuselage, 225-hp engine, a 150-mile range, a top speed of 95 mph and a two-man crew. The R-6 was used in the closing days of WWII for overwater surveillance along coastal areas of the United States. In 1948 the R-6 became the H-6 Helicopter and production totaled 225, counting both designations.

The Bell YR-12 in 1944 was developed from their Model-48 general utility autogyro as Bell's first military helicopter. It had a 600-hp engine, a speed of 100 mph and five seats for a crew of two and three passengers. Only 13 YR-12's were delivered for service trials. In 1948 the survivors were redesignated YH-12 Helicopters for the Air Force.

The Bell YR-13 Sioux Rotating Wing was delivered in 1947 for versatility trials in the roles of trainer, photo-recon, air evacuation, observation, cargo transport, communications and general utility duties. It had a 175-hp engine, three seats, a 220-mile range and a speed of 95 mph. Only thirteen of the YR-13 limited production examples had been delivered when the designation was changed to H-13 Sioux Helicopter in 1948.

H—HELICOPTER: 1948 TO 1962

In 1948 the new autonomous USAF replaced the "R" for Rotating Wing designation with the "H" for Helicopter category in their new aircraft identification scheme.

Vought-Sikorsky R-4 Rotating Wing, first full-production helicopter and Sikorsky's initial entry into the military market. Seen here equipped with pontoons, ready for air-sea rescue duty. (USAF Photo 30563)

The **Vought-Sikorsky H-4 Helicopter** was a direct redesignation in 1948 of all surviving R-4 Rotating Wing aircraft with a Warner R-550 200-hp engine, two-man crew, a range of 170 miles and a speed of 90 mph. No new production orders were submitted for the H-4; therefore, the total ordered was 132 under the designation of R-4.

The **Vought-Sikorsky H-5 Helicopter** was a direct redesignation in 1948 of R-5 Rotating Wing aircraft with a 450-hp engine, a two-man crew, two litters, a 360-mile range and a speed of 105 mph. Subsequent improved variants added a rescue hoist and wheeled pontoons for amphibious operations. During the Korean conflict H-5's were used as rescue choppers and for general utility purposes. Production totaled 131 aircraft, including all variants of the R-5 and H-5.

The **Sikorsky H-6 Helicopter** was a direct redesignation in 1948 of all R-6 Rotating Wing aircraft with a 240-hp engine, a two-man crew, a range of 160 miles and a speed of 100 mph. Counting all variants of the R-6 and H-6, production totaled 225 aircraft.

The **Bell YH-12 Helicopter** in 1948 was a redesignation of the YR-12 Rotating Wing aircraft with a 600-hp engine, a two-man crew and a speed of 100 mph. During trials the cabin was expanded to accommodate up to eight passengers. A production order for 34 additional H-12's was cancelled; therefore, only the original 13 aircraft were delivered for experimental work.

The **Bell H-13 Sioux Helicopter** in 1948 was redesignated from the YR-13 Rotating Wing aircraft that appeared in 1947. The H-13 had a single Lycoming VO-435 200-hp engine, mountings for two fixed litters outside, a 240-mile range and a speed of 100 mph. The Sioux served honorably in Korea as a general utility and rescue chopper with two or three seats and either an open or closed fuselage. In 1957 the plush H-13J Ranger was delivered to the USAF for use by the President of the United States. A total of 1,133 aircraft were produced, including 13 as YR-13's and 1,120 as H-13's. In 1962 all surviving H-13's were redesignated UH-13's to classify them in the new tri-service Utility Helicopter category.

The **Sikorsky H-19 Chickasaw Helicopter** first flew for the USAF on October 21, 1949, and subsequently became a workhorse for the Air Rescue Service, MATS. The SH-19 was the Search and Rescue Helicopter version. Operating from land or aircraft carrier decks during the Korean war, Chickasaws performed the duties of air evacuating litter patients, pulling ditched airmen out of the water, picking up combat crewmen downed behind enemy lines, transporting combat troops to forward areas, carrying passengers and hauling cargo. During a typical rescue operation the power-operated winch was maneuvered into hover position above the pinpointed area; next the horsecollar harness was lowered for fitting around the survivor's back and under his arms; then he was swiftly hoisted into the copter for flight to safety. In Korea H-19's were dubbed ''Angels of Mercy'' for their heroic work behind enemy lines in areas of dense underbrush where American troops often found themselves stranded.

The H-19 had a 3-bladed main rotor driven by a single Pratt & Whitney R-1340 700-hp engine to give it a forward speed of 105 mph. Twin tubular-shaped pontoons with two wheels each were mounted below and just outside the cabin. Manned by up to three crewmen, arrangements could be varied to accommodate six fixed stretchers, eight am-

Sikorsky SH-19 Chickasaw, eminent rescue and utility Helicopter of the Korean war. Photographed while landing at Eglin AFB, Florida on April 1, 1958. (USAF Photo 159257-AC)

bulatory patients or 8 to 10 fully equipped combat troops. A capacity payload of 2,500 pounds could be hauled on a range of 400 miles. Production ended in March 1961 after 1,280 H-19's had been built for military usage. In 1962 all surviving Air Force SH-19's became HH-19 Search and Rescue Helicopters in the new tri-service scheme of aircraft designations. The CH-19 was for the Marines and the UH-19 was used by the Army and Navy.

The **Piasecki (later Vertol) H-21 Workhorse Helicopter** was delivered to the USAF in 1952 for use as a troop carrier and cargo transport. The H-21B was first delivered to the Tactical Air Command in November 1953 for use as a tactical assault transport in Korea. The SH-21 joined the Air Rescue Service with MATS in October 1953 as a rescue chopper along the Distant Early Warning (DEW) line extending from Alaska through Canada to Greenland near the 70th parallel. The H-21C Shawnee was the Army version. The Work-horse was the first tandem-rotor gyroplane in Air Force history. It had two main rotors, one on each end of a flat-V fuselage, that appeared at first sight to be pulling in opposite directions.

The H-21 had a single 1,250-hp engine pulling the two divided main rotors to produce a top speed of 125 mph. The landing gear was of a tricycle type with two wheels at the center juncture and a nose wheel beneath the front cockpit where the 2-man crew sat side by side. The rudder configuration looked like a football goalpost (H). The cargo transport version had a large door on the left side and could carry 14 fully equipped combat troops or 4,000 pounds of freight on a range of 300 miles. The airborne ambulance variant accommodated 12 stretcher patients. A total of 217 H-21's were built for the Air Force. In 1962 all surviving H-21's became CH-21 Cargo Transport Helicopters and SH-21's became HH-21 Search and Rescue Helicopters in the tri-service designations.

The **Kaman H-43 Huskie Helicopter** was delivered in 1958 as the first airborne fire-fighter and crash rescue aircraft in USAF history; it was also the first Kaman aircraft built for the Air Force. Developed from the civilian Model-600, the H-43 looked like something from Mars with twin booms, two counter-rotating rotors on the same shaft, no tail rotor, four up-and-down pointing tailfins and four wheels positioned one at each corner of the rectangular shaped cabin. Delivered to bases of all commands, Huskies were operated by crews from Air Rescue Service detachments spread out all over the country. On alert status the H-43 can get airborne in less than a minute, speed to the crash scene, discharge the firefighting crew and hover over the site while downwash from the rotor protects survivors from smoke and flames. On December 9, 1959, Walt Hodgson climbed to 30,100 feet in an H-43 and set a new altitude mark for helicopters.

The H-43 had a Pratt & Whitney 600-hp engine driving the 4-bladed main rotor to produce a top forward speed of 115 mph. Pilot and observer sat tandem along one side while the rest of the cabin could be arranged for 2 to 4 firefighters, or four litters with a medic, or seven to ten passengers. As a cargo carrier it had a capacity payload of 3,970 pounds which could be hauled on a range of 400 miles. A total of 254 H-43's were delivered to the Air Force. In 1962 the H-43 became the HH-43 Search and Rescue Helicopter.

H—HELICOPTER: 1962 TO THE PRESENT

The Bell family of H-1 Helicopters evolved through a wide variety of versions for all branches of the military service. Modeled after the Army's UH-1B Huey, the USAF **UH-1F Iroquois Utility Helicopter** made its first flight on February 20, 1964. Delivery for trials as an armed combat chopper began in March 1964 to the 4486th Test Squadron at Eglin AFB, Florida. Subsequent production deliveries were made to the Strategic Air Command and Tactical Air Command for use in Vietnam flying Top Secret psychological warfare and surveillance missions with a downward folding radar scanner under the cabin. a distinguishing feature was the stabilizing tailplane located in the fuselage forward of the tail section. The TH-1F was a Helicopter Trainer and the HH-1H is the Search and Rescue variant.

The UH-1F had a Lycoming T-58 1,100-hp engine driving a two-bladed main rotor spinning at 320 revolutions per minute to develop a forward speed of 125 mph. It had a 2-bladed tail rotor for hover stability and tubular skid-type landing gear with a wheel at the rear of each skid. Pilot and copilot sat in the forward cockpit and nine combat soldiers occupied bench seats across the center and rear of the cabin. A payload up to 4,500 pounds could be hauled on a combat range of 320 miles. Armament came in a variety of configurations including rocket packs on either side, electrically controlled machine guns on both sides, a nose mounted 40mm grenade launcher or two side-mounted 30mm

Bell UH-1F Iroquois gunship of the 14th Air Commando Squadron takes off on a strike against the Viet Cong in 1967. A 7.62mm minigun, capable of firing 6,000 rounds of ammo per minute, is seen protruding from the doorway. (USAF Photo 103051)

cannon. Production totaled 146 UH-1F's and TH-1F's when the contract was completed in 1967.

The **Bell HH-1H Iroquois Search and Rescue Helicopter** was delivered to the Air Rescue and Recovery Service (ARRS) of the Military Airlift Command (MAC) beginning in 1971. The ARRS was formerly ARS, the acronym being changed after MAC succeeded MATS in 1962. Developed as a local base rescue chopper, the HH-1H has basically the same airframe as the UH-1F, but additional equipment includes inflatable float bags, a cargo hook, a rescue hoist and auxiliary fuel tanks. Access is via two jettisonable crew doors up front and two compartment doors in the rear.

The HH-1H has a Lycoming T-53 engine driving the 2-bladed rotor at 320 rpm to generate a top speed of 125 mph. The spacious 220 cubic foot cabin accommodates the pilot and 14 fully equipped combat troops. As an airborne ambulance it can handle six stretcher patients and one medical attendant. The cargo capacity is 3,900 pounds, which can be hauled on a combat range of 340 miles; the ferry range extends out to 700 miles. When production ended in 1973 a total of 30 HH-1H's had been delivered to the Air Force at a total cost of about ten million dollars.

The Sikorsky series of H-3 Helicopters first showed up in the Air Force inventory as the CH-3B Cargo Transport Helicopter, of which six were delivered for trials in September 1961. The **HH-3E Jolly Green Giant Search and Rescue Helicopter,** delivered in December 1963, was a redesignation of the CH-3E in the new tri-service scheme of aircraft categories. The Jolly Green Giant served with MAC, TAC, SAC, PACAF and USAFE in a wide variety of duties including recovery of NASA space capsules from the ocean. Distinguishing features are short stubby wings with sponsons for carrying jettisonable fuel tanks, five-bladed main and tail rotors, a half-tailplane on top of the fin, armor plating, self-sealing fuel tanks, and aerial refueling capability using the probe and drogue technique to hook up with Lockheed HC-130 Hercules Tankers. A large hydraulically operated ramp in the rear of the fuselage can be lowered for loading vehicles or other heavy freight packages. The Navy version is named the Sea King and the Coast Guard variant is called the Pelican.

The HH-3 has two General Electric T58-GE 1,500-hp engines producing a top speed of 165 mph. Pilot and copilot sit side by side in the cockpit located on the nose. The large cabin with four windows on each side can accommodate 30 fully equipped combat troops or 15 stretchers with medical technicians. A cargo load of 5,000 pounds can be carried on a combat range of 485 miles, or 2,500 pounds can be hauled for 800 miles. The armament consists of two 50-caliber machine guns mounted in front of the door and at the loading ramp.

On May 31, 1967, two Jolly Green Giants became the first Helicopters in history to fly non-stop across the Atlantic Ocean, being refueled nine times each on the 4,240-mile trip from New York to Paris that took 30 hours and 43 minutes. HH-3's transported men and equipment between Otis AFB, Massachusetts, and the Texas Tower radar sites off the east coast. Operating as combat rescue copters behind enemy lines in Vietnam, HH-3's rescued downed flyers, oftentimes right from the grasp of communist troops. If rugged terrain prohibited landing, the Giant could hover motionless overhead while a sling-hoist was lowered to the stranded men. Thousands of wounded ground troops owe their lives to

Sikorsky HH-3 Jolly Green Giant, famed rescue Helicopter and mercyship in the Vietnam war. Snapped on a mission in 1965. (USAF Photo 178636)

speedy evacuation from the battlefield by Jolly Green Giants. Bob Dyberg, a pioneer member of the Air Rescue Service in World War II, was a Rescue Forces Commander of aircrews in Vietnam who made headlines with such daring exploits as the dangerous raid on a North Vietnamese Prisoner Of War (POW) camp deep inside enemy territory. Production totaled 83 H-3 Helicopters for the USAF.

The **Bell UH-13 Sioux Utility Helicopter** was a redesignation in 1962 from the H-13 Helicopter, which was itself a redesignation from the YR-13 Rotating Wing that made its first flight in 1947. The UH-13 had a Lycoming 250-hp engine and a forward speed of 105 mph with a combat range of 250 miles. It had three seats inside the cabin and attachments for two litters outside. Production totaled 1,133 Helicopters in the Sioux family. When the going got rough during the early days of the Vietnam war, UH-13's got the battle call to support COunter-INsurgency (COIN) operations throughout the jungle areas assigned to Corps I, Corps II, Corps III and Corps IV.

The **Sikorsky HH-19 Chickasaw Search and Rescue Helicopter** was a redesignation in 1962 from the SH-19 Helicopter of Korean war fame. The HH-19 had an 800-hp engine and a top speed of 115 mph. Pilot and copilot rode in the cockpit and eight fully equipped troops or eight litters and a medic could be accommodated in the cabin. The normal combat range was 500 miles.

In Vietnam when a call for "Mayday" was received, a Command Control chopper orbited above the downed airmen while A-1 Skyraiders provided interdictory fire in the surrounding area by strafing and dropping 500-pound bombs on any approaching enemy troops. Then the HH-19 would swoop in, drop its hook, and hoist the survivors aboard. Occasionally the Viet Cong tricked our crews by feigning downed pilots and calling for help in fluent English, using the correct frequency. When the Helicopter landed, it was caught in a trap. American communicators discovered that Charley (the VC) couldn't pronounce our four-letter words clearly because his English class hadn't taught that kind of language. That's how the American air-to-ground communications jargon came to include quite a bit of profanity—and it worked! Production totaled 1,280 of all the Chickasaw variants for the military services.

The **Piasecki-Vertol HH-21 Workhorse Search and Rescue Helicopter** was a redesignation in 1962 from the SH-21 tandem-rotor chopper of Korean war days. The HH-21 had a single Wright R-1820 1,425-hp engine, giving a top speed of 135 mph. Pilot and copilot sat side by side in the cockpit up front and 14 combat troops or 12 stretchers with a medical attendant could be accommodated in the main compartment. In Vietnam the Workhorse was used as a combat and rescue Helicopter. Production totaled 217 of the H-21 planes, including all the derivatives.

Piasecki-Vertol YH-21 Workhorse pictured on October 7, 1953, at Thule Air Base, Greenland. HH-21's later joined the war in Vietnam serving as both combat and rescue Helicopters. (USAF Photo 150805-AC)

The **Kaman HH-43 Huskie Search and Rescue Helicopter** was a redesignation in 1962 of the H-43 crash rescue and airborne firefighter. The HH-43 had a single Lycoming

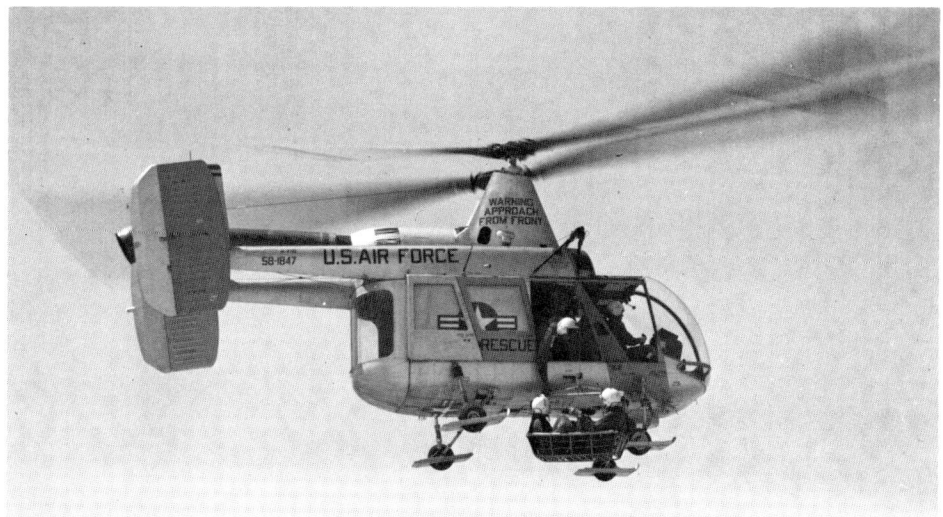

Kaman HH-43B Huskie demonstrates use of the litter basket, picking up two crewmen from the ground at Edwards AFB, California on April 21, 1960. In Vietnam, Huskies filled the roles of crash rescue, firefighter, ambulance and troop carrier. (USAF Photo 171228)

860-hp engine and a forward speed of 125 mph. In Vietnam the ten-place Huskie was alternately arranged to carry four firemen, four litters, or 8 troops, when used in the roles of firefighter, ambulance, rescue craft or troop carrier. A capacity payload of 4,000 pounds could be carried on a range of 300 miles. A total of 254 H-43 and HH-43 Huskies were built for the Air Force.

The **Sikorsky HH-53B Sea Stallion** heavy-lift Search and Rescue Helicopter made its initial flight for the USAF on March 15, 1967, and deliveries to the Air Rescue and Recovery Service (ARRS-MAC) began in June 1967. Basically an enlarged version of the HH-3 Jolly Green Giant, equipment included a rescue hoist, a cargo hook, sophisticated electronics devices, a flight refueling probe and auxiliary fuel tanks. The HH-53C delivered in August 1968 was an improved version with more powerful engines and modernized survival equipment. Other sub-numbers are employed by the Navy and Marines. All of the HH-53's retained the stubby wings, half-tailplane configuration and aerial refueling capability characteristic of the Jolly Green Giant.

The HH-53 has two General Electric T64-GE 3,080-hp turboshaft engines driving a 6-bladed main rotor to achieve a top speed of 185 mph with a range of 540 miles. Some later variants for the other services had 7-bladed main rotors. A 4-bladed tail rotor provides hover stability and balance. Pilot and copilot ride in the forward cockpit, and accommodations are provided in the cabin for 38 combat troops, or 24 stretchers with four medics, or 8,000 pounds of freight including several varieties of large packages. For defense purposes two 50-caliber machine guns can be mounted at the doors. Production totaled 72 HH-53's for the Air Force.

On April 24, 1980, eight Navy RH-53 Minesweeper Helicopters took off from the aircraft carrier Nimitz in the Gulf of Oman and participated in ''Project Delta,'' an unsuccessful attempt to rescue 50 American captives who were being held as hostages by

Sikorsky HH-53C Sea Stallion, heavy-lift rescue and utility Helicopter used by the ARRS (MAC). Photo taken in 1968. (USAF Photo 178699)

Iranian militants in Teheran. One of the RH-53's returned to the Nimitz, two developed mechanical problems, one crashed into an Air Force C-130 Transport after "drop-dead time" was declared in the staging area, and four were abandoned at the "Desert One" staging area, together with the bodies of 8 airmen from the all-service "Blue Light" team who died in the inferno caused by the collision between the RH-53 and the C-130.

OBSERVATION AND LIAISON AIRCRAFT

CO—CORPS OBSERVATION: 1919 TO 1924

Copied from the French "Corps d'Armée" aircraft of World War I, the "CO" for Corps Observation category was introduced into the Air Service scheme in 1919 to designate two-seated aircraft, armed with flexible guns in the back seat, for use in the roles of scout, enemy surveillance, liaison between the front line and command headquarters, communications wire stringing, artillery spotting, aerial reconnaissance, photography, command transport and general utility duties.

The **Fokker CO-4 Observation** aircraft in 1922 was the only "CO" designation to have more than three planes built. General Billy Mitchell took a personal liking to the airplane when he first saw it at the Fokker plant in Holland while on a tour of Europe during 1921. The CO-4 set a pattern for the general design of Observation planes in the 1920's and 1930's; all of them were similar in appearance, differing only in airframe design and engine efficiency among the various manufacturers. The CO-4 was a single-engine biplane with a standard single-finned tail section, two fixed wheels up front and a landing skid at the rear. Pilot and observer had separate open cockpits arranged tandem style on top of the fuselage. The Wright 180-hp engine gave it a top speed of 130 mph. A single swing gun was mounted on a ring around the back seat; some of these aircraft had the gun removed for use as VIP taxis on short hops. A total of seven CO-4's were built for the Army Air Service.

The **Engineering Division XCO-5 Corps Observation** plane in 1924 was a redesignation of the TP-1 Two-seat Pursuit. Only one XCO-5 was used by the Army Air Service. On January 29, 1926, Lt. John MacReady established a new altitude record of 38,704 feet when the XCO-5 was equipped with an engine supercharger.

AO—ARTILLERY OBSERVATION: 1919 TO 1924

The "AO" for Artillery Observation category was on the list of designations in the 1919 scheme for classifying airplanes.

The **Fokker-Atlantic AO-1 Artillery Observation** plane in 1923 was almost an exact duplicate of the Fokker CO-4 Corps Observation aircraft equipped specially for the job of spotting enemy battle locations, directing artillery fire, evaluating artillery accuracy and recommending sighting adjustments. Two of the AO-1's were built for the Air Service and they turned out to be the only aircraft to carry an "AO" designation.

O—OBSERVATION: 1924 TO 1942

After WWI the Army Air Service used their plentiful DeHavilland DH-4 Liberty Planes to fill needs for both Observation aircraft and Attack light bombers. In May 1924 the entire designation system was overhauled and the "O" for Observation category replaced the "CO" and "AO" classifications. In 1942 the "O" was replaced by "L" for Liaison. Then in 1962 the "L" was abandoned in turn and the "O" for Observation was restored. Observation planes between WWI and WWII often came in families, with each successive model using basically the same airframe and differing only in minor revisions and in experiments with new engines.

The **Curtiss O-1 Falcon** in 1925 was the first aircraft ordered in the new Observation category; it was also the first of the Curtiss Falcon family. The O-1 was a conventional biplane of the day with two tandem seats, standard tail section, two fixed wheels up front and a tailskid at the rear. A single Curtiss D-12 435-hp engine drove a 2-bladed prop to produce a top speed of 135 mph with a range of 600 miles. Armament consisted of two forward-firing guns on the fuselage front and one or two flex guns in the rear cockpit.

The Curtiss O-11 and XO-12 Falcons in 1927 were like the O-1 except for engine changes. The Curtiss O-13 Falcon had the same airframe, but was fitted with a Curtiss V-1570 600-hp engine to participate in the 1927 National Air Races. The Curtiss XO-16, XO-18 and YO-26 Falcons in 1928-29 were experimental aircraft using a variety of engines and cooling systems. The O-39 in 1932 was the last of the Curtiss Falcons with a smaller radiator and reduced rudder area. One O-39 was rigged with streamlined britches and another example tried out an enclosed canopy for the pilot.

Curtiss O-1 Falcon, first in a large family of Curtiss two-seat Observation biplanes. Shown here is the O-1C assigned to Major General James Fechet. (USAF Photo 154633-AC)

Other offshoots from the Curtiss O-1 Falcon family were the A-3 and XA-4 Attack planes in 1927 and the XBT-4 Basic Trainer in 1931. Production totaled 310 of all the Curtiss Falcon variants that used the O-1 airframe.

The **Douglas O-2 Observation** plane in 1924 was the first aircraft delivered in the new "O" category, beating the O-1 to the punch by a few months. The O-2 was the first in a long line of Douglas Observation planes that were to span eighteen years. The O-2 was a standard biplane with a Liberty 400-hp engine, top speed of 130 mph and a 500-mile range. Two fixed guns were mounted on the fuselage and the observer fired a flexible gun from his rear cockpit. The O-2A was equipped for night flying, the O-2B had dual controls for use as a Trainer, and the O-2J was a staggerwing with plush seats for use by VIP officials.

Douglas O-2 Observation plane, first in a long line of Douglas two-seated Observation biplanes. Caught inflight is an O-2H on August 21, 1930. (USAF Photo 39278)

The Douglas O-7, O-8 and O-9 in 1926 had new engines and minor revisions, but stuck to the O-2 airframe. The XO-14 in 1928 had staggered wings and a narrower span. The O-22 in 1929 had a slightly swept-back upper wing and introduced a tail wheel in place of the skid. The O-25 in 1930 had a new engine, new nose and dual controls for training. On August 27, 1933, Capt. Ernest Harmon was killed in the crash of an O-25 near Stamford, Connecticut. The O-29 and O-32 in 1930 were minor advances, and the O-34 in 1930 had a new engine with a swept wing like the O-22. Last of the Douglas O-2 airframe family was the O-38 in 1931 with an enclosed canopy covering both cockpits. A few O-38's were still hanging around when Japan attacked Pearl Harbor on December 7, 1941, and were immediately put to work towing targets for aerial gunnery training.

Other variants to have airframes similar to the O-2 were the Douglas C-1 Cargo Transport in 1925, the XA-2 Attack plane in 1926, the BT-1 Basic Trainer in 1930 and the BT-2 Basic Trainer in 1931. A total of 649 Douglas aircraft were built, using the O-2 airframe as a foundation for each member of the family.

The **Douglas O-5 Observation Seaplane** in 1924 was referred to as the Douglas World Cruiser (DWC) which grew out of the 1923 Douglas Observation Seaplane (DOS). Both the DOS and DWC acronyms were official designations in the early 1920's before the Air Service consolidated its Observation planes into a single category. The O-5 was a standard biplane fitted with twin floats for operating on water surfaces. Pilot and observer sat tandem style in open cockpits. A single Liberty V-12 420-hp engine gave it a top speed of 100 mph. Two fixed guns fired through the synchronized two-bladed prop and a swing gun was operated from the rear seat. A total of six O-5's were delivered for active duty. One example was fitted with two wheels up front and a landing skid at the rear for trials as a landplane at McCook Field, Dayton, Ohio.

On April 6, 1924, four Army Air Service DWC Seaplanes departed from Seattle, Washington, on a 26,345-mile flight around the world. On September 28, 1924, after 175 days of pressing on against many hardships, two of the planes completed the flight when they landed again at Sand Point Field in Seattle. The DWC flown by Lt. Erik Nelson on that historic flight, and named "New Orleans," was presented for exhibit to the Air Force Museum, Wright-Patterson AFB, Ohio. The other DWC, named "Chicago," flown by Lt. Lowell Smith was turned over to the Smithsonian Institute at Washington, D.C.

Douglas O-5 DWC (Douglas World Cruiser), photographed with pontoons on January 10, 1924. The O-5 was immortalized for its flight around the world during April-September of 1924. (USAF Photo 11286-AS)

The **Thomas-Morse YO-6 Observation** plane in 1925 was an all-metal copy of the Douglas O-2 two-seater with a 400-hp engine, top speed of 130 mph, two fixed guns, one flex gun and a range of 500 miles. Only one YO-6 was built, and it was unsuccessful. However, it served as a pioneer in all-metal aircraft construction and encouraged the people at Thomas-Morse to continue their research.

The **Consolidated O-17 Observation** plane in 1928 was a converted PT-3 Primary Trainer with increased fuel capacity, a 4-hour endurance and a back seat equipped with either a 30-caliber machine gun for use as an Observation plane or dual controls for training. It was a standard biplane with a Wright R-790 220-hp engine and a top speed of 100 mph. One version nicknamed "Courier" was redesignated the XPT-8 Primary Trainer. A total of 32 O-17's were delivered to the Air Corps and most of them were transferred to the Air National Guard.

The **Thomas-Morse O-19 Observation** plane in 1928 was the firm's first successful all-metal product after nine long years of experiments with metallic technology at their main plant. The O-19 was a standard biplane with 3-wheel undercarriage and tandem open seats. A single Pratt & Whitney R-1340 450-hp engine drove the two-bladed prop to develop a top speed of 135 mph. A forward-firing cowl gun carried 350 rounds of ammo and the rear cockpit held 600 rounds for use with a ring-mounted flex gun. The O-19D had no armament; it was equipped with plush seats for VIP passengers. Production totaled 175 O-19's for the Air Corps. The Thomas-Morse YO-20, XO-21 and YO-23 were all single examples built for service trials using the O-19 airframe with different engines and new cooling systems.

The **Fokker YO-27** in 1931 was the first limited production twin-engine Monoplane in Air Corps history. It had a high wing with a standard tail section, 3-wheel landing gear and enclosed cockpits for the 3-man crew. The two wing-mounted engines were located close to the fuselage and drove 3-bladed props to develop a top speed of 160 mph. Twelve YO-27's were delivered to the Air Corps as Fokker's only contribution to the "O" category. One YO-27 was modified into the XB-8 prototype Bomber for service trials.

Fokker YO-27 Observation plane, first Air Corps Observation monoplane. Snapped on the ground is the XO-27A after General Aviation became the Fokker representative in America. (USAF Photo 15869-AS)

The **Douglas O-31** in 1931 was the first full-production model Observation Monoplane. It was an all-metal butterfly-wing two-seater with standard tail, 3-wheel landing gear and sliding canopies over the cockpits. The single Curtiss V-1570 675-hp engine gave it a top speed of 190 mph. A fixed gun was mounted in the wing and a swing gun was fired by the observer in the back seat. The Douglas O-43 in 1933 was an improved O-31 with a parasol wing, enclosed cockpits and a new tail surface. The Douglas O-46 in 1935 was like the O-43 except for the Pratt & Whitney R-1535 725-hp engine and a razorback rear fuselage. Production totaled 125 O-31, O-43 and O-46 planes.

The **Douglas O-32** Observation plane was developed from the O-2K airframe in 1931 as a standard biplane with 3-wheel undercarriage, no armament and a Pratt & Whitney R-1340 450-hp engine giving it a top speed of 135 mph. A total of 31 O-32's were built and all of them were converted to BT-2 Basic Trainers in 1931.

The **Douglas YO-35** part-metal high-birdlike-wing Monoplane in 1932 was the first twin-engine Observation plane with retractable landing gear. Two Curtiss V-1570 600-hp engines drove 3-bladed props to develop a top speed of 180 mph. The 3-man crew operated two guns located at the front and rear. One YO-35 was converted into the XO-36, which became the prototype for the Douglas YB-7 Monoplane Bomber. Production totaled 14 YO-35 and XO-36 Observation planes.

The **North American O-47 Observation** plane in 1937 introduced many radically modernized advances compared to old-style Observation planes that had changed very little since WWI. The O-47 was a midwing Monoplane with standard tail assembly and retractable front wheels. Pilot, observer-photographer and gunner made up the crew, who were seated in tandem cockpits under a long canopy. The Wright R-1820 975-hp engine drove a 2-bladed prop to produce a top speed of 220 mph with a range of 450 miles. One fixed gun was mounted in the wing and a flex gun was fired from the rear seat. Windows on the bottom and along each side gave the photographer an excellent downward view for aiming his cameras. Some of the O-47's were caught overseas in the Pacific during the Japanese raids of December 7-8, 1941, and never made it back home. Those in the States were immediately put to work as trainers, as target tugs, and on general utility duties. Production totaled 239 O-47's for the Air Corps.

North American O-47 Observation plane, most-powerful of all the Observation monoplanes. Pictured inflight is an O-47A over Elko, Nevada, February 10, 1939. (USAF Photo 36095-AC)

The **Stinson O-49 Vigilant** in 1940 was designed to satisfy a drastically revised role for Observation planes. The new specs called for an unarmed aircraft with less horsepower and a slower speed for direct support of ground forces. The O-49 was a high-wing monoplane with enclosed cabin. Pilot and observer sat in tandem seats having dual controls. The bubble-type windows provided good downward vision, even when in level flight. The Lycoming R-680 295-hp engine gave it a top speed of 120 mph and a range of

280 miles. In 1942 the "O" designation was abolished in favor of "L" for Liaison, and all surviving O-49's were redesignated L-1 Vigilants. Production totaled 324 O-49's and L-1's combined.

The **Curtiss O-52 Owl** in 1940 was already on order when the specifications for an Observation plane were changed to require a lighter unarmed aircraft. The O-52 was an all-metal high-wing monoplane with standard tail section and retractable 3-wheel undercarriage. Pilot, observer and gunner sat in tandem under a sliding canopy. The Pratt & Whitney R-1340 600-hp engine drove a 3-bladed prop to produce a top speed of 210 mph. Curtiss Owls were never used in combat, but helped to fill the need for Trainers in the early days of WWII. A total of 203 O-52's were delivered to the Air Corps.

The **Taylorcraft O-57 Grasshopper** in 1941 was the first small, light, unarmed cabinplane designed for Liaison duty in close support of front line ground troops. It was a high-wing monoplane with enclosed cabin, a standard tail section, three fixed wheels and a gross weight of 1,300 pounds. The 65-hp engine gave it a top speed of 90 mph. Pilot and observer sat in tandem with a two-way radio, dual controls and special equipment for artillery spotting. In 1942 all O-57's were redesignated L-2 Grasshoppers. A total of 1,911 O-57's and L-2's were delivered for active duty.

The **Aeronca O-58 Defender** in 1941 was another of the Grasshoppers designed for light Liaison duty. It was almost exactly like the other unarmed high-wing cabinplanes with a 65-hp engine, a two-man crew, a top speed of 85 mph, and a 190-mile range. In 1942 all O-58's were redesignated L-3 Defender-Grasshoppers. A total of 1,487 O-58's and L-3's were delivered for service.

The **Piper O-59 Cub** in 1941 was another unarmed Grasshopper high-wing light cabinplane. Pilot and observer were seated tandem style with dual controls. The enclosed cabin provided a clear all-around view. The Continental O-170 65-hp engine gave it a top speed of 85 mph with a 190-mile combat range. In 1942 all O-59's were redesignated L-4 Cub-Grasshoppers. Production totaled 5,413 of the O-59's and L-4's.

The **Kellett YO-60** gyroplane in 1942 was the first and only Observation Helicopter delivered to the Air Forces for World War II. It was an advanced development from the XR-2 Rotating Wing aircraft that was itself a direct conversion from the YG-1 Gyroplane of 1935 vintage. The YO-60 had an overhead rotor for vertical lift, no tail rotor, a tractor propeller for forward flight and two tandem seats on top of the fuselage. The Jacobs 225-hp engine gave it a top speed of 125 mph. Only seven YO-60's were delivered to the USAAF.

The **Stinson-Vultee O-62 Sentinel** in 1942 carried the double-designer name as result of Vultee having bought Stinson out in 1940. The O-62 was a two-seated high-wing light cabinplane with standard tail section and fixed 3-wheel landing gear. The 185-hp engine produced a top speed of 130 mph and gave it a range of 420 miles. The Sentinel carried the designation of O-62 for only a few months before the classification was changed to L-5 in 1942. Production totaled 3,284 O-62's and L-5's for the USAAF and Army.

The **Interstate XO-63 Cadet** ordered in 1941 was the highest numbered aircraft when the "O" category was abolished in 1942. The designation was changed to XL-6 at the factory and deliveries were made in 1943 as the L-6 Cadet-Grasshopper. Therefore, no examples were actually delivered as the O-63. Although Liaison aircraft replaced Observation planes for World War II, the "O" for Observation category was reinstated by the Air Force when the 1962 tri-service scheme of aircraft designations became effective.

FOREIGN-BUILT OBSERVATION AIRCRAFT

The **Morane-Saulnier MS-234** of 1932 was purchased from the French Air Force for use by the United States Military Attache Office in Paris. It was a parasol high-swept-wing monoplane with standard tail section, two fixed wheels up front, a tailskid in the rear and two open tandem seats. The Wright Hispano-Suiza R-975 330-hp engine gave it a top speed of 127 mph with a flight endurance of two and a half hours. Only one MS-234 was procured by America. It was not given an Army designation, but carried an Air Corps serial number, color scheme and markings.

L—LIAISON: 1942 TO 1962

On March 2, 1942, the Army Air Forces became a parallel Command to the Army Ground Forces. However, because of inter-service communications problems, on June 6, 1942, the War Department authorized light Liaison airplanes for Field Artillery Units of the Ground Forces. In 1942 the "L" for Liaison category replaced "O" for Observation planes on the list of USAAF designations. The planes and parts under the "L" designation were supplied by the USAAF, but pilots were recruited and trained by the Ground Forces. The job of a Liaison plane involved observation of enemy movements, artillery spotting, communications wire laying, coordinating air support, air evacuation, camouflage checking, courier and station taxi duties. Approximately 13,558 light Liaison planes were built for use in World War II. Also, in 1942 some of the "L" planes were converted to "TG" for Training Gliders by removing the engine, installing a new nose and shortening the front wheel legs.

Most popular of the WWII Army Liaison planes were the L-2, L-3, L-4 and L-6 models, all referred to as "Grasshoppers." Liaison planes filled the Army's requirements until they were supplemented later by more sophisticated Helicopters.

The **Stinson L-1 Vigilant** in 1942, a redesignation of the O-49 Observation plane, was the first aircraft in the new "L" category. Exactly like the O-49, the L-1 was put to work immediately in the Pacific and European theaters for use as an airborne ambulance, float-fitted amphibian rescue craft, and trainer in the art of glider-pickup. A total of 324 L-1 and O-49 planes were delivered to the USAAF.

The **Taylorcraft L-2 Grasshopper** in 1942 was a redesignation of the O-57 Observation plane chosen to participate alongside the Aeronca O-58 and Piper O-59 during a huge Army field exercise in 1941 for testing in the proposed light Liaison role. Because of their ability to flit from bush to bush quickly, all three aircraft were referred to as "Grasshoppers." The L-2 had a Continental O-170 65-hp engine and was specially equipped as an artillery spotter, but could also be used as a utility transport and glider pilot trainer. A total of 1,911 L-2 and O-57 planes were built for the war. Seven of them served briefly as UC-95 light Cargo Transports before being converted back to L-2's in 1943. The TG-6 Training Glider in 1942 changed the landing gear and had a new front fuselage to accommodate a third seat in place of the engine; otherwise it was like the L-2. A total of 250 TG-6's were built for training glider pilots who were slated to fly the Waco CG-4A Assault Glider.

The **Aeronca L-3 Defender-Grasshopper** in 1942 was a redesignation of the O-58 Observation plane with a Continental O-173 65-hp engine and two-way radio for specializing in communications duties, but it was used in other roles too. A total of 1,487 L-3 and O-58 planes were delivered during the war. The TG-5 Training Glider in 1942 had a new landing gear and was lengthened by two feet in order to substitute a glider pilot seat where the engine used to be; otherwise it was like the L-3. A total of 250 TG-5's were built for the USAAF in WWII.

The **Piper L-4 Cub-Grasshopper** in 1942 was a redesignation of the O-59 Observation plane for use as an artillery spotter, trainer and transport plane. The L-4 first went into battle as a spotter in November 1942, being flown from the deck of an aircraft carrier during the invasion of Northwest Africa by Allied forces. One version of the L-4 was rigged with Brodie launching gear for quick takeoffs and landings. A total of 5,413 L-4 and O-59 aircraft were built, making it the most-produced airframe of all the Observation-Liaison planes. One example was tried out as a UC-83 light Cargo Transport.

Piper L-4 Cub Liaison plane, most-produced "Grasshopper" of WWII. Shown on runway at Minter Field, California, February 27, 1943. (USAF Photo 41048)

The TG-8 Training Glider in 1942 had a different landing gear and substituted another seat for the engine; otherwise it was like the L-4. A total of 250 TG-8's were built. The YL-14 in 1945 was similar to the L-4, but that contract was cancelled at war's end after five examples had been delivered, too late for use in World War II.

The **Stinson-Vultee L-5 Sentinel** in 1942 with a Lycoming O-435 185-hp engine was a redesignation of the O-62 Observation plane, which was developed from the civilian Model-105 Voyager. Sentinels served as airborne ambulances in both World War II and the Korean conflict. One version was equipped with a K-20 camera for use as a Photo-Recon plane. Another variant was temporarily labeled AT-19 Advanced Trainer for shipment to Britain, but was commandeered from the factory as the L-9 Voyager. A total of 3,284 Sentinels were built, second only to the popular Piper L-4 Cub for most-produced honors.

The **Interstate L-6 Cadet-Grasshopper** in 1943 was originally ordered as the O-63 Observation plane, but the designation was changed while in production. The L-6 was a high-wing cabinplane with standard tail section, three-wheeled landing gear and two tandem seats. The Franklin 0-200 115-hp engine gave it a top speed of 105 mph. During WWII the L-6 was used as a utility transport, communications plane and trainer aircraft. Production totaled 251 L-6 planes, least-produced of all the Grasshoppers.

The **Stinson-Vultee L-9 Voyager** in 1942 was commandeered at the factory and stopped before being shipped to England as the AT-19 Advanced Trainer. Originally designated YO-54 in 1941, the L-9 was a three-seated high-wing monoplane with a Franklin C-199 90-hp engine and a top speed of 100 mph. It was used as a light utility transport in WWII. Only twenty L-9's were procured by the USAAF.

The **Consolidated-Vultee XL-13** prototype made its first flight in 1945 and deliveries of the production model L-13A to the USAAF began in 1947. It was designed as a versatile triphibian with wheels, long floats or skis for performing all the duties expected of a Liaison plane while operating on any type of surface. The L-13A was a high-folding-wing monoplane with a plus-shaped tail section, conventional landing gear, three to six seats and cameras when needed for the photography role. The Franklin O-425 245-hp engine gave it a top speed of 115 mph. The range extended to 750 miles when an extra tank was carried. Production totaled 302 L-13 planes, and some of them were still around during Korean war days.

The Canadian **DeHavilland YL-20** made its first flight for the USAF in 1951 and deliveries of the production model L-20A Beaver began to arrive in 1952. The Air Force used Liaison planes in the 1950's to ferry men and supplies to isolated detachments at sites with short, unimproved landing strips. The Army version was used in Korea where it was kept busy hauling field commanders to forward areas and evacuating casualties to safety. Used by Mark Clark and Matt Ridgeway to assess battle situations, the Beaver was affectionately referred to as "the General's Jeep." The L-20 was a high-wing monoplane with standard tail section and three non-retractable wheels. A Pratt & Whitney R-985 450-hp engine gave it a top speed of 160 mph and a combat range of 450 miles. The cabin accommodated a crew of two and up to six combat troops or four litters. The seats could be rearranged to make room for 1,000 pounds of freight. The undercarriage was rigged with either floats, skis or wheels for versatile operations as a triphibian. Production ended in 1960 after 975 Beavers had been built, including 212 for the USAF and 763 for the Army. In 1962 all surviving Air Force L-20's were redesignated U-6 Utility Beavers.

DeHavilland L-20 Beaver, triphibian Liaison monoplane of the Korean war. Caught here on water skis in Alaska, July 12, 1951. (USAF Photo 151762-AC)

The **Piper L-21 Super Cub,** an advanced model of the L-4 Grasshopper, was first delivered in 1951 to the Air Force for service trials and to the Army for use as a utility plane in Korea. The L-21 was a standard high-wing monoplane with a Lycoming O-290 125-hp engine, top speed of 125 mph and a maximum range of 750 miles. Pilot and copilot-observer sat tandem style with dual controls inside the glassed-in cabin. A 50-pound baggage rack was located behind the rear seat. Production totaled 702 L-21 planes, mostly for the Army. In 1962 all surviving L-21's were redesignated U-7 Utility planes for use by the Air Force and Army.

The **Aero Design YL-26A Commander** was delivered to the USAF in 1956 for service trials as a general utility plane. The L-26B was a staff transport and the L-26C was a Presidential VIP transport. The L-26 and Cessna L-27 were the only two twin-engine Liaison planes utilized by the Air Force. The L-26 was a highwing cabinplane with standard tail section and retractable landing gear. Two Lycoming O-540 350-hp engines drove 3-bladed reversible propellers to give it a top speed of 250 mph with a range of 1,700 miles. Pilot and copilot sat up front and five to nine passengers could ride in the cabin. With some of the seats removed, a cargo load of 500 pounds could be carried. Production totaled 18 L-26 planes for the USAF and ten for the Army. In 1960 the Air Force L-26's were redesignated U-4 Utility planes and in 1962 the Army L-26's became U-9's in the tri-service scheme.

The **Cessna L-27A Administrator** was first delivered to the USAF in 1957 and promptly got redesignated U-3A in 1958. The L-27 was a twin-engine low-wing cabinplane with retractable tricycle landing gear and a standard tail section having a swept-back tailplane. Pilot and copilot sat up front and four passengers rode in the cabin. Up to 600 pounds of cargo could be hauled. Two Continental O-470 260-hp engines drove 2-bladed full-feathering props to produce a top speed of 240 mph and a maximum range of 1,300 miles. Production totaled 195 aircraft, of which 160 were U-3A's followed by 35 U-3B's in 1960.

The **Helio-GAC L-28A Super Courier** was delivered in 1958 with large flaps covering three-fourths of the wing's trailing edge to furnish Short Take-Off and Landing (STOL) capability. It was a high-wing cabinplane with standard tail section and non-retractable landing gear. Pilot and copilot sat up front and four seats were fitted in the cabin. The powerful Lycoming O-480 295-hp engine supplied the thrust needed to leap the light plane off the ground. The maximum speed was 160 mph and the 60-gallon fuel tank gave it a range of 670 miles. Five airborne paratroopers or 1,000 pounds of light cargo packages could be transported in the cabin. A total of only three L-28A's were delivered to the Air Force, and in 1961 they were redesignated U-10 Super Couriers in the new Utility category.

Helio-GAC L-28 Super Courier, last of the "L" designated Liaison planes. Seen on the ground at the California USAF Flight Test Center, March 26, 1959. (USAF Photo C-162885-AC)

FOREIGN-BUILT LIAISON AIRCRAFT

The **Fiesler FI-156 Storch (Stork)** was one of the German aircraft captured when the Afrika Korps was cornered near Tunisia and forced to surrender on May 12, 1943, thus bringing an Allied victory in the North African campaign. General Erwin Rommel had used the Storch as his personal transport to run up and down the front lines and assess battle situations. The FI-156 was a high-wing cabinplane with standard tail section, three wheels with long, skinny front legs, and two tandem seats inside the clear-view cabin. An Argus AS-10C 240-hp engine gave it a top speed of 110 mph. For defense a 7.9mm swing gun could be mounted behind the back seat. On September 12, 1943, Benito Mussolini was rescued from imprisonment atop Gran Sasso in the Apennine mountains by SS Colonel Otto Skorzeny in a FI-156 flown by German Ace Hans Gerlach. For that act of daring, Adolf Hitler awarded the Knights Cross to Skorzeny.

Captive aircraft were reconditioned, repainted, given Army serial numbers and marked with USAAF insignia for shipment back to research laboratories in the States; they were not assigned USAAF letters or designations.

O—OBSERVATION: 1962 TO THE PRESENT

The "O" for Observation category was revived in the 1962 scheme for tri-service aircraft designations after having been dropped in favor of "L" for Liaison by the USAAF in 1942.

The **Cessna O-1 Bird Dog Observation** plane was a direct redesignation in 1962 of all surviving Korean war L-19 Bird Dog Liaison planes built strictly for the Army and delivered beginning back in 1950. The O-1F was the first USAF version, deliveries of which began in 1963 to PACAF and the Tactical Air Command where they were used as target markers for Strike aircraft. The O-1F was a high-wing cabinplane with standard tail section and three fixed wheels. Pilot and copilot-observer in tandem were protected from enemy ground fire by flak curtains and armored seats. The Continental O-470 213-hp engine produced a top speed of 115 mph and the semi-selfsealing gas tanks gave it a range of 530 miles. Armament consisted of two 50-caliber machine guns and six air to ground rockets, or four target-marking smoke bombs could be carried on underwing inboard pylons. Optional arrangements provided for two 250-pound bombs underneath, a chemical spray tank or an aerial container for dropping supplies. Bird Dogs served in Vietnam from 1964 until they were phased out during 1967-68 in favor of Cessna O-2 Skymasters and North American OV-10 Bronco (LARA) VSTOL aircraft. Production totaled 3,600 Bird Dogs for all services, including 310 for the Air Force.

Cessna O-1E Bird Dog, all-purpose Observation plane, shown flying convoy escort over South Vietnam, September 19, 1963. (USAF Photo 99256)

The **Cessna O-2A Skymaster** was delivered to the USAF in 1967 to help replace aging Cessna O-1 Bird Dogs whose active duty career had spanned 17 years. With backup push-pull engines, exchangeable fuel pumps and other advanced features, the Skymaster earned a reputation as one of the safest planes in the world. The O-2 can remain aloft for about four hours with both engines running, or can fly on one engine while switching the fuel tanks and stay in the air about seven hours. In Vietnam O-2's were equipped with

Cessna O-2 Skymaster, psy-war plane used on COIN operations in Vietnam. Seen here is an O-2A parked on the flight line at Edwards AFB, California, May 7, 1968. (USAF Photo 178957)

electronics gear for use on COunter-INsurgency (COIN) missions, psychological warfare, air to ground broadcasting, leaflet dropping and Forward Air Control (FAC). The O-2B was used as both an aircrew trainer at Eglin AFB, Florida, and a combat plane in Southeast Asia (SEA), where it was lauded as the most agile FAC plane of the war. The O-2 is a twin-engine, twin-boom, high-wing cabinplane with retractable tricycle landing gear and two tail fins connected at the booms by a long tailplane. Two Continental O-360 210-hp engines located fore and aft of the cockpit drive two-bladed airscrews to give it a top speed of 205 mph with a ferry range of 1,400 miles. Maximum range with a full 830-pound payload is 770 miles. Pilot and copilot sit up front with dual controls, and four passengers with 365 pounds of luggage can ride in the cabin. Armament consists of a 7.62mm Minigun pack, four rockets, flares or napalm canisters on underwing racks. Production ended in 1970 after 546 O-2's had been built for the Air Force.

The **North American-Rockwell International YOV-10 Bronco Light Armed Reconnaissance Airplane (LARA)** first flew on July 16, 1965, and delivery of the production model OV-10A began to the Tactical Air Command and Pacific Air Forces (PACAF) in 1968 as replacement for the Cessna O-1F Bird Dog in Vietnam. Designed specifically for COunter-INsurgency (COIN) guerilla warfare in jungle terrain, Broncos served in the roles of fighter-bomber, close support attack, helicopter escort, forward air control, reconnaissance, observation, troop carrier and light cargo transport. A special landing gear and thick wings with double-slotted flaps allow it to take off after a run of only 750 feet. Twin props turning in opposite directions help stabilize the aircraft when it slows down to 55 mph for loitering over a target. The turning radius of only 500 feet emulates a cat chasing his own tail. The rear of the fuselage is a cargo compartment with removable door for air-dropping supplies. One version was equipped with laser beams for night missions, and another example was assigned to the National Aeronautics and Space Administration (NASA) for STOL experiments. Bronco-Laras earned the distinction of being the most versatile combat planes of that time.

Rockwell OV-10 Bronco-LARA, versatile Light Armed Reconnaissance Aircraft of the Vietnam war. Photo taken over Southeast Asia on December 6, 1968, by Sgt. Russell Parrish. (USAF Photo 106627)

The OV-10 is a twin-engine high-wing monoplane with retractable tricycle landing gear and twin tailfins connected at the top by a long tailplane. The one or two-man crew sit tandem style in ejection seats within the buy-eyed cockpit that provides straightdown vision when in level flight. Two Garrett-Airresearch T76-G 715-shp (shaft horsepower) engines give it a top speed of 280 mph. The combat radius is 125 miles with a full weapons load of 2,400 pounds; the ferry range reaches out to 1,430 miles. Six fully equipped combat troops or 3,200 pounds of freight can be carried. Armament consists of four 7.62mm Gatling guns and two Sidewinder missiles. A bomb load of four 500-pounders and/or an assortment of firebombs (napalm) can be carried underneath. Production ended in 1969 after 164 OV-10's had been delivered to the USAF.

OBSERVATION AMPHIBIANS — 1919 TO 1962

COA—CORPS OBSERVATION AMPHIBIAN: 1919 TO 1925

The "COA" for Corps Observation Amphibian was introduced in the Air Service designation system of 1919.

The **Loening COA-1** in 1923 was the first Corps Observation Amphibian in the Army's history. It was a single-engine biplane with a parasol upper wing and standard tail section. The main float was the bottom of the hull, extending from the tail to a position forward of the propeller. Two outboard floats for balance were attached to the wing tips. The landing gear had two retractable wheels up front and a tailskid at the rear of the main float. The Liberty V-1650 400-hp engine drove a 3-bladed prop to produce a top speed of 120 mph and a range of 400 miles. Pilot and observer sat in open tandem cockpits. The pilot operated a fixed forward-firing fuselage gun and the one-man-band observer operated one or two 30-caliber swing guns when he wasn't busy navigating, operating the radio equipment or snapping pictures with the fuselage camera. A total of eleven COA-1's were built for the Army Air Service. In 1925 three of them participated in MacMillan's Arctic Expedition. The COA-1 turned out to be the only aircraft in the "COA" category, and in 1925 all remaining COA-1's became OA-1's in the new Air Corps designation of Observation Amphibian when "Corps" was dropped.

OA—OBSERVATION AMPHIBIAN: 1925 TO 1948

In 1925 the "OA" category for Observation Amphibian replaced "COA" when the Army Air Corps designation scheme was implemented.

The **Loening OA-1 Observation Amphibian** in 1925 was a direct redesignation of all surviving COA-1 biplane amphibians with no change in the aircraft. On December 21, 1926, Herb Dargue, Ira Eaker, Muir Fairchild and Ennis Whitehead departed from Kelly Field, San Antonio, Texas, in OA-1 Amphibs for a 5-month goodwill tour through Central America, returning to Bolling Field, D.C., on May 2, 1927. The OA-1 named "City of San Francisco" was presented to the Smithsonian Institute. One OA-1 was delivered for service trials as the XO-10 Observation prototype, but it never reached production. A total of 42 OA-1's were delivered to the Air Corps, including 8 redesignated from COA-1's and 34 built under the new designation.

Loening COA-1 (OA-1) Corps Observation Amphibian, the Army's first full-bodied Amphibian and the only aircraft to carry a "COA" designation. Photograph shows the plane in November 1926 after it had been redesignated OA-1. (USAF Photo 1628-AC)

The **Leoning OA-2** in 1929 was like the OA-1 except for a Wright V-1460 480-hp engine, top speed of 125 mph, 585-mile range and the fixed gun mounted in the upper wing instead of on the fuselage. A total of eight OA-2's were delivered to the Air Corps.

The **Douglas OA-3 Dolphin Observation Amphibian** was a direct redesignation in 1933 of all C-21 Amphibian Transports to help with new coastal patrol responsibilities taken over from the Navy. The OA-3 was an unarmed high-wing monoplane with a plus-type tail section and twin engines mounted on braces above the wings. The interior was modified to carry a 4-man crew, but arrangements could be made for three more persons. Two Wright R-975 300-hp engines drove 2-bladed props to give it a top speed of 140 mph with a range of 550 miles. The forward bottom of the hull served as the main float and two outboard floats were hung under the wingends. Two fixed wheels and a tail skid were installed for land-based operations. Conversions from the C-21 to OA-3 totaled eight aircraft—no additional contracts were submitted.

The **Douglas OA-4 Dolphin Observation Amphibian** in 1934 was a direct redesignation of all C-26 Amphibian Transports with twin Pratt & Whitney 350-hp engines and a top speed of 150 mph; otherwise it was similar to the OA-3. A total of seven OA-4's were delivered to the Air Corps for use on coastal patrols and overwater flights. One advanced OA-4 with engine change was to have been the XOA-7, but the contract was cancelled.

The **Douglas XOA-5** prototype in 1935 was a redesignation of the YO-44 Observation plane, which was itself a conversion from the Douglas XB-11 Amphibian Bomber, but no further development followed.

On June 29, 1936, Frank Andrews and John Whiteley set a new Amphibian distance record flying the XOA-5 non-stop on a 1,430-mile trip from San Juan, Puerto Rico, to Langley Field, Virginia.

The **Sikorsky YOA-8 Observation Amphibian** in 1937 was a high-wing monoplane with twin engines welded into the wing and a tailplane fitted about one-fourth the way up the fin. The belly float extended the full length of the hull and twin balancing floats were suspended about halfway to the wing tips. Internal arrangements could accommodate eleven people; eight seats were removable for hauling cargo. Two Pratt & Whitney R-1690 750-hp engines drove 3-bladed props to produce a top speed of 185 mph with a range of 775 miles. Three retractable wheels were mounted for land-based operations. A total of five YOA-8's were delivered to the Air Corps. The XOA-11 in 1941 was similar to the YOA-8, but it crashed on a flight to Trinidad. The JRS-1 was a Navy version.

The **Grumman OA-9 Goose Observation Amphibian** in 1938 was a highwing monoplane with twin engines mounted on top of the wings and a tailplane about one-third the way up the fin. The main float covered three-fourths of the hull's bottom, and two outrigger floats were slung near the wingtips. Three retractable wheels served as the land-based undercarriage. Two Pratt & Whitney R-985 450-hp engines drove 3-bladed props to produce a top speed of 195 mph and a range of 700 miles. Pilot and copilot sat in the cockpit and four seats could be installed in the cabin. A total of 31 OA-9's were procured for the Air Corps. The Navy versions were designated JRF-5/6. In 1948 all surviving OA-9's became A-9 Amphibians in the new USAF designation scheme.

The **Consolidated-Vultee OA-10 Catalina Observation Amphibian** in 1942 was a direct transfer of the Navy PBY-5A Patrol Bomber to serve as an air-sea Search and Rescue plane. During WWII the Catalina was used in a variety of roles including patrol bomber, long-range reconnaissance, anti-submarine patrol, convoy escort, torpedo carrier, glider tug, cargo hauler, personnel transport and mailcarrier. The OA-10 was a high-wing monoplane with twin engines mounted in the wings and a plus-type tail section. The underside of the hull served as the only float, and the tricycle landing gear was retractable into the fuselage. Two Pratt & Whitney R-1830 1,200-hp radials gave it a top speed of 190 mph. Sufficient fuel could be loaded for a 2,550-mile range. Pilot and copilot sat side-by-side with dual controls. The bombardier sat in a plexiglass nose cone

Consolidated-Vultee OA-10 (PBY-5) Catalina, versatile WWII Observation Amphibian-Mercy-ship-Rescue Plane-Patrol Bomber. (USAF Photo 32693-AC)

and operated a flex gun through the gun port. The navigator sat in the cabin, and two gunners operated Brownings from the side windows. A bomb load of 4,000 pounds could be carried, including torpedoes and four 325-pound depth charges hung on underwing pylons. On May 26, 1941, Tuck Smith, a young American Navy pilot flying a British Catalina on a classified reconnaissance mission, spotted Germany's most powerful battle-ship, the *Bismarck,* in the North Atlantic off the coast of Ireland. British warships closed in and sank the "unsinkable" *Bismarck,* which carried over 2,200 German seamen to the bottom. Baron Burkard von Mullenheim-Rechberg was the highest ranking survivor; he spent the rest of the war in a Canadian POW camp. Later in the war, Tuck Smith was airborne in a Catalina out of Hawaii when Pearl Harbor was hit by Japanese planes on December 7, 1941. He completed the mission and survived the war to remain on active duty until retirement. Production of the OA-10 ended in 1945 after 367 had been delivered to the USAAF. In 1948 all surviving OA-10's were redesignated A-10 Amphibians for the USAF. PBY-5, PBY-6, PBN-1, PB2B-1, OA-10, A-10, Catalina, Canso and Nomad all refer to the same airplane.

The **Grumman OA-13 Goose** in 1942 was exactly like the OA-9 except for engine change. The OA-13 was a high-wing monoplane with 450-hp engines, six seats, a range of 700 miles and a top speed of 195 mph. A total of five OA-13's were procured for the USAAF, including three bought from private owners and two JRF-5's transferred from the Navy. They served on coastal patrols and air-sea rescue missions.

The **Grumman OA-14 Widgeon** in 1942 was similar in appearance to the other Grumman Observation Amphibians except for its smaller size and lighter weight. The OA-14 was a high-wing monoplane with twin engines embedded in the wings and a plus-type tail section. The underside of the hull served as the main float and two outboard floats suspended under the wing ends prevented capsizing. Three retractable wheels were provided for operations on land. Two Ranger L-440 200-hp engines gave it a top speed of 150 mph and a range of 650 miles. Pilot and copilot sat up front, and three seats in the cabin were for additional crewmen or passengers. A total of 16 OA-14's were adapted to military service with the USAAF for use on coastal patrols and overwater rescue missions.

A—AMPHIBIAN: 1948 TO 1962

In June 1948 the new USAF abolished the "A" for Attack category and redesignated all Attack planes into "B" for Bombers. In the same designation scheme the "OA" category was abolished and "A" for Amphibian substituted. When the prefix "S" was used the "SA" designation stood for Search Amphibian.

The **Grumman A-9 Goose Amphibian** was a direct redesignation in 1948 of all surviving OA-9's with no change to the aircraft. No new production orders were submit-ted. The A-9 had two 450-hp engines, six seats, a 700-mile range and a top speed of 195 mph. The Grumman YA-12 prototype in 1948 was adapted for trials from the Navy J2F-6, but that project was cancelled. Only three of the YA-12's were built.

The **Consolidated-Vultee A-10 Catalina Amphibian** was a direct redesignation in 1948 of all OA-10's carried over from the USAAF to the USAF for air-sea rescue duties. There was no change in the aircraft and no new production contracts were made.

Grumman SA-16 Albatross, Korean war JATO triphibian search and rescue plane. Snapped inflight is a workhorse of the Air Rescue Service. (USAF Photo 155150-AC)

The **Grumman SA-16 Albatross Search Amphibian** was delivered to the USAF in 1949 as the only Air Force aircraft to carry an "SA" designation. Modern electronics, radar and communications equipment provided the capability of thorough scanning along either side of the plane's flight path. The SA-16A in 1953 was a triphibian with wheeled float-skis allowing operations from land, sea, ice or snow. The SA-16B in 1957 was a stretch job with increased wingspan, longer range and faster speed. During the Korean conflict SA-16's were credited with saving the lives of over 900 combat troops.

The SA-16 was a high-wing monoplane with twin engines mounted in the wings near the fuselage. The underside of the hull served as the main float, and two outboard floats were suspended near the wingtips. The tail section was standard and the landing undercarriage was of the retractable tricycle type. Two Wright R-1820 1,425-hp engines had two-speed superchargers driving three-bladed props to produce a top speed of 265 mph. The range was 2,700 miles with a 23-hour endurance time when carrying a full load of fuel and flying under optimum conditions. Pilot and copilot sat in the cockpit, and positions were provided in the hull for navigator, radio operator, engineer and two medical or rescue technicians. Up to twelve stretchers could be accommodated in the cabin. Ten to fifteen survivors could be carried, and an Airdrop Rescue Kit (ARK) could be floated down by parachute to provide temporary shelter and medical aid to as many as 40 stranded survivors. A total of 395 SA-16's were delivered to the Air Force. In 1962 all surviving SA-16's were redesignated HU-16 Search and Recovery Utility aircraft in the new tri-service designation system.

PHOTOGRAPHY AIRCRAFT

F—PHOTOGRAPHY: 1930 TO 1948

The "F" for Photography (Reconnaissance) category was added to the Air Corps list of designations in 1930. All the "F" aircraft in World War II were conversions or derivatives of other aircraft already in production. In 1945 the "F" became a prefix, but continued to stand for Photography. In 1948 the new USAF added the prefix "R" for Reconnaissance to replace "F" for Photography in order to use the letter "F" as a designator of Fighter airplanes. For example: Photo-Recon derivatives of the Boeing B-29 Superfortress Bomber were designated F-13 Photography planes in 1944, became FB-29 Photography-Bombers in 1945, and finally were redesignated RB-29 Reconnaissance-Bombers in 1948.

The **Fairchild F-1 Photo** plane was the forefather of Photography aircraft for the Air Corps. It was ordered in 1929 as the seven-seated C-8 Cargo Transport, but the designation was changed to F-1 when the first production examples were delivered in 1930. It was a high-wing monoplane uniquely designed so that the wings folded on each side for storage in hangars with narrow doors. Downward looking cameras were installed in the fuselage. The three-man crew included pilot, copilot and photographer. A total of fifteen F-1's were delivered to the Air Corps, and later in 1930 all of them were redesignated C-8 Cargo Transports, but they continued to be used in the Photography role. Similar models were commandeered for service in World War II under the designation of C-96 Cargo Transports with Photo capability.

The **Beechcraft F-2 Expediter-Bugsmasher** in 1941 was a conversion from the C-45 light Cargo Transport. The F-2 had multiple-lens cameras for mapping large areas, the F-2A had four cameras in the fuselage and the F-2B had a trimetrogon arrangement of cameras for taking wide-angle shots. (Trimetrogon means taking one shot straight down and two at oblique angles simultaneously.) A total of 70 F-2's were built for the USAAF.

The **Douglas F-3 Photo-Havoc** in 1940 was a conversion from the A-20 Havoc Attack plane. The F-3 had the distinction of being America's first Photography plane used in combat. It had a T-3A camera in the bomb-bay and the F-3A added more cameras in the back of the fuselage. A total of fifty A-20's were converted to F-3's for the war.

The **Lockheed F-4** and **F-5 Lightnings** in 1942 were derivatives from twin-boomed P-38 Pursuit planes. They were America's first high-speed Photography aircraft. The F-4 had four K-17 cameras. The F-5 had varying camera arrangements and a longer range; the F-5D added a piggy-back seat to accommodate a photographer as the second crew member. The Lightning series turned out to be the most produced of all the Photo planes. A total of 1,373 were delivered, including 119 F-4's and 1,254 F-5's for the USAAF in World War II.

The **North American F-6 Mustang** in 1942 was the USAAF's first combat variant from the P-51 Mustang Pursuit project. The F-6A had two K-24 fuselage cameras, and the F-6C had one K-17 and one K-22 camera in the bay. A total of 482 F-6's were delivered for WWII. In 1945 all F-6's became FP-51's and in 1948 all surviving FP-51's were redesignated RF-51's, which were subsequently used on Photo-Recon missions in Korea. The TRF-51 was a Recon-Trainer for the USAF.

The **Consolidated F-7 Liberator** in 1942 was a conversion of the B-24 Heavy Bomber into a long-range Photo plane by adding extra fuel tanks and modifying the fuselage rear into a camera bay. Arrangements varied from six to eleven cameras at positions in the nose, bomb-bay and camera bay. A total of 90 aircraft were rebuilt into F-7's, but many more field-modified B-24's were used as Photo-Recon planes without officially changing their designation.

The Canadian **DeHavilland F-8 Mosquito** in 1943 was a descendant of the British Mosquito Fighter-Bomber that made its first flight for the RAF in November 1940. The F-8 was a twin-engine high-wing monoplane with standard tail section and conventional landing gear having oversized front wheels that retracted into engine nacelles. Pilot and navigator-photographer sat side by side in the enclosed cockpit above the wing's leading edge. Two Packard-Merlin 1,300-hp engines driving 3-bladed props gave it a top speed of 365 mph, and the range stretched out to 2,700 miles when all tanks were full. British Mosquitos were used on quick-hitting missions, such as bombing the building in Berlin, Germany, where Reichsmarshall Goering was making a speech, and a tree-top raid on Gestapo headquarters in Oslo, Norway. The British Mosquito PR-XVI Photo-Recon variant was used by the 8th Air Force in England as a dual-control trainer and on Weather-Recon missions; those planes received USAAF markings and serial numbers but were not given American designations. A total of 40 Canadian-built F-8's were procured for the USAAF in WWII. An additional batch of Mosquitos was acquired on a Reverse Lend-Lease arrangement with Britain.

The **Boeing F-9 Flying Fortress** in 1942 was a modification of the B-17 Heavy Bomber into a long-range Photography plane with extra fuel tanks, trimetrogon (wide-angle) cameras mounted in the nose and downward aimed cameras in the bomb-bay and rear fuselage. The F-9B had different camera arrangements. In 1945 the F-9 Photography plane became the FB-17 Photo-Bomber, and in 1948 the FB-17 was redesignated RB-17 Recon-Bomber for the Air Force.

The **North American F-10 Mitchell** in 1941 was a conversion of the B-25 Medium Bomber with extra tanks and no armament. Trimetrogon cameras were housed in a fairing underneath the nose cone and a camera nook replaced the tail gun in the rear fuselage. Only ten F-10's were built for the USAAF.

The **Hughes XF-11** and **Republic XF-12 Photo** planes were ordered in 1944 but didn't fly until 1946, too late for WWII. Both contracts were cancelled.

The **Boeing F-13 Superfortress** in 1944 was a derivative of the B-29 Heavy Bomber with extra fuel tanks in the bomb-bay. Variations of K-18 and K-22 cameras were installed in the nose, fuselage and tail. These aircraft took surveillance pictures all over the western Pacific, including photographs over Japan that led to the selection of Hiroshima and Nagasaki as the first atomic bomb targets. F-13's tagged along with flights of B-29 Bombers to photograph the results of raids throughout the Japanese islands. Flying a F-13A named ''Tokyo Rose'' out of Tinian in the Mariana Islands on November 1, 1944, Captain Ralph Steakley made the first flight over Tokyo by an American land-based aircraft. In 1945 the F-13 Photo plane became FB-29 Photo-Bomber, and in 1948 the FB-29 was redesignated RB-29 Recon-Bomber.

The **Lockheed F-14 Shooting Star** in 1944 was a conversion of the P-80 all-jet Pursuit with a new nose cone for housing the cameras. The F-14 was the first and only jet-propelled aircraft to carry an ''F'' designation for Photography, but that didn't last long. In 1945 the F-14 Photo-Jet became the FP-80 Photo-Pursuit and in 1948 the FP-80 was redesignated RF-80 Recon-Fighter. A total of 55 Shooting Stars were delivered for the Photo-Recon role.

The **Northrop XF-15 Reporter** was ordered as a Photography version of the P-61 Black Widow Night-Pursuit. It was the last of the ''F'' for Photography aircraft. By the time of delivery in 1946 the designation had been changed to FP-61 Photo-Pursuit. It had a tandem-seated cockpit for pilot and navigator-photographer, who operated the camera systems. In 1948 all surviving FP-61's were redesignated RF-61's to become Recon-Fighters in the brand new USAF. The name of Reporter distinguished it as the only Photography plane to have a name different from that of its progenitor. A total of 36 Reporters were delivered for service in the Photo-Recon role.

PURSUITS OF WWI

In July 1917 Congress authorized $640 million for development of military airplanes—too little and too late! No amount of money could compensate for having fallen so far behind other major nations in aeronautical technology. As a result of a visit to the Western Front by Major Ray Bolling in June 1917, it was decided that Trainers could be built in America, but that Pursuit-Fighters should be purchased from our Allies. The Air Service of the American Expeditionary Force (AEF) fought the entire war without a single American designed and built dogfighter.

On August 13, 1917, the 1st Aero Squadron Commanded by Major Ralph Royce made its debut at the Western Front to become the first unit sent for flying duties with the AEF. In March 1918 Major Raoul Lufbery and his soon-to-become-famous 94th Aero Squadron arrived at the front. Proudly displaying their "Hat in the Ring" insignia, members of the old 94th joined with fifteen other Aero Squadrons to begin tilting air superiority in favor of the Allies.

During the war America received a total of 1,371 combat Fighters designated Nieuport 28C, SPAD XIII, Sopwith Camel and SE-5 Scout. Our gallant young men flew those planes to fame, downing 781 enemy aircraft while producing 88 "Aces" in only seven months of air-to-air combat. The romantic, chivalrous Frenchmen originated the title of "Ace," which was bestowed for being credited with five unassisted air victories against enemy aircraft. America was the only other Ally to officially adopt that title as a distinguisher of air heroes.

NIEUPORT 11

France claimed to have scored the first aerial victory in history when a Voisin Fighter shot down a German Taube on October 14, 1914. Kaiser Wilhelm retaliated by hiring the Dutch airplane designer, Anthony Fokker, to develop a warplane with machine guns firing through a synchronized propeller. The resulting Fighter aircraft were so superior to Allied planes of the day that light scouting planes used in combat came to be known as "Fokker fodder."

To counteract the challenge of those superior Fokker warplanes, France developed their first really successful Allied Fighter, the Nieuport 11, which came into the battle in August 1915. Nicknamed the *"Bébé"* because of its comparative small size, the Nieuport 11 was armed with a Lewis machine gun, carrying 47 rounds of ammo, and eight le Prieu incendiary rockets for bursting low level kite balloons hoisted aloft by the enemy for artillery spotting. Flying his Nieuport 11 Bébé, Captain Tricornot de Rose, a pioneer of French Fighter tactics, became one of the first pilots in the war to attain the title of Ace.

The Nieuport 11 was never owned or operated by American Military units; however, on April 20, 1916, the Bébé became the first Fighter flown into battle by American civilian soldiers who made up the famous Lafayette Escadrille. On April 24, 1916, taking off on a Dawn Patrol from the airport at Luxeuil les Bains, Kiffen Rockwell scored America's first air victory when he downed a Boche plane with a single burst of gunfire and became the first American to be awarded the Médaille Militaire. On September 23, 1916, Kiffen Rockwell was killed by a dum-dum bullet. On June 18, 1916, Clyde Balsley of the Lafayette Escadrille became the first American to be wounded in air combat when he and his Bébé were shot down near Verdun. Victor Chapman of the Lafayette Escadrille, the son of the well-known author, John Jay Chapman, was the first American to be killed in WWI; he was shot down on June 23, 1916, at Verdun, and died the next day.

NIEUPORT 17E

Sometimes referred to as the most famous of the Nieuports, the Nieuport 17 made its combat appearance along the Western Front in March of 1916. It had no throttle—therefore it had no cruise speed; it flew wide open all the time unless ignition in the cylinders was somehow suppressed. Fortunately, our pilots used the Nieuport 17E only as a trainer and never had to fly it in combat. It was underpowered and outclassed in comparison to German Fighters when Americans went into battle. The ''E'' suffix was a symbol for ''*école*'' (school), meaning a trainer version. On April 24, 1917, Major Billy Mitchell became the first American Army Officer to fly over enemy lines when he went along on a sunset reconnaissance mission in a French twin-engine, 3-seat Observation plane piloted by François Lafonte; cover was provided by a formation of Nieuport-17 Fighters.

Doing most of his combat flying in the double-barreled Nieuport 17 equipped with one Vickers and one Lewis gun, Lt. Charles Nungesser of the French Air Force scored 45 victories, earned the Médaille Militaire and the Crois de Guerre, and survived the war to emerge as France's third-ranking Ace. René Fonck, with 75 air victories to his credit, was the French—and also the Allied—Ace of Aces in WWI. Georges Guynemer ranked second with 54 kills. On May 8, 1927, long after the war, Captain Nungesser met his death somewhere between Europe and America on the first solo attempt at a westward crossing of the Atlantic Ocean. Taking off from le Bourget Field, Paris (the airport at which Charles Lindbergh was to land 13 days later), Charles Nungesser nosed his plane toward the west and disappeared over the horizon, never to be heard from again.

NIEUPORT 28C

America's first combat dogfighter was the diminutive Nieuport 28C which made its initial flight on June 14, 1917. Beginning in March 1918, Nieuport 28C Fighters were assigned to four Aero Squadrons of the Air Service, AEF. They were accepted for the time being by Colonel Billy Mitchell, not because of any outstanding characteristics, but for the reason that they were the only aircraft available. In fact, if the Nieuport 28 had any special talent, it was for shedding fabric and vital parts during steep dives and other strenuous maneuvers. The French Escadrilles had already firmly rejected it for front line service. The ''C'' suffix stood for ''*chasse*''—''hunt'' or ''chase''—that's where the Army got the name ''Pursuit'' for its dogfighters.

French Nieuport-28C. First Fighter flown into combat by Army pilots in WWI. (USAF Photo 152232-AC)

The devil-may-care American airmen took that castoff relic and flew it to fame in only three months. Lt. Doug Campbell, flying a Nieuport 28C-1 on a dawn patrol with the 94th Aero Squadron, shot down a Pfalz Albatros D-III over Toul Airdrome on April 14, 1918. That was the Army's first air victory of the war and also America's first Military air victory in history. At the time, Lt. Campbell's plane had only one machine gun installed because the Squadron had just received its first shipment of Vickers .303's and there weren't enough for two per plane. Doug Campbell became America's first Ace in history by downing his 5th German plane on May 31, 1918. In July 1918, when the Aero Squadrons were fully stocked with SPAD-XIII Fighters, most of the replaced Nieuport 28's had their guns removed and were used as combat trainers with the 213th Squadron. The Air Service received a total of 297 combat-ready Nieuport 28C's during the war.

SPAD XIII

Produced by the Société des Productions Armand Deperdussin Aircraft Company, the SPAD XIII first flew on April 4, 1917, and proceeded to rack up a set of statistics that led to its being regarded as being the finest Fighter developed by the Allies in WWI. Deliveries of the SPAD began to dribble in to Aero Squadrons during March 1918, and as production was spurred by vitally needed American dollars, sufficient quantities had arrived in July 1918 to replace the outclassed Nieuport 28C in fifteen of the sixteen Aero Squadrons. Armed with two Vickers machine guns mounted on the cowl, those highly maneuverable speedsters had a climb rate of 6,500 feet in five minutes—mighty adept for those days.

Flown by the famous Aces René Fonck, Georges Guynemer, Eddie Rickenbacker and Frank Luke, the SPAD XIII contributed enormously toward winning the battle of the skies

French SPAD-XIII. Captain Eddie Rickenbacker's famous AEF dogfighter of WWI. (USAF Photo 155364-AC)

and helping to silence the guns on the ground. Eddie Rickenbacker, flying first in Nieuport 28C's and later in SPAD XIII's, was credited with shooting down 26 enemy planes in seven months to be acclaimed as America's Ace of Aces. Captain Rickenbacker scored his first dogfight victory on April 29, 1918. Frank Luke was the second leading Ace with 21 air victories to his credit. Lt. Luke was the first American Military Aviator to be awarded the Congressional Medal of Honor for shooting down 18 enemy planes in 17 days while participating in the 1,481-plane air battle at St. Mihiel during September 1918. General Billy Mitchell, leader of that Sky War, read the citation posthumously after Lt. Luke was shot down and killed on September 29, 1918. Raoul Lufbery, former member of the Lafayette Escadrille and America's first civilian war Ace, ranked third with 17 air victories.

The AEF received 893 combat-ready SPAD XIII Fighters during the hostilities, and after the war dust had settled, 435 of them were in good enough shape to be shipped back home for use in the peacetime Army Air Service.

SOPWITH CAMEL

Armed with two Vickers machine guns and capable of carrying four 25-pound Cooper bombs under the fuselage, the Sopwith Camel was introduced in July 1917 by the Sopwith Works of England. Its nickname was derived from the appearance of the gun breeches that were enclosed by a distinctively "humped" cover. The men who flew the Camel scored 1,294 air victories, which turned out to be the record for Fighters flown by the Allies in WWI. Practically all of those kills were scored by the men of the British Royal Flying Corps (RFC). The Camel was a tricky, dangerous plane to fly and an extensive training period was required for learning all of its idiosyncrasies—but the pilots who drove it bragged that it was the best dogfighter in the war. An excellent advance in its effective firing power was that the belt-fed guns didn't require reloading as frequently as did the magazine-fed guns on most of the other planes.

On April 21, 1918, Roy Brown, a Canadian pilot leading a routine flight of Camels near the Somme, confronted and shot down Baron Manfried von Richthofen, Germany's leading war Ace who claimed eighty victories against Allied airmen—the "Red Baron" was dead. Ernst Udet and Erich Loewenhardt ranked second and third among Germany's Aces with 62 and 53 air victories respectively. Von Richtofen was given the nickname of Red Baron because he always flew in a triplane painted bright red except for its markings. On August 10, 1918, Lt. Stuart Culley was riding his Camel when he outclimbed and shot down the last enemy Zeppelin to fall during the war.

Because the SPAD XIII had already been adopted as America's favorite Fighter, only one Aero Squadron was fully equipped with the pride of Sopwith; as a result only 143 of the Sopwith Camels were delivered to the AEF.

SE-5 SCOUT

The Scouting Experiment Number 5 was produced by the British Royal Aircraft Factory and appeared at the front during "bloody April" of 1917, the blackest month of the war in terms of casualties for the British RFC. Armed with either two Vickers guns or a Vickers and a Lewis gun, the Scout soon proved a match for the German Albatross, which up until that time had ruled the skies. Four 25-pound trench bombs were carried on racks under the fuselage for use in ground attacks in support of infantry down on the ground. That tough little Fighter was so rugged that a pilot once flew it through the side of a house and emerged unhurt. Flying with the famous English "High Hat" Squadron, Major Edward Mannock, Britain's leading Fighter pilot, scored 50 of his 73 victories while flying in the SE-5 Scout. William Bishop was a close second with 72 kills and Ray Collishaw downed 60 of the enemy to rank third.

While serving as an American civilian volunteer with the RFC, Elliot Springs became an Ace early in the war flying SE-5 Scouts. When the 103rd Pursuit Squadron was formed in February 1918 with surviving civilian airmen, mostly from the abandoned Lafayette Escadrille, Elliot W. Springs was commissioned a Captain and appointed its Commander. He went on to score a total of 12 kills to rank in a fifth place tie on America's list of Aces; George Vaughn was just ahead of him with 13 victories.

In August 1918 one Aero Squadron was assigned 38 SE-5A's, which turned out to be the only ones received by the Air Service. Meanwhile, a few of the hulls were shipped to America for assembly, but the American Liberty engine was too powerful—and by the time an engine had been developed that was compatible with the hull, the war was over.

LE PERE LUSAC-11

The Packard Le Pere Lusac-11 Fighter-Pursuit was delivered in September 1918, but only three of them got to France before WWI ended. The Le Pere was a standard two-seater biplane with a Liberty-12 400-hp engine and a top speed of 133 mph. Armament consisted of two nose guns and two swing guns in the rear seat. In 1920 the Army Engineering Laboratory at McCook Field, Ohio, provided a Lusac-11 with a gear-driven turbo-supercharger attached to the engine—the first Army aircraft so equipped. One example had two spoked wheels under the tail and was fitted with eight guns. On February 27, 1920, Major Rudolph Schroeder flew the Le Pere out of McCook Field to establish a world altitude record of 33,113 feet. On September 18, 1921, Lt. John MacReady flew a Le Pere while raising the mark to 34,510 feet above Dayton, Ohio, and on February 22, 1924, he pushed it up to 41,000 feet. Production totaled 27 Le Pere Lusac-11 planes.

HANRIOT-MACCHI HD-1

The most popular Fighter in Italy during WWI was the French Hanriot HD-1, of which 831 were built by the Italian Macchi factories. The first Italian-designed Fighter was the Ansaldo A-1 Balilla (Hunter), but it was too late for the war. The Ansaldo SVA-5 was their most popular long-range Reconnaissance plane. Italy also imported French Nieuport and SPAD Fighters for use against Austria-Hungary Aces Godwin Brumowski (37 kills), Julius Arigi (29 kills) and Frank Crawford (28 kills). The leading Italian Ace of WWI was Francesco Baracca, who got the first of his 34 kills in a Nieuport-11 Fighter on April 6, 1916—that was also Italy's first air victory of the war. Baracca was shot down and killed in his ''Black Pony'' emblazoned SPAD-XIII Fighter on June 19, 1918. Silvio Scaroni won 26 air duels to rank second on Italy's list of WWI Aces, followed by Ruggiero Piccio with 24 victories and Flavio Baracchini with 21 to his credit.

PURSUITS — WWI TO WWII

POST-WWI FOREIGN-BUILT FIGHTERS

During WWI America procured 893 SPAD XIII Fighters for use along the Western Front. When the war ended in November 1918, about 435 of them were in good enough condition to be shipped back home for use in the post-war Air Service. Those aircraft were used for a variety of purposes including laboratory research, training, engine test bed, altitude experiments and public exhibits. The SPAD XIII was requested by engineers in the States because it was respected as the finest dogfighter produced in World War I; at that time America had not yet succeeded in developing a military Fighter that measured up to foreign standards.

In April 1918 the Fokker D.VII Fighter made its appearance on the Western Front as the best German airplane of WWI. The D.VII was considered by America to be such a major part of the German military capability that it was the only airplane listed by name in the Armistice agreement as a war machine that was specifically not to be produced any longer inside the geographical boundaries of Germany. America received approximately 350 enemy aircraft as war reparations, of which 142 were Fokker D.VII Fighters. Most of them were used in exhibits throughout the United States, but a few were sent to McCook Field, Dayton, Ohio, for study and evaluation. Army engineers were impressed with the aircraft's sturdy structure and adopted some of its features in later aircraft designs.

ORENCO-D FIGHTER

The aeroplane was invented in the United States in 1903; the Army's Aeronautical Division was organized in 1907; the Aeronautical Division received its first Trainer aircarft in 1909; the Air Service of the American Expeditionary Force went to war in 1917; but it wasn't until 1919 that the Army received its first all-American designed and constructed Fighter combat plane.

In 1918 the ORdnance ENgineering COmpany (ORENCO) of Baldwin, Long Island, New York, contracted to build four wooden prototypes of their Model-D military plane for use in the air-to-air Dogfighting role. Most of the world referred to their air combat planes as "Fighters." The French scheme of designations placed their dogfighters in the "C" for *Chasse* category. Literally translated, *Chasse* means hunt or "chase," and the Army later substituted the synonym "Pursuit" in the 1919 scheme of designating Air Service airplanes. Built in 1918, those prototypes were referred to as Orenco-D Fighters.

Ordnance Engineering Company (ORENCO) Pursuit-Fighter, the Army's first all-American de-signed and built dogfighter of post-WWI days. (USAF Photo 164961-AC)

With occupation funds left over from WWI, the Army submitted an order for bids on a production model of the Orenco-D Fighter, and Curtiss won the first contract with the lowest bid. Deliveries of the Curtiss Orenco-D began in January 1919 and the Army Aviators boasted, "At last we are in the race!" The Orenco-D was a single-engine, one-seater biplane with standard tail assembly, fixed front wheels and a landing skid at the rear. The Wright-Hispano 330-hp engine gave it a top speed of 136 mph. The pilot sat in an open cockpit with two Vickers machine guns mounted on top of the fuselage for synchronous firing through the two-bladed propeller.

A total of 54 Orenco-D Fighters were built, including the four prototypes and 50 Curtiss production models.

MORSE-BOEING FIGHTERS

The Thomas Morse MB-3 Fighter was ordered in 1918, made its first flight on February 21, 1919, and deliveries began to arrive at Army fields in 1920. The MB-3 was a standard biplane with spoked front wheels and a landing skid at the rear. The single Wright H-3 300-hp engine produced a top speed of 140 mph. The pilot sat in an open cockpit under the trailing edge of the top wing. Two forward-firing 30-caliber machine guns were fitted on the fuselage to shoot through the synchronized propeller.

The Morse-Boeing MB-3A in 1922 used basically the same airframe as the MB-3 except for an enlarged tailfin and either a 2-bladed or 4-bladed prop. Boeing became the manufacturer, having submitted the lowest contract bid for a production model. First delivery of the MB-3A was made in 1922 to the much-decorated 94th Pursuit Squadron at Selfridge Field, Michigan—during WWI the 94th Aero Squadron had been commanded by Capt. Eddie Rickenbacker. Selfridge Field was named in honor of Lt. Tom Selfridge who was killed in a crash of the Wright Model-A Trainer, the Army's first aeroplane. One disarmed MB-3A had a hand-carved four-bladed propeller and a plush cockpit for use by General Billy Mitchell as his own personal aircraft.

Another design from the Morse-Boeing MB-3 family was the MB-6 in 1921 with a Wright 400-hp engine for entry in the National Air Races. It was redesignated the R-2 Racer for participation in the 1921 Pulitzer Race. The MB-7 was a high-wing monoplane built for the 1922 Pulitzer Race. The MB-9 in 1922 was an all-metal monoplane Fighter, and the MB-10 in 1923 was a Trainer. None of these offshoots from the MB-3 was ever put into quantity production.

The MB-3/MB-3A family was the first Army Fighter to be ordered in substantial quantity. It was the standard combat dogfighter in the Army Air Service throughout the early 1920's. Production totaled 244 aircraft delivered to the combat units.

PURSUITS: AIRCOOLED AND WATERCOOLED, ETC.

In the designation scheme used by the Air Service during the early twenties, Pursuit planes were given suffixes to further clarify their type. ''PA'' meant Pursuit Aircooled, ''PG'' indicated Pursuit Ground-attack, ''PN'' stood for Pursuit Night and ''PW'' represented Pursuit Watercooled.

The **Curtiss PN-1** in 1921 with a WWI Liberty engine was the first air combat fighter to be officially designated as a ''Pursuit.'' Two of them were built for trials and they turned out to be the only aircraft to carry the ''PN'' designation.

The **Loening PA-1** in 1922 was the only aircraft developed in the ''PA'' category. Only two of them were delivered for tests.

The **Aeromarine PG-1** in 1923 was the only ''PG'' aircraft to serve on active duty. Three of them were fitted with a 37mm cannon and a verification camera for experimental work.

The **Army Engineering Division PW-1** was designed and built in 1921 at McCook Field, Dayton, Ohio, using as a model the 1919 vintage Verville-Clark VCP-2 Experimental Pursuit. Only two PW-1's were built for performance testing. The Loening PW-2 in 1922 had aerodynamic and structural problems and was cancelled after eight examples had been flown.

The **Fokker PW-6** and **PW-7** were single-example biplane Pursuits purchased in 1922 from the Fokker Company, which had moved to Holland after WWI because the Versailles Treaty did not allow any further aircraft construction inside Germany.

The **Curtiss PW-8** was delivered to the Army Air Service on May 27, 1923, as the first successful aircraft to be called a ''Pursuit.''

It was a standard biplane of that day with a single Curtiss D-12 440-hp engine producing a top speed of 170 mph. The pilot sat in an open cockpit with two fixed forward-firing machine guns on the fuselage. The XPW-8A pre-Hawk had a set of completely redesigned wings to become a prototype for the P-1, soon to become the first in a long line of Curtiss Hawk Biplane Pursuits. Production totaled 28 of the PW-8 Pursuits. On June 23, 1924, Lt. Russ Maughan, winner of the 1922 Pulitzer Trophy, flew a PW-8 on the first coast-

to-coast, dawn-to-dusk flight from Mitchel Field, New York, to San Francisco, California, in 21 hours and 48 minutes, barely able to see the runway when he landed.

The **Boeing PW-9** was first flown by Boeing's test pilot, Frank Tyndall, in April 1923
and delivery began in 1924—Tyndall Field at Panama City, Florida, was named in honor
of Frank Tyndall. The PW-9 was a standard biplane that formed an important link in
aerodynamic progress during the twenties when the public was more concerned with the
making of "bathtub gin" than with the development of Army airplanes. The 435-hp
engine gave it a top speed of 155 mph. Experimental developments from the same
airframe were the P-4 Pursuit, the P-7 Pursuit with a 600-hp engine and the AT-3
Advanced Trainer, but none of them were highly successful. The PW-9 was the last
Pursuit in the "PW" category, which was phased out beginning in 1925. Production
totaled 114 PW-9 Pursuits delivered to the Army for use with the Pursuit Squadrons.

Boeing PW-9 Pursuit-Watercooled, Boeing's first Pursuit and last of the "PW" designated aircraft.
(USAF Photo 162455-AC)

P-1, P-2, P-3, P-5 HAWKS

In 1925 the Army adopted a new scheme for assigning designations to aircraft. The "P"
for Pursuit category became the only classification for air-to-air combat aircraft built after
that time; however, there was an overlap transition period during which some of the older
"PW" Pursuits retained their designations until retired from service.

The **Curtiss P-1 Hawk** was delivered to McCook Field, Dayton, Ohio, in August 1925
as the first aircraft in the new "P" for Pursuit category. Developed from the Curtiss
XPW-8A prototype, the P-1 had a 425-hp engine and a top speed of 160 mph. Otherwise,
it was a standard one-seated biplane of the type popular in the "decade of the flapper."
The Army received 148 P-1 Hawks as their first-line Pursuit during the phasing out of the
aging "PW" designated airplanes. An additional 75 aircraft were supplied to the Navy as
F6C Fighters, and sixteen were exported to South America. One of the P-1 derivatives
was sold to Japan in 1930, but World War II was not even the remotest of possibilities
during those days of peace and depression before Adolf Hitler came to power in Germany.

On July 15, 1930, Lt. Frank Tyndall was killed when his P-1 Hawk crashed near Mooresville, North Carolina.

Other Curtiss Hawks to use the same airframe as the P-1 with engine change and a few minor modifications were the P-2 in 1926 with a 500-hp engine, the P-3 in 1927 with an aircooled engine, and the P-5 in 1927 with a turbo-supercharger for high-altitude performance. Production totaled sixteen P-2, P-3 and P-5 Hawks.

P-6 HAWK

Although the Curtiss P-6 Hawk in 1928 had basically the same airframe as its predecessor Hawks, it was regarded as an entirely different series. The P-6 featured many improvements, including better all-around performance, fancy pants on the wheels, and machine guns mounted on the fuselage sides—rather than on top—to provide better visibility. Two original XP-6 prototypes in 1927 were conversions from P-1 and P-2 Hawks with 600-hp engines. They were entered in the 1927 National Air Races and one took the trophy at 201 mph while the other placed 2nd at a speed of 189 mph.

The P-6E Hawk in 1932 was delivered to the 17th Pursuit Squadron displaying the Diving Owl insignia at Selfridge Field, Michigan. It had a Curtiss V-1570 700-hp engine and a top speed of 198 mph. A tail wheel was introduced at the rear to replace the tailskids on previous models. Production totaled 75 P-6 Hawks.

Other variants of the Curtiss Hawk Biplanes were the YP-11, XP-17, YP-20, XP-21, YP-22 and the YP-23 in 1932 that turned out to be the last Biplane Pursuit delivered to the Army.

The Curtiss Hawk Biplanes spanned ten years of active duty before giving way to the Boeing P-26 Peashooter monoplanes. Final tabulations showed a total of 239 Hawks built, counting P-1, P-2, P-3, P-5 and P-6.

Curtiss P-6 Hawk, speedy biplane Pursuit of the peacetime era between WWI and WWII. (USAF Photo 15332-AC)

P-12 BOEING PURSUIT

The Boeing P-12 Pursuit made its first flight on April 11, 1929, and first deliveries to the Air Corps began in October 1929, coinciding almost exactly with the Stock Market crash ushering in the Great Depression. The Boeing F4B Navy version was developed simultaneously in a joint Army-Navy program limited by a tight budget. On April 12, 1930, nineteen Army Pursuits led by Capt. Hugh Elmendorf set a formation-flying altitude record of 30,000 feet in the California skies. Capt. Elmendorf was killed in the crash of a Consolidated YP-25 Pursuit at Patterson Field, Ohio, on January 13, 1933. The P-12/F4B family was distinguished as one of only a few aircraft designs effective as both a land-based Army Pursuit and a carrier-based Navy Fighter.

The P-12E was delivered in 1931 with an all-metal fuselage. It was a typical biplane with a 500-hp engine producing a top speed of 187 mph. The pilot sat in an open cockpit, but one P-12E experimented with a canopy covering. A tail wheel replaced the tailskid used on most Fighter aircraft since WWI. Two machine guns were fitted on the fuselage and two 116-pound bombs could be slung on underwing racks. One P-12K had a turbo-supercharger installed for high altitude performance trials. The P-12E was flown by the 95th Pursuit Squadron, boasting a history of honorable service dating back to WWI.

During February-June 1934 President Franklin Roosevelt ordered P-12 Pursuits to participate in airmail-carrying flights. There were severe snowstorms over the Rockies, and before the operation was over the Air Corps had lost 14 planes while proving conclusively that military aircraft should be left to military duties. Politicians in Washington should have learned what Air Corps officials already knew: our Pursuits were inferior to civilian planes that had successfully carried the mail for more than a decade.

Production ended in March 1932 after 366 P-12 Pursuits had been built for the Air Corps. An additional 170 F4B's were delivered to the Navy. Twenty-six examples were

Boeing P-12 biplane Pursuit, popular between-the-wars Air Corps Pursuit and Navy Fighter (F4B). Photo taken over Air Mail Route 18 between Oakland and Salt Lake City. (USAF Photo 28977)

exported, bringing the total built to 562 aircraft. All P-12's were retired by 1935, being replaced in service by Boeing P-26 Peashooters. Hasty exit of the P-12 was brought about by the appearance in 1934 of the Martin B-10 Bomber, which could outrun any Biplane Pursuit then in service.

YP-16 (PB-1) TWO-SEAT PURSUIT

The Berliner-Joyce XP-16 prototype made its first flight in October 1929 and delivery of the limited production model YP-16 began in 1932. Reporting for duty with the 1st Pursuit Group, the YP-16 was the first two-seated Pursuit since the Engineering Division TP-1 in 1924. The YP-16 had a 600-hp engine and a top speed of 175 mph. Pilot and copilot-observer rode in open tandem cockpits equipped with dual controls. Two nose guns fired forward through a 3-bladed prop and a flexible gun could be mounted in the rear seat. A bomb load of 250 pounds could be carried on underwing hardpoints. A distinguishing feature was mounting of the top wing into the fuselage rather than elevating it on struts in the conventional "parasol" fashion.

Production of the YP-16 totaled 25 aircraft, which represented the last gasp of the Biplane Pursuits before they were pushed into the background by newly developed monoplanes. The Berliner-Joyce Company over-invested in their biplane projects at a time when two-winged combat planes were going out of style; consequently they went out of business in 1934.

In 1935 the Air Corps introduced the "PB" for Pursuit Biplace (two-seated) category into their aircraft designation scheme. The Berliner-Joyce PB-1 in July 1935 was a direct conversion of all surviving YP-16 Pursuits, the first aircraft assigned to that new classification. The PB-1 turned out to be the only Biplane Pursuit that carried a "PB" designation.

P-26 PEASHOOTER

The Boeing XP-26 prototype made its first flight on March 20, 1932, and delivery of the production model P-26A Peashooter began in January 1934 to the 34th Pursuit Squadron. Far in front of its contemporaries, the P-26 was the Army's first all-metal Pursuit, the first production model Monoplane Pursuit, the first Pursuit to exceed 225 mph, and it turned out to be the last open-cockpit Pursuit accepted by the Air Corps.

The P-26 was a low-wing monoplane with a prominent engine cowling. It had an open cockpit centered above the wing, a standard tail section, and conventional three-wheeled landing gear fitted with streamlined fairings on the front assembly. The 500-hp engine produced a top speed of 234 mph, allowing it to barely keep up with the speedy Martin B-10 Bomber. Two machine guns were fitted on the fuselage sides and the bomb load could be varied among two 100-pounders or five 30-pounders on racks underneath.

Production of the P-26 ended in 1935 and phasing out began in 1937, coinciding with arrival of the Seversky P-35 and Curtiss P-36 Monoplanes. By 1940 all Stateside P-26's had been grounded and some were used for training mechanics at Tech Schools. All P-26's located overseas had been transferred to the governments of China, Panama, Guatemala and the Philippines for use as part of their independent self-defense forces.

Boeing P-26 Peashooter, the Air Corps' first all-metal monoplane Pursuit and the oldest dogfighter to enter combat in WWII. Pictured inflight over Oahu, Hawaii on March 6, 1939. (USAF Photo 38361-AC)

Peashooters had the distinction of being the oldest Pursuits to enter combat during WWII. The P-26's exported to China were the first to see action, being used against the invading Japanese on July 7, 1937. The Filipino Peashooters saw action against the Japanese and made a good showing during the turbulent days between the initial attacks on December 8, 1941, and the Death Marches to Prisoner of War camps that came after Americans surrendered at Bataan on April 9, 1942. American and Filipino pilots courageously pitted their outdated Peashooters against modern Japanese Fighters and managed to destroy a few of the enemy before being completely wiped out.

Production totaled 139 P-26 Peashooters delivered to the Air Corps, plus twelve that were exported.

P-30 (PB-2) TWO-SEAT PURSUIT

The Consolidated P-30 was a development from the YP-25 by way of the XP-24 prototype that originated in 1931. The production model P-30 was delivered for trials in August 1934 as the only two-seated Monoplane Pursuit ever accepted by the Army. The 675-hp engine produced a top speed of 270 mph. Pilot and observer sat in tandem seats under a long sliding canopy. Two fixed guns fired forward through the three-bladed propeller and a swing gun was fitted in the rear cockpit. The tail section was standard and the 3-wheeled landing gear had retractable front wheels. The Consolidated YA-11 Attack plane in 1934 was an offshoot from the P-30.

Production ended in May 1935 after 54 P-30 Pursuits had been delivered to the Air Corps.

In July 1935 all surviving P-30's were redesignated PB-2 Pursuits to put them in the new "PB" for Pursuit Biplace category. The PB-2 turned out to be the only Monoplane 2-seated Pursuit to carry the "PB" classification. Fifty-two of the original P-30's were converted in this manner. Production was never resumed because the flexible gun was rendered ineffective by the fast speed of the aircraft. It was decided to concentrate on development of Attack aircraft for that role. On May 25, 1936, Major Hez McClellan was killed while test-flying a PB-2 out of Wright Field, Ohio.

P-35A GUARDSMAN

The Seversky P-35 made its first flight in August 1935 and deliveries of the plane to the Air Corps began in 1937 as the first one-seated Pursuit with an enclosed cockpit and retractable landing gear. The P-35 was a low-wing monoplane with a 1,050-hp engine producing a top speed of 290 mph. Two machine guns were fitted on top of the fuselage and a 210-pound bomb load could be carried on underwing racks. In September 1937 Jackie Cochrane-Odlum set a women's speed record of 292 mph at Detroit, Michigan, in a P-35 Pursuit. On September 3, 1938, she flew a Seversky Pursuit to beat an all-male field and take home the Bendix Trophy.

Production totaled 77 P-35's built for the Army. In 1939 Seversky was absorbed by the Republic Aircraft Company, at which time the aircraft were sometimes referenced with the double-designer name of Seversky-Republic. One P-35 was equipped with a super-charged engine and modified into the XP-41, which became the prototype for the Republic P-43 Lancer Pursuit of 1940.

The Republic P-35A Guardsman was orginally designed by Seversky as the EP-1 Export Pursuit in 1938. In 1940 when restrictions were imposed on the export of arms to certain nations, the Army received 60 EP-1's that had been stopped on their way to Sweden. Designated P-35A, that aircraft had an airframe similar to the original Seversky P-35. The P-35A engine generated 1,200-hp to produce a top speed of 320 mph. Two 50-caliber fuselage guns fired synchronously through the 3-bladed prop and two 30-caliber guns were installed in the wings to share pioneering of wing-mounted guns with the Curtiss P-36 Hawk. Ten 35-pound bombs could be carried on underwing racks. Of the sixty P-35A's supplied to the Air Corps, twelve were sent to Central America and 48 were shipped overseas to the Philippines. During the Japanese invasion of the Philippines that began on December 8, 1941, all 48 of those P-35A's were either damaged or totally destroyed; the fighting career of the Guardsman ended almost as suddenly as it had begun.

Final tabulations showed that 125 P-35 and P-35A variants served on active duty with the Air Corps. Export examples added up to 72 for a total of 197 produced. Just for the record, the P-35 is commonly referred to as the Seversky P-35, and the P-35A is generally referred to·as the Republic P-35A.

P-36 HAWK (HAWK-75, MOHAWK)

The Curtiss XP-36 prototype first flew in May 1935; the YP-36 limited production model was delivered for trials in March 1937; and the production P-36A Hawk began to arrive in April 1938. Distinguished as a pioneer in wing-mounted machine guns, the P-36 was an all-metal low-wing monoplane with a tail wheel and retractable front landing gear. The 1,200-hp engine produced a top speed of 300 mph. Two guns were mounted on top of the cowl and two guns could be fitted in the wings. Underwing bomb racks were optional equipment for carrying 300 pounds of bombs on a combat range of 800 miles. The pilot sat in an enclosed cockpit above the wing's trailing edge.

One Squadron of P-36's was sent to Hickam Field, Hawaii, in February 1941. Within 45 minutes after the attack at Pearl Harbor on December 7, 1941, P-36's and P-40's were scrambled and got airborne to shoot down the first Japanese aircraft in WWII.

Production totaled 214 P-36 Hawks for use by the Air Corps. Other variants with similar airframes were: the YP-36 in 1939 as the first to exceed 300 mph in level flight; the XP-40 in 1938 as a prototype for the yet-to-become famous WWII P-40 series; the XP-42 in 1939 and the XP-46 in 1940, two designs that never arrived at the production stage.

The Curtiss Hawk-75 was an export version very similar to the P-36. Armament varied among two cowl guns, four 7.5-inch wing guns, or a six-gun mixture. Some models had underwing racks for 300 pounds of bombs. Top speed was from 280 mph to 320 mph, and the ferry range extended to 1,060 miles.

Hawk-75 variants earned the dubious honor of having fought on both sides for more countries than any other combat plane in history. One batch went to China in 1938 for use against the invading Japanese. Shipments went to France beginning in 1938, and on September 8, 1939, a P-36 armed with four 7.5-inch guns was flown by a pilot of the Armée de l'Air to shoot down a German Messerschmitt, the first Allied WWII air victory over France. Some Hawk-75's went to Norway in 1939, but they were captured by the Germans while still in their crates. When France was overrun in June 1940, undelivered Hawk-75's were diverted to England, where they received the name of Mohawk for rerouting to India, South Africa and Portugal. The Indian Mohawks fought in the CBI (China-Burma-India) theater until they were withdrawn in 1943. The South African and Portuguese Mohawks were used against the Italians in East Africa and in the Medittera-nean. Germany gave some of the captured Hawk-75's to Finland where they were used against the Russians; then when Finland broke relations with Germany, they turned the same planes against the Luftwaffe. Other captured Hawks were used against American forces by Rommel in North Africa. Some Hawks went to Siam where they were used

Curtiss P-36 Hawk, WWII Pursuit that fought on both sides for more countries than any other combat plane in history. (USAF Photo 198-AC)

against Japanese invaders of Indo-China (Southeast Asia) in 1941. A few that went to Iran were taken over by the British for use in the Far East. One boatload slated for Holland was diverted to the Dutch East Indies for use by Netherlands pilots against Japanese forces. Canada used their batch of Hawk-75's to train Norwegian refugees who had escaped the Nazi onslaught. The Hawks that went to South America saw service as homefront defensive Interceptors.

Construction of Hawk-75 variants added up to 775 aircraft produced for export in America and 205 built under license by Allies for a total of 980 aircraft, nearly all of which were at some time or other used in actual combat during the early months of WWII. Add the 214 P-36's built for the Air Corps, and the grand total comes to 1,194 Curtiss Hawk Monoplanes in the P-36 family.

YFM-1 AIRACUDA

In 1935 a category called "FM" for Fighter Multiplace, meaning more than two seats, was added to the Air Corps scheme of aircraft designations. The Bell XFM-1 prototype made its first flight on September 1, 1937, and the limited production model YFM-1 Airacuda began to arrive for trials in September 1939. It was a low-wing monoplane with twin "pusher" engines mounted in nacelles on top of the wings. Two 1,150-hp engines produced a maximum speed of 270 mph. Two flexible guns were mounted in plexiglass-covered cones at the front of each engine nacelle. Underwing bomb racks were optional equipment, depending on the mission. The crew of five consisted of pilot, copilot and navigator in the central cockpit and one gunner in each of the nacelle cones.

Production totaled only 13 of the Airacudas, built for experimenting with pusher propellers again, giving wing gunners a clear view over a wider area on Attack sorties. The YFM-1 turned out to be the sole aircraft manufactured in the "FM" category because of its similarity to twin-engine multi-place Attack planes already on production orders.

PURSUITS OF WWII

P-38 LIGHTNING

The Lockheed XP-38 prototype first flew on January 27, 1939; the YP-38 limited production model took to the air in September of 1940; and deliveries of the production model P-38A Lightning began to the 1st Pursuit Group at Selfridge Field, Michigan, in June of 1941. The Lightning was a mid-wing monoplane with retractable tricycle landing gear. The pilot's cockpit was located atop the short center fuselage. Twin booms housed the engines, 3-bladed propellers, super-chargers, radiators and landing gear. The tail assembly had twin fins, one at the rear of each boom, connected by a long tailplane. Two Allison V-1710 engines with 1,425-hp each produced a top speed of 420 mph. Four 50-caliber machine guns and one 20mm or 37mm cannon were mounted in the nose. A maximum bomb load of 3,200 pounds could be carried on a range of 300 miles, or 2,000 pounds for a distance of 450 miles. Maximum endurance was 12 hours with auxiliary tanks. Some variants had provisions for rockets on underwing racks. One model of the P-38 was called a Pathfinder; it had a radar bombsight in the plexiglass nose section. Another version, called a Droop Snoot Fighter-Bomber, had a bombardier position with a Norden bombsight in the clear-view nose cone. A further variant called the Night Lightning had an observer-radar operator seat in the cockpit behind the pilot and an elongated radar cannister under the nose. The F-4 and F-5 Photo-Recon versions carried four K-17 cameras in the nose; some Photo-Recce's were armed and some were unarmed.

The P-38 Lightning got its first taste of combat on August 7, 1942, when Lt. Elza Shahan of the 27th Fighter Squadron shot down a Focke-Wulf FW-200C Fighter over Iceland to claim America's first air victory against the Nazis in WWII. Arriving in the British Isles, P-38's flew their first sortie on August 28, 1942, escorting B-17 and B-24 Bombers on a daylight raid over Europe. Appearing for the first time at the North African front on November 13, 1942, Lightnings were referred to by members of Rommel's Afrika Korps as "der Gabelschwanz", meaning "fork-tailed." In the final months of the war in Europe, P-38 Fighter-Bombers led by Pathfinders and Droop Snoots hit targets on the continent to hasten the downfall of the Third Reich. During their combat tour in Europe from August 1942 until May 1945, P-38's destroyed 2,540 enemy aircraft while losing 1,750 to all causes. The P-38 was a warm-weather Fighter, therefore it was much more successful in the Pacific than in Europe.

A Squadron of F-4 Recon-Fighters led by Karl Polifka arrived at Port Moresby, New Guinea, in April 1942, the first Lightnings in the Southwest Pacific. The first air battle was fought out of Port Moresby on December 27, 1942, when a flight of twelve P-38 Pursuits engaged a formation of Japanese aircraft, downing eleven of the enemy without suffering a single loss. On April 18, 1943, Lightnings led by Tom Lanphier out of

Lockheed-Vega P-38 Lightning, USAAF twin-boomed Pursuit renowned for its exploits in the Pacific during WWII. (USAF Photo 23363-AC)

Guadalcanal shot down the G4M Betty Bomber carrying Admiral Yamamoto from Rabaul to Bougainville. Yamamoto, mastermind of the Pearl Harbor attack, was killed. Subsequent victories in the Solomons secured the Southwest Pacific, thwarting the planned invasion of Australia and New Zealand by Japanese forces.

Richard Bong, flying P-38's with the 49th and 475th Groups in the Southwest Pacific, scored 40 victories to be acclaimed as America's Ace of Aces for all time. His first kill was a Japanese Bomber over Dobodura, New Guinea, during the first air battle of December 27, 1942. Dick Bong received the Congressional Medal of Honor for his actions during General MacArthur's return to the Philippines in October 1944. On August 6, 1945, after surviving 146 combat missions, Dick Bong was killed when the engine of his P-80 Shooting Star Jet Fighter failed during a test hop for Lockheed at Burbank, California.

Tommy McGuire, also with the 49th and 475th Fighter Groups, was credited with 38 air victories to rank as America's 2nd leading Fighter Ace. Mickey McGuire was awarded the CMH for gallantry in battle on Christmas day of 1944. He died in a dogfight on January 7, 1945. McGuire AFB, New Jersey, is named in his honor.

Charles MacDonald, serving first with the 348th Fighter Group and later as Commander of the 475th Fighter Group, scored 27 air victories while flying 204 missions and 688 combat hours in the New Guinea-Solomons-Philippines air war.

In the summer of 1944 Charles Lindbergh went to the Southwest Pacific as a civilian on "technician status" to evaluate single-engine and double-engine Fighters in combat. After he had spent a few weeks of flying Marine Corsairs in the Solomons, George Kenney put him to work teaching cruise control techniques to pilots of Charles MacDonald's 475th Fighter Group. On July 28, 1944, while tagging along in a P-38 out of Biak, West Irian, to observe an Escort mission, Mr. Lindbergh was credited with shooting down a Japanese Fighter near the oil refineries at Boela, Ceram.

The Navy's leading Ace of WWII in the Pacific was David McCampbell, flying F6F Hellcats off the aircraft carrier *Essex* to destroy 34 enemy aircraft. Winner of the CMH, Dave McCampbell shot down nine enemy aircraft in a single day to lead all the services in that department. Joe Foss, winner of the CMH, led the Marine Aces in the Pacific with 26 air victories while flying F4F Wildcats out of Henderson Field, Guadalcanal, Solomon Islands.

Japan's leading Ace was Hiroyoshi Nishizawa with 104 air victories. Shoichi Sugita, the pilot responsible for Tommy McGuire's death, ranked 2nd with 80 victories, followed by Saburo Sakai, participant in the initial attacks on the Philippines, with 62 victories. Saburo Sakai shot down the first American P-40 Warhawk over the Lingayen Gulf at Luzon in the Philippines when those outnumbered Pursuits rose to meet the enemy on December 8, 1941.

Production of the P-38 evolved through the P-38M, ending in August 1945 with cancellation of all outstanding contracts at war's end. A total of 10,423 P-38's, including 500 Photo Recon planes, were delivered to the USAAF. An additional 134 examples were exported, adding up to a grand total of 10,557 aircraft built.

P-39 AIRACOBRA

The Bell P-39 was ordered as the P-45, but the designation of the prototype was changed to XP-39 at the factory. It made its first flight on April 6, 1939. All of the early examples were built for England to bolster its defenses against the feared German invasion of the British Isles. Deliveries of the first Air Corps production model P-39C Airacobra began to arrive at Selfridge Field, Michigan, in January 1941. Distinctions of the P-39 were: the first single-engine Pursuit with tricycle landing gear, the first Fighter to carry a cannon, and the first to locate the engine behind the cockpit. An 8-foot shaft connected the engine to a 3-bladed prop in the nose. The TP-39 was a two-seated Operational Trainer.

The P-39 was a low-wing monoplane with standard tail section and retractable tricycle landing gear. The pilot sat in an enclosed cockpit above the wing's leading edge. The 1,200-hp engine generated a top speed of 400 mph. Four machine guns in the wings, two on top of the fuselage, and a shell-firing 37mm cannon in the nose, gave the Airacobra lethal firepower. A 500-pound bomb could be carried under the fuselage for a combat range of 600 miles, or a maximum range of 750 miles was attained with a fuel tank in place of the bomb.

Immediately after the attack at Pearl Harbor, P-39's were ordered to Hawaii, Midway, Alaska and the Aleutian Islands. The 8th Pursuit Group arrived at General George Brett's Southwest Pacific Air Force in March 1942 and scored their first victory out of Port Moresby, New Guinea, on April 6, 1942. Led by Boyd "Buzz" Wagner, Airacobras destroyed 40 enemy aircraft in their first month of combat. Lt. Buzz Wagner had already achieved the distinction of being America's first WWII Ace when he shot down his fifth enemy aircraft over the Philippine Islands on December 18, 1941. Wagner came home for a rest with 8 air victories to his credit, only to die in a plane crash at Eglin Field, Florida.

P-39 "Klunkers" joined the Cactus Air Force at Henderson Field on August 22, 1942—"Cactus" was the code name for Guadalcanal. On August 15, 1943, Airacobras out of the forward landing strip at Marilinan, New Guinea, destroyed 14 planes out of a formation of 24 Japanese bombers and fighters while losing only four aircraft in the action. When 7th Air Force B-24 Liberators launched the Central Pacific campaign with their first raid at Tarawa in the Gilberts on November 13, 1943, Airacobras flew cover out of Canton in the Phoenix Islands.

Airacobras joined the Atlantic and European war in July 1942. Flying out of Keflavik, Iceland, on August 14, 1942, Bill Shafter recorded the first P-39 air victory by shooting down a Focke-Wulf reconnaissance patrol plane. P-39's went into Northwest Africa with the American invasion forces in November 1942 for use by the 12th Air Force against Erwin Rommel's Afrika Korps. Airacobras exported to Russia were used effectively as dive-bombers against German tanks along the Eastern Front. Alexander Pokryshkin, Russia's 2nd ranking Ace with 59 air victories, was regarded as the Soviet's most successful P-39 pilot. Russia's leading Ace in WWII was Ivan Kozhedub with 62 air victories.

Production ended with the P-39Q in 1944, after 4,750 had been delivered to the USAAF. An additional 4,808 went to America's Allies, making a grand total of 9,558 built for the war. Phasing out of the P-39 began in 1943, and by August 1944 they had all been withdrawn from combat. The P-39 did its job—helping to hold the line until more efficient Pursuits came along in plentiful quantities.

Bell P-39 Airacobra, heroic WWII Pursuit that helped hold the line alongside the Curtiss P-40 Warhawk in the crucial early days. Shown here on an AAF Training Command flight along the Texas coast above Matagorda Island near Foster Field. (USAF Photo 28072-AC)

P-40 WARHAWK

The Curtiss XP-40 made its first flight on October 8, 1938, as a reengined P-36A; the Hawk-81 export model appeared in May 1940; and the P-40B production model was delivered to Air Bases beginning in February 1941. The P-40 turned out to be the last in a long line of Curtiss Hawks that had begun with the P-1 Biplane Hawk back in 1925.

The P-40 was a low-wing monoplane having a single-finned tail section, 3-wheeled retractable landing gear, and a canopy-covered cockpit with good forward visibility. The 1,200-hp engine drove a 3-bladed prop at a top speed of 375 mph. Armament varied up to six wing guns carrying 280 rounds per gun, or four wing guns and two cowl guns. The bomb load ranged from a 500-pounder under the fuselage for a distance of 375 miles, to 1,500 pounds as a Fighter-Bomber with underwing racks on a combat range of 250 miles. The ferry range stretched out to 1,400 miles with a drop tank.

The name of Warhawk was officially adopted by the Air Corps with delivery of the P-40F. The P-40N was the most-produced of all the variants. One batch of Warhawks was

Curtiss P-40 Warhawk, WWII Pursuit immortalized by the Flying Tigers in China and Burma.
(USAF Photo 10786)

used in the Advanced Operational Trainer role, and one example had two tandem seats.
Other experimental prototypes had a 4-bladed prop and a bubble-type canopy with clear
all-around vision. A souped-up version attained a speed of 420 mph. Export models went
to England, Australia, New Zealand, China and Russia—the P-40 was the first American
Fighter to be used by the Russians on the Eastern Front.

P-40 Fighters joined the Allied Forces in the Middle East during June 1941, and later
they became a part of General George Brereton's 9th Air Force in Egypt. In October 1942,
P-40's of the 9th Air Force in Northeast Africa took part in the Allied counter-offensive at
El Alamein near Alexandria, Egypt, to begin the rout of Erwin Rommel's German Afrika
Korps in Libya. On April 17, 1943, Warhawks of the 57th Fighter Group out of El Djem,
Tunisia, shot down 77 German transports and fighters in a sky battle over the Gulf of
Tunis while losing only six planes in the action. After Rommel's defeat at Tunis on May
12, 1943, P-40's participated in the invasion of Sicily two months later.

The first American overseas unit of P-40's in the Atlantic was the 33rd Pursuit Squadron sent to Iceland on July 25, 1941. Warhawks joined the European war in April 1942,
flying out of bases in England until being gradually phased out by more modern Fighters
in 1943 and 1944. In the European Theater of Operations, American Warhawks destroyed
520 enemy aircraft while losing 553 to all causes—not a bad record for an airplane that
was considered vastly inferior to Fighters of the German Luftwaffe. The P-40 performed
nobly alongside the P-39 Airacobra in the early days of WWII.

In August 1941 a freighter unloaded 100 P-40's at Rangoon, Burma. In September
1941 Claire Chennault's American Volunteer Group of 100 pilots arrived in China on
promises of $600 a month and a bonus of $500 for every enemy plane destroyed. The first
air battle of those ''Flying Tigers'' occurred on December 20, 1941, when P-40's flying
out of Kunming, China, shot down six enemy planes. On Xmas day of 1941 Duke
Hedman became an Ace by claiming his fifth victory when he and 18 buddies flying
P-40's, characterized by a shark's mouth painted at the intake, downed a third of an

enemy formation without losing a single plane. Greg "Pappy" Boyington became a civilian Ace by downing six enemy aircraft while flying with the Tigers. When the Flying Tigers were disbanded, Pappy Boyington joined the Marines, flying F-4U's in the Solomons to score 22 more victories for a total of 28. Boyington became a POW on January 3, 1944, when his plane went down in the Pacific. He was one of the lucky ones to make it back home when the war ended. During their seven months of combat, the Flying Tigers never lost a battle, destroying 286 of the enemy while losing 26 Warhawks and 23 pilots. After the Flying Tigers were dissolved in July 1942, many of the surviving pilots joined the 23rd Fighter Group to continue combat flying in the CBI theater under Command of Bob Scott, who had earned a reputation as a "one-man wave of destruction" while with the Tigers along the Burma Border.

The two waves of attacks by Admiral Nagumo's Japanese task force at Pearl Harbor destroyed about 75 Warhawks. George Welch and Ken Taylor took off from Wheeler Field to shoot down eight enemy aircraft. For his part in that action George Welch was invited to the White House by President Franklin D. Roosevelt; Lt. Welch scored 18 air victories during WWII.

George Whiteman became the first pilot to die in WWII when his Pursuit was destroyed on takeoff at Wheeler Field by the initial batch of enemy raiders.

Immediately after the first wave of Japanese planes hit the Philippines, P-40's of the 20th Pursuit Squadron scrambled from Clark Field at Manila to shoot down three aircraft. About ninety P-40 Warhawks were destroyed by the enemy during the first week of fighting in the Philippines.

When the Japanese began to reach out across the Pacific in all directions, P-40's were rushed to Alaska, Midway, Hawaii, the Netherlands East Indies (Indonesia), New Guinea and the Solomon Islands.

Production ended in December 1944, and by July 1945 only one USAAF unit was still equipped with P-40's. A total of 13,733 were built, including 8,410 for the USAAF and 5,323 for our Allies. Warhawk, Tomahawk, Kittyhawk, Tigerhawk, Tiger Shark, Tiger Mouth, Flying Tiger, Hawk and Gypsy Rose Lee were names that referenced the P-40. "Gypsy Rose Lee" applied to the P-40L that was "stripped" down to lighten the load and increase its range. After the war P-40's were transferred to the Air National Guard where they served until they were retired in 1950.

P-43 LANCER

The Seversky XP-41 prototype developed in 1938 was simply a P-35 with a turbo-super-charged engine. When Seversky was absorbed by Republic in 1939, the XP-41 became the Republic XP-43. The limited production YP-43 was delivered in 1940, and the Republic P-43 Lancer production model began to arrive in 1941. The P-43 was a low-wing monoplane with enclosed cockpit, 3-bladed prop, standard tail section and retractable landing gear. The 1,200-hp engine produced a top speed of 360 mph. Armament consisted of two wing guns and two fuselage guns. Six 20-pound bombs could be hauled on underwing racks for a combat range of 800 miles.

America never used the P-43 in combat. Export versions shipped to China in 1942 were flown in action with satisfactory results by pilots of Chiang Kai-Shek's Chinese Air Force against Japanese invaders.

Production totaled 282 P-43 Lancers, of which 124 went to China, leaving 157 for the USAAF. The P-43 project was abandoned in order that Republic could devote full time to development of the vastly superior P-47 Thunderbolt.

MK-V SPITFIRE (BRITISH)

The British Supermarine Spitfire Fighter was a single-engine, low-butterfly-wing monoplane having standard tail assembly and a 3-wheel landing gear with retractable front legs. A single Merlin 1,470-hp engine drove a 3-bladed prop to generate a top speed of 370 mph. The pilot sat in an enclosed cockpit at the wing's trailing edge. Wing-mounted armament consisted of either 8 machine guns or 4 guns and two 20mm cannon with a combat range of 470 miles.

In October 1940 the Royal Air Force formed the 71st Fighter Squadron, thereafter known as the "Eagle" Squadron, which was manned by American civilian volunteers and equipped with British Hurricane Fighters. In early 1941 two more Eagle Squadrons were formed, the 121st and 133rd. In August 1941 all three Eagle Squadrons received their first Spitfire Fighters. In September 1942 the three Eagle Squadrons were transferred to the USAAF to become the 334th, 335th and 336th Fighter Squadrons of the 4th Fighter Group, which was Commanded by 23-year-old Chesley Peterson, the youngest Colonel in the USAAF. Their Spitfires went along with the deal on a Reverse Lend-Lease arrangement. In 1943 the 4th Fighter Group, operating out of Debden, London, replaced their Spitfires with P-47 Jugs, and in February, 1944, under the Command of Don Blakeslee, they traded the P-47's for P-51 Mustangs.

The first American Eagle to be killed in the war was Bill Fiske, who was shot down over Tangmere, England, on August 16, 1940, and died the following day.

Meanwhile, on June 18, 1942, the 52nd Fighter Group became operational in General Carl Spaatz's newly organized 8th Air Force that had its Headquarters at Bushy Park, London. The 52nd was stocked with Spitfires, which were the first ones to wear USAAF markings, but did not receive USAAF designations. In November 1942 the 52nd Fighter Group and its Spitfires went to Gibraltar to become a part of Jimmy Doolittle's brand new 12th Air Force, formed to support operation "Torch," the invasion of Northwest Africa at Casablanca, Oran and Algiers.

Flying a Spitfire on August 19, 1942, Frank Hill was credited with being one of the first Americans to score an air victory over Western Europe when he shot down a Focke-Wulf FW-190 at about the same time that Sam Junkin downed a FW-190 over Dieppe, France, during the British attempt to establish a Second Front.

On February 9, 1944, Bob Hoover, flying his Spitfire with the 4th Fighter Squadron out of the island of Corsica in the Mediterranean, was jumped by a flight of FW-190's. During the dogfight that ensued he shot down two of the enemy planes and was shot down himself. Bob Hoover was captured by the Germans and spent the duration in a Prisoner of War camp. He was freed near the end of the war and returned home safely.

James Johnson, a Spitfire pilot, was the British Ace of Aces in WWII with 42 victories to his credit. Marmaduke Pattle ranked 2nd with 41 victories, mostly in Hurricanes. The "Duke" was killed in a dogfight with a German BF-110 on April 20, 1941.

Approximately 600 British Spitfires were procured by the USAAF.

P-47 THUNDERBOLT—JUG

The Republic XP-47B prototype first flew on May 6, 1941, after both the XP-47 and XP-47A experiments were abandoned. Therefore, the first production model was the P-47B Thunderbolt delivered to the 56th Fighter Group in March 1942 as the heaviest one-seated Pursuit in American history. On November 15, 1942, Hal Comstock and Roger Dyar thrilled the designers at Republic by power-diving their P-47B's at a speed of 725 mph from 35,000 feet. The XP-47J flew at 504 mph, the fastest sustained level flight speed ever achieved with a piston engine. The P-47M had a top combat speed of 475 mph for use against the German V-1 Buzz Bombs over Britain. The TP-47G was a two-seated Combat Operational Trainer.

The P-47 was a low-wing monoplane with standard single-finned tail and conventional retractable landing gear. The pilot sat in an enclosed cockpit that formed a straight line with the top of the rear fuselage in what was referred to as the "Razorback" configuration. Later versions had a "Bubbletop" canopy that provided better all-around pilot vision. The engine varied from 2,000-hp to 2,800-hp driving a four-bladed propeller at a top speed of 430 mph. Eight 50-caliber machine guns were fitted in the wings and a maximum bomb load of 2,500 pounds could be carried on some versions; the P-47N was rigged with racks for three 500-pound bombs and ten rockets. Combat radius varied from 475 miles with a heavy load to 1,150 miles and an endurance of nine hours when carrying three drop tanks.

P-47B Jugs reported for duty in the European war with the 56th and 78th Fighter Groups in January 1943. Flying out of England they went into action on April 8, 1943, carrying out Fighter sweeps and escorting Heavy Bombers over the continent. P-47's made their first appearance over Berlin on March 6, 1944, flying escort for B-17 and B-24 Bombers.

Republic P-47 Thunderbolt, celebrated WWII Pursuit of the European and Pacific campaigns. This photo shows the most-produced Fighter in American history caught in the air on August 3, 1944. (USAF Photo A-40909-AC)

Colonel Hub Zempke, Commander of the 56th Fighter Group "Wolfpack," scored 18 air victories while leading his unit to glory as the top P-47 outfit in the 8th Air Force until he was shot down and taken prisoner when his plane crashed over enemy territory on October 30, 1944. Flying with the Wolfpack, Dave Schilling, successor in Command to Hub Zempke, scored 23 air victories. Colonel Schilling was one of the most decorated men in the war.

Frank Gabreski, Commander of the 61st Fighter Squadron, 56th Fighter Group, was the leading American Ace in Europe with 31 air victories. He was captured as a POW when his plane went down on July 20, 1944. Frank Gabreski survived prison life and remained in the Air Force to become a Jet Ace in Korea by downing seven Russian MIG-15 Jet Fighters in his F-86 Sabre Jet. Hub Zempke, Dave Schilling and Frank Gabreski were given a title of respect, "The Terrible Three," by pilots of the German Luftwaffe.

Bob Johnson, flying with the Wolfpack, was the 2nd ranking P-47 Ace with 28 victories. His first kill was a Focke-Wulf FW-190 over Belgium on June 13, 1943. The 56th Fighter Group destroyed 1,006 enemy aircraft, including the all-time record of 672 in the air.

P-47 Jugs and P-38 Lightnings began combat operations in Italy with the 15th Air Force before the fall of Rome in June 1944, being used mostly in the Fighter-Bomber role. John Voll, who got his first victory on June 23, 1944, over the oil fields at Ploesti, Rumania, led the 15th Air Force Aces with 21 air victories; Herschel Green was a close second with 18 victories.

During two years and one month of combat missions over Europe, Thunderbolts destroyed 6,300 enemy planes and lost 3,500 in the process.

The 348th Fighter Group, Commanded by Neel Kearby, reported to General George Kenney in the Southwest Pacific during June 1943. On September 5, 1943, Neel Kearby got his first kill over Lae, New Guinea. Colonel Kearby was awarded the Congressional Medal of Honor for downing six enemy aircraft over Wewak, New Guinea, on October 11, 1943. His luck ran out over Wewak on March 4, 1944, when he was shot down and killed after scoring 24 air victories.

Joining the Burma campaign in April 1944, Jugs loaded with 500-pound bombs conducted around-the-clock attacks against a force of 18,000 Japanese troops attempting to escape across the Sittang river into Siam. Over half of the enemy force was wiped out in that action.

Production ended in December 1945 after 15,660 P-47's had rolled out of the factory, the largest number for a Fighter plane in American history, and second only to the Consolidated B-24 Liberator for most-produced honors on the part of a combat plane. Exports totaled 1,120 examples, and the 14,540 USAAF airplanes flew one and a third million combat hours.

After WWII P-47 Jugs were retained to form the core of USAAF Fighter capability together with P-51 Mustangs. Some Air National Guard units were equipped with Jugs, and many were sold to friendly nations abroad. In March 1946 P-47's became a part of three newly organized Major Commands: the Strategic Air Command, the Tactical Air Command, and the Air Defense Command. In 1948 all surviving P-47 Pursuits were redesignated F-47 Fighters in the new, autonomous USAF, where they served until retired with honors in 1955 after thirteen years of service. The last combat appearance of the P-47 was by those shipped overseas that participated in the Guatemalan revolution of 1953. The last known Thunderbolt survivor was withdrawn from service by Brazil in 1967.

P-51 MUSTANG—KITE

The North American XP-51 prototype first flew on October 26, 1940, and the export deliveries of the Mustang I to England began in November 1941. The first P-51 variant delivered to the USAAF was the F-6 Photo-Recon Mustang in July 1942, followed in September 1942 by the A-36 Attack Mustang. The production model P-51A Pursuit began to arrive at USAAF bases in March 1943, but it wasn't used in battle. The first combat-ready version was the P-51B supplied to Fighter units in England during Autumn of 1943.

The P-51 was a low-wing monoplane with standard single-fin tail section, conventional retractable landing gear and a four-bladed propeller. The pilot sat in an enclosed cockpit that blended into the fuselage; on later models a bubble-type canopy was substituted for better pilot visibility. The Allison 1,100-hp engine was later changed to a Merlin 1,380-hp package, producing a top speed of 440 mph. Armament varied among four, six, or eight machine guns, or four 20mm cannon. A ton of underwing stores consisted of bombs, rockets or drop tanks. Combat radius varied from 400 miles with a heavy load to 950 miles with spare fuel tanks.

The 354th Fighter Group, 9th Air Force, flew their P-51B's into combat for the first time on December 13, 1943, escorting B-17 Bombers for a record distance of 490 miles. Donald Blakeslee, Commander of the 4th Fighter Group "Eagles," 8th Air Force, traded his P-47 Jugs for P51 Mustangs in February 1944. On March 3-6, 1944, members of the Luftwaffe were startled to confront Mustangs escorting B-17 and B-24 Bombers over the heart of Berlin under the leadership of Colonel Blakeslee.

North American P-51 Mustang, WWII Pursuit that flew to glory in the European theater. Seen inflight is the P-51D with a bubbletop cockpit. (USAF Photo 27977-AC)

Jim Howard of the 354th Fighter Group, 9th Air Force, was awarded the Congressional Medal of Honor for shooting down four enemy Interceptors on June 11, 1944, while single-handedly flying cover for a formation of B-17 Flying Fortresses over Oschersleben, Germany.

Glenn Eagleston was the leading Fighter Ace in the 9th Air Force with 19 air victories. Flying with the 354th Fighter Group he got his first kill on December 13, 1943, downing a Messerschmitt BF-110 twin-engine Fighter over Kiel. Glenn Eagleston survived the war and remained on active duty to fly F-86 Sabre Jets in the Korean war. Bruce Carr was the over-all leader in the 9th Air Force with 26 aircraft destroyed in the air and on the ground.

George Preddy flying with the 352nd Fighter Group, 8th Air Force, shot down 26 enemy aircraft to lead the 8th Air Force. On August 6, 1944, over Hamburg he became the first American pilot to shoot down six German Fighters in a single day. George Preddy was killed in a dogfight over Belgium on Christmas day of 1944.

John Meyer, Commander of the 487th Fighter Squadron, 352nd Fighter Group, 8th Air Force, was the over-all leader with 37 aircraft destroyed in the air and on the ground. He got his first kill on October 25, 1943, and went on to fly 200 combat missions in the war.

One of the best remembered Mustangs of the war was the P-51B named "Shangri-La" flown by Fighter Ace Don Gentile with the 336th Fighter Squadron, 4th Fighter Group in Jimmy Doolittle's 8th Air Force. Don Gentile, former volunteer with the British "Eagle" Squadrons, and his wingman, John Godfrey, combined to destroy thirty enemy aircraft during their first six weeks of combat to become the most-respected team by the Germans at the front. Don Gentile got his first three kills with the Eagle Squadrons and followed with 20 in the 8th Air Force for a total of 23 victories. He survived 182 combat missions, but was killed after the war in a plane crash at Andrews AFB, Maryland, on January 28, 1951. John Godfrey downed 18 enemy aircraft before being shot down himself and taken POW in August 1944. He survived the war, but died of natural causes on June 12, 1958.

The 4th Fighter Group "Eagles" amassed the all-time record for an American Fighter unit of 1,016 enemy aircraft destroyed, including 550 total air victories. Final statistics for P-51 Fighters in the ETO showed 4,950 enemy planes shot down and 4,130 destroyed on the ground for a total of 9,080 aircraft while losing 2,520 in combat.

Erich Hartmann, flying his Messerschmitt BF-109 Gustav single-engine Fighter with the Luftwaffe under Reichsmarschall Hermann Goering during WWII, was credited with 352 air victories, mostly along the Soviet front, to go down in history as the world's all-time leader in aerial combat. Only a few P-51 Mustangs were included in that total of air kills. For that incredible feat he was convicted as a war criminal and spent ten years in the salt mines of Siberia.

The F-6 Reconnaissance Fighter was the first version to join the war in the Pacific. On January 11, 1945, Bill Shomo, flying his F-6K (P-51) with the 82nd Tactical Recon Squadron out of Mindoro in the Philippines, earned the Congressional Medal of Honor when he shot down seven Japanese aircraft in his first encounter over Luzon to establish the all-time record for kills in a single day by a USAAF pilot. His wingman, Paul Lipscomb, downed three enemy Fighters in the same air battle to account for ten destroyed out of the original formation of one Bomber and twelve escort Fighters.

In February 1945 the P-51D joined the Pacific war at Iwo Jima right in the middle of fierce hand-to-hand fighting for possession of that flyspeck on the map where 27,000 brave men from both sides lost their lives. In March 1945, when Iwo Jima was secured, those Mustangs began flying Escort for B-29 Superfortresses over the Japanese mainland. On April 7, 1945, 7th Air Force P-51D's made their combat debut over Tokyo and Nagoya in a sky filled with flak.

Joe Myers from the 78th Fighter Group, 8th Air Force, got the first P-51 kill of a German ME-262 Jet Fighter-Bomber over Europe on August 28, 1944.

Production ended with cancellation of contracts at the close of World War II after 15,576 P-51 Mustang variants had rolled out of the factories. That total came within a hundred planes of the record number for a Fighter set by the Republic P-47 Thunderbolt. Included were 482 F-6's, 500 A-36's and 11,644 P-51's, all for the USAAF, plus 2,950 for the British Royal Air Force.

During the post-WWII period Mustangs became the standard Pursuit for the USAAF, being used overseas with American Occupation Forces. Some were trnasferred to National Guard and Reserve units and others were sold at a bargain price to friendly nations. In 1948 the new United States Air Force (USAF) abolished the ''P'' for Pursuit category and redesignated all P-51 Pursuits as F-51 Fighters.

In July 1950 the F-51H became the only veteran WWII piston engine Fighter to enter combat in Korea, being used on ground attack and close-support missions. Then, to round out a triple-war career alongside the Douglas B-26 Invader and Douglas C-47 Gooney Bird, the F-51D was remanufactured by the Cavalier Aircraft Corporation in 1967 for use on COunter-INsurgency (COIN) operations in Vietnam. That version of the F-51D had sophisticated electronics devices and required a second crewman seated in tandem behind the pilot. Six 50-caliber machine guns were fitted in the wings, three on each side. Underwing hardpoints had a weapons capacity of 4,500 pounds including bombs, rockets and/or HVAR missiles. The last Mustang was retired in 1969 after 27 years of hard work and loyal service.

P-61 BLACK WIDOW

The Northrop XP-61 prototype first flew on May 21, 1942, the limited production YP-61 was delivered in September 1942 and the production model P-61A Black Widow Nightfighter began to arrive in October 1943 as a much-needed replacement for the Douglas P-70 Night Havoc. The P-61 was a twin-engine, twin-boom, mid-wing monoplane with 4-bladed props, retractable tricycle undercarriage and a long tailplane connecting twin fins at the rear of each boom. The three-man crew consisted of pilot, radar operator and radio operator-gunner, all seated in the central fuselage. Engines varied from 2,000-hp to 2,800-hp with a top speed of 430 mph. Four 50-cal machine guns were fitted in the top turret, and four 20mm cannon fired downward from a belly turret. A maximum bomb load of 6,400 pounds could be carried at four hardpoints on a combat radius of 415 miles—the range stretched out to 1,900 miles when fitted with four 300-gallon drop tanks instead of the bombs. Sophisticated radar equipment developed at the Massachusetts Institute of Technology (MIT) was housed in the central nacelle. Early examples had a top turret, which was first deleted and later restored when buffeting problems were solved. The F-15A was a Photo-Recce version with the top turret deleted and a new two-seated tandem cockpit substituted for the old configuration.

Black Widows reported for duty in the Southwest Pacific with the 18th Fighter Group in May 1944 and shot down their first enemy aircraft on July 7, 1944. Carroll Smith was the leading USAAF Nightfighter Ace of WWII with 7 air victories, his 5th kill coming on December 29, 1944. Flying with the 418th Nightfighter Squadron in the Southwest Pacific, he scored two victories in a P-38 Night Lightning and added 5 more victims while teamed up with his radarman, Phil Porter, in their Black Widow. P-61's arrived in England during June 1944 and on their first raid of July 14, 1944, destroyed five enemy aircraft. Paul Smith and Bob Tierney, crewmates with the 422nd Nightfighter Squadron, 9th Air Force, were the first P-61 Nightfighter team to reach Ace status over Europe, shooting down their 5th victim, a German Junkers JU-188 Bomber on December 26, 1944. During its short tour in the European war, the Black Widow shot down 58 enemy aircraft while losing only 25 to all causes.

By late 1944 P-61 Nightfighters had replaced all P-70 Night Havocs to become standard equipment for all Nightfighter units. The last blow of the war was struck on August 15, 1945, when a P-61 Black Widow shot down a Nakajima KI-43 Fighter, code-named "Oscar" in the Intelligence Agency's system of identifying enemy aircraft.

After the war some P-61's were used as scientific research aircraft. On April 22, 1946, pilotless P-61 drones were flown into severe thunderstorms to obtain vital stress and remote-control statistics. On August 17, 1946, Larry Lambert came flying out of a P-61 to become the first human being ejected from an airplane by automatic emergency escape equipment. He lived to tell the tale.

When the Air Defense Command (ADC) was formed in March 1946, Black Widows became their Night Interceptors until replaced in 1948 by North American P-82 Twin Mustangs. The surviving F-15 Photo-Recon Black Widows were redesignated RF-61C in the new USAF scheme of aircraft categories and served until being retired in 1952. In retrospect, the P-61 was one of the finest aircraft ever built for the Military service—it had the misfortune of coming along at exactly the wrong time.

When production ended in 1946 a total of 742 P-61 variants had been built for the USAAF, including 36 as the F-15A (RF-61C). Contracts for many more examples were cut back due to limitation of postwar funds.

Northrop P-61 Black Widow, twin-boomed Nightfighter that struck the last blow of WWII in Japan.
(USAF Photo 28521)

P-63 KINGCOBRA

Developed as an enlarged P-39 Airacobra, the Bell XP-63 prototype first flew on December 7, 1942—exactly one year after the attack at Pearl Harbor—and the production model P-63A Kingcobra was delivered beginning in October 1943. The P-63 was a low-wing monoplane with standard tail assembly and retractable tricycle landing gear. The pilot sat in an enclosed cockpit centered above the wing's leading edge. The 1,325-hp engine drove a 4-bladed prop to achieve its top speed of 410 mph. Two 50-caliber machines guns and a 37mm cannon carrying 58 shells were fitted in the nose. Two 50-caliber guns were installed in underwing housings, one on each side of the aircraft. The 1,500-pound bomb load was carried on racks under the wings and fuselage. The range with a full load of ordnance was 390 miles, which could be extended to 450 miles by substituting a 175-gallon drop tank for the bomb under the fuselage. One version could be fitted with six rockets, three under each wing.

Intended as a Fighter-Bomber for use in the ground attack role, the P-63 came along six months after the P-47 Thunderbolt began doing an outstanding job in Europe. Since the Russians had put their P-39's to good use as tank busters, the Soviet Union immediately requisitioned a majority of the improved P-63's for use at the Eastern Front. Therefore, the USAAF received only a few P-63 Pursuits and they were never used in combat.

The RP-63 Restricted Pursuit was delivered in 1943 as—now, listen carefully,—a manned target that could be shot at by armed aircraft during aerial gunnery practise! It had an extremely tough armor-plated hide that caused the brittle practise bullets to shatter upon contact. The RP-63 was nicknamed "Pinball" because every time it was hit a red light on each wing tip "lit up like a pinball machine," thereby providing instant verification to trainees during the shooting. A tough canopy protected the pilot from getting hit. The "R" prefix stood for "Restricted" which was affixed to aircraft being put to use as something other than what they were designed for. In 1948 under the USAF scheme, the RP-63 was redesignated QF-63 meaning Aerial Target-Fighter.

Production ended in 1946 after 3,300 variants of the P-63 had been built. The USAAF employed 500, mostly as Aerial Targets. The Soviets procured 2,500 P-63 Fighter-Bomber models and 300 of the Fighter version went to the Free French. The P-63 was a good airplane, but it came along at a time when America had just completed two combat planes that were a little bit better, the P-47 Thunderbolt and the P-51 Mustang.

YP-64 LEND-LEASE PURSUIT

The North American YP-64 limited production Pursuit was an outgrowth of the AT-6 Texan family of combat planes and Advanced Trainers. In 1939 North American began developing a series of Fighters and Attack planes for export to small friendly nations: one of those designs was the Model NA-68. Siam (later Thailand) ordered six of the armed NA-68 Fighters in 1940 to bolster their defenses against the expanding Japanese military forces. In November 1940 those aircraft were not delivered by America for fear that they might fall into Japanese hands. Assigned the designation of P-64 because they were one-seaters, the aircraft were not used as Pursuits, but were put to work as Advanced Trainers alongside AT-6 Texans already in service.

The P-64 was a low-wing monoplane having a standard single-fin tail section and conventional 3-wheeled landing gear with retractable front wheels; the front legs were covered with fancy streamlined fairings. The pilot sat in an enclosed cockpit just above the wing's trailing edge. The 775-hp engine drove a 2-bladed prop to produce a top speed of 270 mph. Two 30-caliber machine guns were mounted on the cowling, and two 20mm cannon could be fitted in the wings. Racks for 550 pounds of bombs could be slung underneath.

When WWII ended, the surviving P-64's were disarmed and sold as salvage to private industry. A couple of them were rigged with dry ice mechanism for experiments in cloud-seeding to produce rain. They were affectionately referred to as ''Rainmakers.''

P-66 VANGUARD

The Vultee P-66 Vanguard was derived from the 1939 Model-48 Fighter developed for export to friendly nations. In 1940 Sweden submitted an order for 144 Model-48 Fighters. When the shipment was just about ready to roll, America invoked the Arms Embargo Act of October 1940, prohibiting the export of war materials to certain nations: Sweden was one of those countries standing in the path of the German onslaught. Most of those aircraft were diverted to China for use against the invading Japanese Army, but the Air Corps received a few for use as Target Tugs in combat training programs. At this time, America was building aircraft for export across both oceans, but had not yet begun to gear up for World War II.

The P-66 was a low-wing monoplane with standard tail assembly and three-wheeled retractable undercarriage. The pilot sat in an enclosed cockpit above the wing's trailing edge. The 1,200-hp engine drove a 2-bladed prop to attain a top speed of 410 mph, with sufficient fuel for a range of 850 miles. Two guns were carried on the fuselage and four guns were fitted in the wings.

Production totaled 144 P-66 Pursuits; China procured 129 of them and the Air Corps received fifteen.

P-70 NIGHT HAVOC

The Douglas XP-70 prototype in September 1941 was a direct conversion from an A-20 Havoc Attack plane with British radar equipment installed in the nose. The bomb capability was deleted and the armament was altered; otherwise, all they did was paint the entire aircraft black, including the plexiglass nose cone. The P-70 Night Havoc in March 1942 was distinguished as America's first radar-equipped Nightfighter; it carried the new Airborne Intercept Radar System (AIRS) developed at the Massachusetts Institute of Technology. The job of a Nightfighter was to intercept and destroy enemy Bombers coming in on night raids. Specifications called for a Fighter sneaky enough to ambush the enemy, fast enough to overtake him, and agile enough to out-fox any maneuver a Bomber could make.

The P-70 was a twin-engine mid-wing monoplane with standard tail section and retractable tricycle landing gear. Pilot and copilot-navigator sat in a cockpit forward of the wings, and the radar operator-gunner occupied a separate dorsal cockpit just behind the wings. Two 1,600-hp engines driving 3-bladed props were mounted under the wings. Top speed was 330 mph and the combat range extended out to 1,060 miles. Armament came as either four 20mm cannon in an under-fuselage tray carrying 240 rounds per gun, or six 50-caliber machine guns in the nose.

All Night Havocs delivered to the USAAF were direct conversions from A-20 Attack planes with no alteration of the airframe—no production order was ever submitted. The first batch of P-70's was supposed to have a super-charged engine with a top speed of 390 mph and a high service ceiling, but those power plants were critically needed elsewhere; therefore, the P-70 had to settle for a less powerful engine than called for in design specs. The net result was inefficiency and poor performance at altitudes for which it was intended: enemy Bombers that came in at 25,000 feet were too high for our underpowered pioneer Night Interceptor. Conversions from A-20 to P-70 configuration totaled 269 aircraft, including 184 for use as Nightfighters and 85 as Operational Trainers while awaiting arrival of the Northrop P-61 Black Widows already under development. The phasing out of the P-70 began with delivery of the first P-61's in October 1943, and by the end of 1944 all P-70's had been withdrawn from combat.

P-82 TWIN MUSTANG

The North American XP-82A made its first flight on April 15, 1945, and the production model P-82B Twin Mustang began to arrive in July 1945, a little too late for combat in WWII. Production of Twin Mustang day Fighters and Nightfighters continued into 1948 with several versions equipped for a variety of roles, including Tactical Fighter, Interceptor, long range Escort, Fighter-Bomber, Nightfighter and Reconnaissance. Often referred to as two P-51 Mustangs coupled together, in reality the P-82 incorporated entirely new concepts to increase the range while flying at a faster speed and carrying a heavier load of weapons. The P-82E Fighter and P-82G Nightfighter were delivered to the Tactical Air Command (TAC) and the P-82F Nightfighter was delivered to the Air Defense Command (ADC) as replacement for the WWII Northrop P-61 Black Widow.

North American P-82 Twin Mustang, twin-boom Fighter that scored the first USAF air victory of the Korean war. Photo taken at the Air Force Museum, Wright-Patterson AFB, Ohio, May 1974 by SSgt. Scott Crist. (USAF Photo KKE-67544)

The P-82 was a twin-engine, twin-boom, low-wing monoplane with a cockpit centered above the wings on each boom. The four-wheeled retractable landing gear was unique, having a front and rear wheel under each boom. The pilot occupied the left seat while the right seat accommodated a copilot with dual controls on the day Fighter, or an observer-radar operator on the Nightfighter. A long tailplane connected twin fins at the rear of the booms. Two 1,600-hp engines drove 4-bladed props rotating in opposite directions to produce a top speed of 460 mph. Armament consisted of six 50-cal machine guns on the wing's center section. The bomb load of 4,000 pounds was carried on four underwing hardpoints for a combat range of 1,600 miles. Four drop tanks in place of the bombs increased the range to 2,200 miles. A long pod protruding forward from the center of the wing housed radar or photography equipment, depending on the mission.

On February 28, 1947, a P-82 flown by Bob Thacker and John Ard set a Pursuit distance record of 4,970 miles from Honolulu, Hawaii, to New York City in 14 hours 32 minutes. Under the new Air Force scheme of designations in 1948 the P-82 Pursuit became the F-82 Fighter in a change of numbering but with no alteration to the aircraft. The P-82 turned out to be the highest number assigned to a prop-driven aircraft in the ''P'' for Pursuit category.

In 1950 the 68th and 339th Fighter Squadrons, 8th Fighter-Bomber Wing, went to Itazuke Air Base, Japan, for duty with the Fifth Air Force. Within twenty-four hours after North Koreans invaded South Korea on June 25, 1950, President Harry Truman ordered those units and their Twin Mustangs into combat. On June 26, 1950, Bill Hudson, flying his F-82G with the 68th Fighter Squadron, shot down an enemy YAK-9 over Kimpo to score the first air victory of the war. That was also the first victory credited to a pilot of the USAF since it became an autonomous branch of the Military service back in September 1947.

During the Korean war F-82's were modified with 2,200-hp engines to attain a top speed of 475 mph. Eight guns were added in a special center nacelle, the bomb load was increased to 7,200 pounds, and the range was extended to 2,500 miles. Racks were optional for two 2,000-pound torpedoes or 25 underwing rockets. A blast of firepower from the F-82 Twin Mustang was likened to a broadside from a light cruiser. F-82's flew a total of 1,870 sorties in Korea, being used for any purpose that the situation demanded.

Production ended in 1948 after 257 Twin Mustangs had been delivered for service, including approximately 157 for TAC and 100 for ADC.

RACER PLANES

R—RACER: 1920 TO 1925

In 1920 when the Army Air Service was formed and the Pulitzer Trophy plus financial rewards were offered as stimulants for airplane development by the Pulitzer brothers, Billy Mitchell convinced his bosses to allow Army participation in American Air Races. The Schneider Trophy for Seaplanes was another of those coveted championship cups. In 1920 the ''R'' for Racer category was added to the list of aircraft designations. When General Mitchell was court martialed in 1925, the Pulizer prize was withdrawn and the Army ceased construction of Racer planes as such.

The **Engineering Division R-1 Racer** in 1921 was a redesignation of the VCP-R Verville-Clark Pursuit-Racer in which Lt. Corliss Moseley had won the inaugural Pulitzer Race at Mitchel Field, NY, on November 25, 1920, reaching a speed of 156 mph. Capt. Hal Hartney took second place in a Thomas-Morse MB-3 Pursuit at 148 mph. The Army Engineering Division had its research laboratory at McCook Field, Dayton, Ohio, and the VCP-R was lettered after the design team of Verville and Clark (in those days military planes carried designer designations). The R-1 was a single-engine one-seated biplane with standard tail section, two fixed wheels up front and a landing skid at the rear. The Packard A-2025 12-cylinder engine drove a two-bladed prop to generate 660 horsepower at takeoff. Only one VCP-R was built and it became the only R-1 Racer flown by the Army.

The **Thomas-Morse R-2 Racer** in 1921 was a modification of the MB-6 Morse-Boeing Pursuit, built for the Pulitzer Race held at Omaha, Nebraska, on November 5, 1921. The R-2 lost to Bert Acosta, who took the Silver Plaque in a Navy Curtiss CR-1 Racer at 177 mph. The R-2 was a standard lightweight biplane with a Wright H-1 400-hp engine and a top speed of 175 mph. Only one R-2 was built for the Army.

The **Engineering Division-Sperry R-3 Racer** in 1922 was designed at McCook Field and built by the Sperry Company specially for the 1922 Pulitzer Race, which it lost to the Curtiss R-6 Racer. The R-3 was a standard light biplane with a Wright H-3 400-hp engine and a top speed of 183 mph. The R-3 was re-fitted with a Curtiss D-12 500-hp engine and flown in the Pulitzer Race at Wright Field, Dayton, Ohio, on October 4, 1924, by Lt. Harry Mills, who captured the prized Trophy at a speed of 216 mph. During the same race Capt. Burt Skeel was killed when his Curtiss R-6 Racer crashed; that was the first fatality of the Pulitzer competition. Three R-3 Racers were built for the Army Air Service.

The **Loening R-4 Racer** in 1922 was a Monoplane with a Packard A-2025 660-hp engine, but it was unsuccessful as a racing plane. Only two experimental R-4's were built.

The **Thomas-Morse R-5 Racer** in 1922 was a radically different airplane, about ten years ahead of its time. The R-5 was an all-metal high-parasol-wing Monoplane with standard single-finned tail assembly and a conventional undercarriage having two wheels and a tailskid. The Packard engine drove a 2-bladed propeller to generate 600-hp at takeoff. Designed for use by the Army Air Service as both a Fighter and Trainer, the R-5 had stability problems caused by incompatibility between the engine and the airframe. Only one experimental R-5 was built.

The **Curtiss R-6 Racer** in 1922 was a sleek little souped-up biplane that grossed out at a little over 2,000 pounds. The wingspan was the narrowest and the length was the shortest of all the Army Racers. The tail section had an enlarged fin and the landing gear was standard with two wheels and a tailskid. The Curtiss 550-hp engine drove a 2-bladed prop to produce an absolute maximum speed of 240 mph. Lt. Russ Maughan won the Pulitzer Trophy at Selfridge Field, Mich., on October 14, 1922, flying his R-6 Racer at 206 mph. Lt. Les Maitland made it a one-two sweep for the Army R-6, finishing second at a speed of 201 mph. Billy Mitchell set the world record for a one-kilometer course on October 18, 1922, when he flew the R-6 at 224 mph. On March 29, 1923, Russ Maughan broke that record by flying his R-6 over a one-kilometer circuit at 237 mph. Only two R-6 Racers were built for the Army Air Service.

The **Curtiss R-8 Racer** in 1923 was the designation assigned to a Navy R2C-1 standard biplane acquired by the Army for one dollar on an inter-service transfer. A Navy R2C-1 was flown by Al Williams at Lambert Field, St. Louis, Missouri, in October 1923 to win the Pulitzer Prize at a speed of 244 mph. Hal Brow ran a close second at 242 mph in his Navy R2C-1. Only one R-8 was built; it crashed in 1924 and was never restored. The R-8 turned out to be the last of the ''R'' designated Racers, stepping aside for cooperative developments by the Army and Navy.

The **Curtiss R3C-1 Land-Racer** in 1925 was a joint Army-Navy development that, by agreement, carried a Navy designation. It was a standard biplane with an enlarged tail section and conventional landing gear. A Curtiss V-1400 engine drove a 2-bladed prop to develop 590-hp at full throttle. On October 12, 1925, Lt. Cy Bettis flew an R3C-1 at Mitchel Field, NY, to capture the final Pulitzer Trophy at 249 mph. Three R3C-1 Racers were built for the services.

The **Curtiss R3C-2 Sea-Racer** in 1925 was exactly like the R3C-1 except for pontoons underneath and a Curtiss V-1400 610-hp engine.

An R3C-2 Sea-Racer piloted by Jimmy Doolittle won the Schneider Race for Seaplanes on October 26, 1925, at a record speed of 233 mph. On the next day Lt. Doolittle set a straightaway record of 246 mph in the same airplane. Only one R3C-2 was constructed for Army-Navy shared usage.

The Pulitzer Races accomplished their objectives. Technological advances saw the winning speed increase from 156 mph to 249 mph in only five years. But the pastime was getting too expensive for the apathetic taxpayers of the twenties: the Army Racers died a natural death.

After the Pulitzer Races were discontinued in 1925, Army Air Corps Pursuits participated in most of the annual National Air Races. In September 1926 at Philadelphia, Pennsylvania, the John L. Mitchell Trophy was won by Lt. L.G. Elliott in a Curtiss P-1 Hawk at a speed of 160 mph. At Spokane, Washington, in 1927, Lt. Gene Batten captured the Spokesman-Review Trophy in a souped-up Curtiss P-6 Hawk at 201 mph. In 1929 at Cleveland, Ohio, Lt. R.G. Breene took second place in a P-3 Hawk at 187 mph, losing the Thompson Trophy to Doug Davis at 195 mph in a Beech Travel Air low-wing monoplane.

STRATEGIC-TACTICAL RECONNAISSANCE AIRCRAFT

SR—STRATEGIC RECONNAISSANCE: 1962 TO THE PRESENT

The Lockheed SR-71A Blackbird Strategic Reconnaissance aircraft made its first flight on December 22, 1964, as a progressive development from the YF-12A Blackbird Fighter. Designed as a replacement for the U-2 Gray Ghost Strat-Recon plane, the SR-71A came on the scene as the fastest and most advanced unarmed aircraft in the world. Deliveries to the 9th Reconnaissance Wing, Strategic Air Command, Beale AFB, California, began in January 1966. Equipped with the latest in electronics and photography devices, that triple-sonic aircraft can survey 60,000 square miles of terrain in less than an hour and the cameras can zero in on a gnat. The flattened underside of the fuselage blends smoothly with the wings to give it a tremendous lifting surface. The SR-71B and SR-71C versions are tandem-seated Operational Trainers.

The **U-2 Spyship** should have been given a ''SR'' designation; however, at the time of its development in 1955 the ''U-2'' was deemed necessary for security reasons, and it has never been changed. So, in reality, the Air Force has two famous Strategic Reconnaissance planes, but they carry designations in different categories.

The **SR-71** is a low-delta-wing monoplane with twin tailfins, one on top of each engine nacelle. Two Pratt & Whitney J56-P turbojet engines mounted halfway out on the wings generate 34,000 pounds of thrust each. The retractable tricycle landing gear has eight wheels, and the two-man crew includes pilot and systems operator. The range is approximately 3,500 miles with a full load of fuel.

Existence of the SR-71 was revealed by President Lyndon Johnson on July 24, 1964, and the entire aviation world was soon to be astounded by the news that it had flown from America to England in about half the time of the old record. A few significant operations by Blackbirds included surveillance of the Suez Canal in 1970, vigilance along the China mainland in 1971, overflying North Vietnam in 1973 after cessation of bombing, and aerial reconnaissance over the Middle East during the Yom Kippur war in late 1973. On July 27-28, 1976, Eldon Joersz flew his SR-71A to a new speed record of 2,190 mph and Bob Helt soared to a sustained altitude mark of 85,100 feet.

Lockheed SR-71 Blackbird, the world's fastest Strategic Reconnaissance aircraft, pictured airborne in 1965. (USAF Photo 175876)

TR—TACTICAL RECONNAISSANCE: 1978 TO THE PRESENT

The **Lockheed TR-1 Tactical Reconnaissance** plane was under development in 1979 as a planned successor to the Lockheed U-2 Gray Ghost surveillance aircraft. The TR-1 will be equipped with the latest sophisticated equipment, including all-angle radar and Electronics Counter-Measures (ECM) devices. A single Pratt & Whitney J75-P turbojet engine develops 17,000 pounds of thrust to produce a top speed of 430 mph. The range stretches out to 3,300 miles, and the optimum altitude is about 60,000 feet; the service ceiling is about 72,000 feet. Delivery of the TR-1 is expected to begin sometime in the early 1980's.

THE EARLY TRAINERS

Between August 1909, when the Aeronautical Division received its first flying machine, and July 1916, when the first substantial contracts for WWI aircraft were issued, Army officials used every penny they could steal, beg, or borrow to procure airplanes. Those aircraft scrounged from manufacturers all over America are referred to here as the "Early Trainers." The aeroplane had a difficult time working its way into the military scheme of things. While the Wright Brothers were secretly working on their "Kitty Hawk" flying machine financed from their own personal savings, Samuel Langley was experimenting with his "Aerodrome," backed by $50,000 in Army funds. When the Aerodrome crashed into the Potomac twice in 1903 and ended up a failure, the general public and members of Congress turned apathetic at the mere mention of the word aeroplane. Meanwhile, Alexander Graham Bell was also trying to get his "White Wing" off the ground—like the Wrights', his was a privately funded venture.

After the successful flight of the "Kitty Hawk" on December 17, 1903, the Wright Brothers made many attempts to persuade Congress and the Army of the wisdom of investing in the aeroplane; their appeals fell upon deaf ears because of adverse publicity about the Langley project. Only after the Wright Brothers gained international renown by demonstrating their flying machine in Europe did the Army perk up its ears and begin to show a little interest, mainly because of the insistence of President Teddy Roosevelt.

Wright Model-A Flying Machine piloted by Orville Wright with Brigadier General James Allen as passenger, Fort Myer, Virginia, September 12, 1908. Lt. Tom Selfridge was killed in this aeroplane. (USAF Photo 4613-AS)

Wright Model-B Flyer, the Army's first aeroplane, Fort Myer, VA, August 1909. (USAF Photo 19230-AC)

Although the description and performance statistics of the early planes may sound unexciting, each was a remarkable milestone. Most of them represented a giant stride forward in advancing aeronautical technology. Even the individual planes within a particular Model number differed in minor ways—no two were exactly alike.

Finally, on December 5, 1907, four years after the Kitty Hawk flight, the Board of Ordnance and Fortification ordered a Wright Model-A flying machine. The contract was signed and sealed on February 10, 1908. The Wright Model-A was delivered for trials on September 7, 1908, and crashed during an evaluation flight at Fort Myer, Virginia, on September 17, 1908. Lt. Tom Selfridge was killed in the accident and Orville Wright was injured. The Model-A was later restored to its original configuration for display in the Smithsonian Institute as the Army's first aeroplane. On August 2, 1909, an improved Wright Model-B was accepted by the Army at Fort Myer to become the first official active duty Army Trainer. Ground was cleared in September 1909 for the first Army airfield at College Park, Maryland. It was there on October 26, 1909, that Lts. Fred Humphreys and Frank Lahm became the Army's first pilots by soloing in the Wright Model-B Trainer. On April 11, 1911, the first Army Flying School was inaugurated at College Park with Capt. Charles Chandler as Commandant. On September 26, 1911, Lt. Tom Milling flew a Wright Trainer with two passengers and set a new 3-man endurance record of one hour and 55 minutes at Mitchel Field, New York. On June 7, 1912, Charles Chandler fired a Lewis machine gun for the first time from the righthand seat of a Wright Model-B while Tom Milling and Roy Kirtland handled the controls on alternate flights.

The **Wright Model-A** twin "pusher" had the "empennage" up front—that is to say the elevators were forward of the wings instead of being at the tail. The term "pusher" meant that the propellers were installed behind the wings and pushed the aircraft forward. On the Wright Model-B pusher the elevators were moved back to become the tail section. The twin chain-driven propellers were retained, two landing wheels were mounted under the fuselage, and separate side-by-side positions were arranged so that the instructor and student could share the single set of controls. The Wright Model-C in 1912 was the most popular of the Wright Trainers; it was an improved version having open side-by-side seats with full dual controls and steering wheels to replace the levers. Capt. Charles Chandler, 2nd Lt. Hap Arnold and 2nd Lt. Tom Milling earned the new title of Military Aviator flying the Model-C. The Wright Model-F in 1912 and Model-HS in 1913 were pusher biplanes with enclosed fuselage, chain driven props, twin-fin tail, two front wheels and a landing skid at the rear. Twelve of those early Wright Trainers were delivered to the Aeronautical Division, seven of which were the most-produced Model-C.

The Army's second airplane design was the **Curtiss Model-D Trainer** in 1911. It was a pusher type with elevators in front and a new tricycle landing gear. The instructor and the pilot sat on the wing and shared one set of controls. On May 10, 1911, Lt. George Kelly was killed in a Curtiss Model-D Trainer at Fort Sam Houston, Texas, the first Army pilot to die in an airplane crash. The Curtiss Model-E in 1912 was an improved version, and the Model-F was a pusher type flying boat. The Curtiss Model-G in 1913 was a tractor design; the term ''tractor'' meant that the propellers were mounted up front and acted as pullers rather than pushers.

The **Burgess Model-H** in 1912 was the Army's first ''tractor'' design Trainer. Features introduced were fully enclosed fuselage, tandem open cockpits, standard single-finned tail section and conventional landing gear having two front wheels and a tailskid—tail wheels were not to replace the tailskids until about fifteen years later. The Burgess Model-I in 1913 was a seaplane used on scouting patrols by the 2nd Aero Squadron in the Philippines. A total of ten Burgess early Trainers were built for the Aero Division.
During trials of the Burgess Model-H on June 1, 1912, Lt. Hap Arnold set an Army altitude record of 6,540 feet. On August 12, 1912, Lts. Hap Arnold and Roy Kirtland delivered the first production Model-H from Marblehead, Massachusetts. On May 28, 1913, Lts. Tom Milling and Bill Sherman set a two-man endurance record of 4 hours 22 minutes in a Burgess Model-H out of Texas City, Texas. On March 12, 1915, cadet Lt. Q.B. Jones and crew flew a Burgess tractor plane to set a 3-man duration mark of 7 hours 5 minutes.
Those pioneer Wright, Curtiss and Burgess Trainers were primitive and accident-prone. Their crackups at the Army Pilot Training School, San Diego, California, brought bad publicity in the news media. Army disgust at the lack of Federal funds caused Congress to take a long look at budgetary policies.

Design of the **Curtiss Model-J, Model-N** and **Model-JN** began in 1914 and eventually led to the Army's first large airplane procurement when 94 JN-4 Jenny Trainers were ordered in 1916. Those aircraft were modernized tractor biplanes with fully enclosed fuselage, two tandem seats, standard tail assembly and conventional landing gear having two wheels up front and a landing skid under the tail. On October 8, 1914, Capt. H.L. Muller set an American altitude record of 16,798 feet in a Curtiss-90 Trainer. On January 5, 1915, Lts. Art Christie and J.C. Carberry set an American 2-man altitude mark of 11,690 feet in a Curtiss Jenny Trainer at San Diego, California. The Curtiss Model-R in 1915 was a heavier standard biplane used as an experimental Bomber-Trainer. The Curtiss Model-L was an amphibian Trainer and the Model-S was an armed tandem-seat land Trainer. A total of 74 Curtiss early planes were delivered prior to WWI.

The **Martin Model-T** in 1914 and **Model-S** in 1915 were biplane amphibians and the Martin Model-TT in 1915 was a triplane amphibian that made the first bomb drops in the

Army's history. All of the Martin Trainers were tandem seated tractor planes used for observation and pilot training. The Model-T served with the 2nd Aero Squadron in the Philippines. The Martin (later Wright-Martin) Model-R in 1915 was rigged with either wheels or pontoons for use as a Trainer. It was a standard tractor biplane with tandem open seats. On January 15, 1915, Lt. Q.B. Jones set a one-man duration record of 8 hours and 64 minutes in a Martin Model-R at San Diego, California. A total of 46 Martin early trainers were built for the Army.

The **Lowe-Willard-Fowler LWF Model-V** in 1915 was a standard tractor biplane with tandem open seats. It was used as both a Trainer and an Observation plane. A total of 23 were delivered.

The **Sturtevant Model-S** in 1915 was a standard tractor biplane used as an Observation plane and Trainer. Eleven of them were built for the Army.

The **Sloan** (later **Standard) Model-H** in 1916 was a standard biplane with slightly swept back wings. A total of 12 were used as Army Trainers and Observation planes.

In 1916 Army airplanes flew combat reconnaissance missions along the Mexican border in support of General John J. "Black Jack" Pershing's punitive expedition against Pancho Villa. Aircraft used in that brush-fire war were the Curtiss Model-R, Curtiss Model-JN, Sloan Model-H and Sturtevant Model-S, all serving in combat for the first time.

A total of approximately 200 planes were acquired by the Army during the early years between 1909 and 1916. As of December 31, 1916, a total of 122 Army pilots had been trained since Fred Humphreys soloed on October 26, 1909.

TRAINERS OF WWI

AMERICAN TRAINERS

In 1911 there were 26 aeroplane pilots in all of America; eight of them were in the Army. In March 1913 the 1st Aero Squadron, Aeronautical Division, Signal Corps, was formed to become the first air combat unit in American history. In August 1914 when World War I began, the Aviation Section of the Army Signal Corps had only twenty airplanes. When America declared war against the Central Powers on April 6, 1917, the entire Army inventory contained 55 harmless airplanes. The Aviation Section had a strength of 35 Military Aviators and 1,088 enlisted men spread out among seven squadrons: four in America, one in the Philippine Islands, one in Hawaii and one in Panama. On July 24, 1917, Congress passed an appropriation authorizing $640 million for expansion of the Airplane Division.

When the Presidential Proclamation of 1917 banned private flying in the United States, a lot of privately owned airplanes were commandeered for Military service. Every Army airfield and almost every civilian airport in America joined in a united effort to turn out the best pilots possible within the critically short time at hand. Rapid growth of the Air Service resulted in delivery of approximately 19,000 airplanes of all types. Almost 14,000 of them were built in America and about 5,000 were procured from France and England. At peak strength in 1918 the Air Service had grown to 45 combat-ready Squadrons and boasted some 195,023 personnel, including 20,708 Officers.

To put it mildly, the pilot training program in WWI wasn't very well organized. Young Americans were given primary training in aircraft like the Curtiss JN-4 Jenny and Standard Model-SJ two-seated Trainers with dual controls. Then the lucky ones moved up to Advanced Trainers like the Thomas-Morse Model-S4 single-seater, equipped with one 30-caliber machine gun to get a short check-out in combat tactics before reporting "over there" on the Western Front to be rushed into action.

CURTISS JN-4 JENNY

An outgrowth of the 1915 vintage JN-3 that flew combat reconnaissance with General Pershing during the punitive expedition against Pancho Villa along the Mexican border, the Curtiss JN-4 Jenny was conceived and built in 1916 to assuage the Aviation Section's disgust with the antiquated pusher-type, open-to-the-wind, accident-prone Trainers that had been funded by Congress up to that time. The Military design model of the contract with Glenn Curtiss was the JN-4D, which was destined to become the standard Army Trainer throughout the war and for about seven years thereafter.

Curtiss JN-4 Jenny, noted WWI pilot Trainer and post-war barnstormer. Photo shows a JN-4H over Chicago, Illinois. (USAF Photo 7495-AS)

The JN-4D was a standard two-seat biplane with a Curtiss OX-5 90-hp engine and a top speed of 75 mph. Performing its role as an Army Trainer in relative obscurity, the Jenny went quietly and efficiently about its task of preparing young cadets for their hazardous job in the war.

But after the war things began to liven up a bit for the surplus Jennies. Many pilots-turned-barnstormers purchased surplus JN-4D's from the Government for anywhere from $50 to $600 apiece and set out to earn their fortune flying by the seat of their pants. They lived by the creed that "a pilot is as good as he talks" and were masters at the art of shooting bull. After a period in which haywire was the main component of their maintenance kit for working on the OX engine, the plane became known as an accident looking for a place to happen. One gypsy oldtimer affectionately referred to his Jenny as the "Galloping Dominoes," explaining that he lived by the "7 come 11" philosophy that if he didn't have a forced landing or crash within seven hours of flying time he was sure to have one in eleven.

Stuntmen like Ormer Locklear were called the "Flying Fools" for their wild antics in the air. They were the original pioneers in the science of air-to-air refueling: they merely walked from the wing of one Jenny to another with a can of gas under one arm while holding on for dear life with the other.

Approximately 7,280 Jennies were built, including about 4,800 for the Army Air Service. The world looked back and patted Curtiss on the back for having mass-produced the first commercially successful aircraft in American history. It has been estimated that more than a million aviators at one time or another flew at the controls of a Curtiss Jenny.

STANDARD MODEL-SJ (J-1)

The Standard-SJ primary trainer of 1916 was ordered into mass production when America entered World War I. The Model-SJ was a standard biplane having tandem open seats with dual controls, a 150-hp Wright-Hispano engine and a top speed of 85 mph. The front wheel assembly had a third wheel in the nose to prevent the plane's being tilted forward by novice flyers. Model-SJ trainers worked side by side with Jennies to train young cadets for war duties. A total of about 1,932 Model-SJ and its sister Model-J1 were built for the Army by the Standard Aircraft Company.

THOMAS-MORSE MODEL S-4

The Thomas-Morse Model S-4 solo trainer was originally designed as a scout in 1916. When America entered WWI, the Model-S4 was quickly converted for use in the rapidly expanding program for training young cadets. The S-4 was a standard one-seat biplane with a LeRhone 80-hp engine and a top speed of 100 mph. For use as an advanced combat trainer, the S-4 was fitted with a 30-cal machine gun on the top of the fuselage just ahead of the cockpit. The forward-firing gun was synchronized to shoot through the two-bladed propeller. A total of 597 Model S-4 trainers were delivered for the Army training program. During the twenties, Model S-4's were entered in the National Air Races, and during the thirties they were used in the filming of Hollywood war movies.

FOREIGN-BUILT TRAINERS

In World War I America badly needed men who could pass the prerequisite qualifications for flying. Many of the pilot trainees were selected from foot soldiers aboard ships on their way across the Atlantic. Some trainees were soldiers from the Air Service who volunteered to fly after arriving overseas. None of those young cadets had had any previous flight training in the States. They had to be trained from scratch after arriving in England or in France.

The cadets arriving in England without any previous flying experience were given their primary instruction in trainers like the **Avro Model-504K,** the **Royal Aircraft Establishment Model BE-2,** or the **RAE Model FE-2,** all having two seats with dual controls. Advanced training was taken in aircraft like the **Sopwith 1A-2,** a two-seated former observation aircraft that could be fitted with machine guns. Combat training was received in Fighters like the one-seater Sopwith Camel with two fuselage guns and four practice bombs underwing. After brief training in those planes the cadets were given their "wings" and declared "combat-ready."

In France the AEF employed some two dozen different aircraft as trainers. Primary training was provided in two-seaters like the **Dorand Model-AR**, the **Farman Model-F** and the popular **Nieuport Model-12,** all dual control biplanes. Monoplane primary trainers were the **Caudron Model-G** and the **Morane-Saulnier Model MS-12.** Two-seated combat instruction was given in Attack Trainers like the **Breguet Model-14** and **Salmson Model-2A,** both armed with three machine guns and underwing practice bombs. Advanced combat training was received in single-seated biplanes like the **SPAD-VII** or the family of **Nieuports 17, 21, 23, 24** and **27**, all fitted with one or two machine guns for aerial simulation of dogfights. After brief check-rides in combat trainers, the fuzzy-cheeked pilots were rushed to the Western Front where many of them flew their first combat mission on the day of arrival. After surviving their first "Dawn Patrol" they bellied up to the bar at the club and had a straight shot with the rest of the fully qualified Military Aviators. But many never returned from that first Dawn Patrol.

TRAINERS BETWEEN WWI AND WWII

Designations used for Trainer aircraft between WWI and WWII evolved through six categories. The "TA" for Trainer Aircooled and "TW" for Trainer Watercooled categories were used from 1919 to 1925 for distinguishing between aircooled engines and watercooled engines. The "PT" for Primary Trainer was introduced in 1924 and the "AT" for Advanced Trainer was added in 1925: the Air Corps used Primary Trainers and Advanced Trainers in their two-stage Cadet training program. The "BT" for Basic Trainer was inserted in 1930 for intermediate Trainers sandwiched between Primary and Advanced Pursuit Trainers when the Air Corps experimented with a three-stage training program. The "BC" for Basic Combat Trainer was adopted in 1935 for Transition training in armed aircraft when that stage was necessitated by political events taking place in Europe. The "BC" category was dropped in 1940, and the surviving planes in that group became Advanced Trainers.

TRAINERS OF THE EARLY 1920'S

The **Elias TA-1 Trainer Aircooled** in 1921 was the first experimental "TA" Trainer tested by the newly formed Army Air Service. The Dayton-Wright TA-3 in 1921 was the first successful Trainer in the "TA" category. It was a standard biplane with side by side seats in an open cockpit. The TA-3 turned out to be the last Army airplane with a rotary engine; eleven of them were built. The Dayton-Wright TA-5 in 1922 was unusual in that it was the only aircraft to experiment with three front wheels arranged one under the fuselage and one about three feet from each wingtip. The Huff-Daland TA-6 in 1923 was the last of the "TA" Trainers.

The **Engineering Division TW-1 Trainer Watercooled** in 1920 was the first experiment in the "TW" category. The Dayton-Wright TW-3 in 1921 was the first successful "TW" Trainer for the Air Service. In 1923 the Dayton-Wright Company was absorbed by Consolidated, and subsequent deliveries were referred to as Consolidated TW-3's. The TW-3 turned out to be the last side-by-side Trainer accepted by the Military until the Cessna T-37 Jet Trainer came along in 1955. Thirty of the TW-3's were built. The Huff-Daland TW-5 in 1923 was the last of the "TW" Trainers.

THE PRIMARY TRAINERS

The **Consolidated PT-1 Trusty** in 1924 was the first post-WWI Trainer ordered in quantity as replacement for the aging WWI Curtiss JN-4 Jennies. The PT-1 was an improved version of the TW-3 with a 180-hp engine and a lengthened fuselage to accommodate a second tandem seat. The Consolidated PT-3 Trusty in 1928 differed very little from the PT-1 except for a new 220-hp engine. Production of both the PT-1 and PT-3 totaled 469 aircraft, which formed the nucleus of the Army's Primary Trainers until being replaced in service by the Stearman PT-13 in 1936.

Minor Primary Trainers of the early thirties were the **Fleet YPT-6 Husky Junior** in 1930, the **Stearman YPT-9 Cloudboy** in 1931, and the **Consolidated PT-11** and **YPT-12** in 1932, all standard biplanes with tandem open seats. Production totaled 69 of these four types, some of which were fitted with 300-hp engines and redesignated into the new ''BT'' Basic Trainer category.

The **Stearman PT-13 Kaydet** in 1936 was the first of a long line of Stearman Primary Trainers that were to stretch all the way through World War II. The PT-13 was a standard biplane with tandem open cockpits, a 220-hp engine and a top speed of 120 mph. The PT-13B experimented with a 280-hp engine, the highest powered Primary Trainer in the Army. Production of the PT-13 totaled 691 aircraft when Congress increased the Fiscal budget. Some PT-13's were still around for WWII.

The **Ryan YPT-16** in 1939 was the first all-metal monoplane Primary Trainer in Air Corps history and was the forerunner of Ryan monoplane Trainers that were to be produced in WWII. The YPT-16 was a single-engine, low-wing two seater with open tandem cockpits, streamlined leg fairings and fancy paints on the wheels. Deliveries of the YPT-16 totaled sixteen aircraft.

Consolidated PT-1 Trusty, first Army Pilot Trainer in the ''PT'' category. (USAF Photo 12202-AS)

Huff-Daland AT-1, first Air Corps plane Advanced Trainer assigned an "AT" designation. (USAF Photo 12104-AS)

THE ADVANCED TRAINERS

The **Huff-Daland AT-1 Advanced Trainer** in 1925 was a direct redesignation of surviving TW-5 Trainers as result of a new aircraft numbering scheme. The AT-1 was a standard biplane with 180-hp engine and a top speed of 110 mph. The Huff-Daland AT-2 in 1925 and the one-seat Boeing AT-3 in 1926 were minor single examples used only for experimental purposes.

The **Curtiss AT-4 Hawk** in 1927 was the first Advanced Trainer to be ordered in quantity. It was a single-seat biplane with a 180-hp engine and a top speed of 160 mph. Two fixed machine guns on the cowling fired forward through the synchronized propeller. On February 17, 1928, Bill Randolph was killed during takeoff in an AT-4 at Gorman, Texas. The Curtiss AT-5 Hawk differed only in having a new 220-hp engine. A total of 71 AT-4 and AT-5 Hawks were delivered for use as Trainers at the Army Air Corps 43rd School Squadron, Kelly Field, San Antonio, Texas. In 1929 all of the surviving AT-4's and AT-5's had 435-hp engines installed and were redesignated P-1 Pursuit Hawks, but they remained with the 43rd Squadron as Combat Trainers armed with only one fuselage gun.

The **North American BC-1 Basic Combat Trainer** in 1938 was the first monoplane Advanced Trainer in Air Corps history. Designed to replace Pursuits as combat trainers, the BC-1 had a 600-hp engine and a top speed of 205 mph. It was a low-wing two-seater with instructor and student in tandem under an enclosed canopy. The landing gear was conventional with retractable front wheels. The BC-1A and BC-2 in 1939 were improved BC-1's by North American. Production totaled 355 of all the "BC" variants. In 1940 all of the surviving BC-1A's were redesignated as the AT-6 Texan to become the first batch of that famous WWII Advanced Trainer.

THE BASIC TRAINERS

The **Douglas BT-1** in 1930 was the first to carry the new "BT" for Basic Trainer designation. The BT-1 was a direct conversion from the Douglas O-2 Observation biplane with a 450-hp engine, tandem open seats and a top speed of 135 mph. One fuselage gun and one flexible gun could be mounted if desired. The BT-1 was the last Air Corps Trainer to use the pioneer WWII Liberty engine. Forty-nine BT-1's were converted from O-2's for service.

The **Douglas BT-2A** in 1931 was a direct conversion from the Douglas O-32 Observation biplane. It was very similar to their BT-1. The BT-2B and BT-2C in 1931 came out of the factory as unarmed standard biplanes with a few minor improvements. Conversions and new productions totaled 197 BT-2's which formed the core of Basic Trainers for the Army.

Minor Basic Trainers included the **XBT-3** and **XBT-5**, both 1931 conversions from the Stearman YPT-9 Primary Trainer for tests with 300-hp engines. The Consolidated XBT-6 and YBT-7 in 1932 were converted from the YPT-12 Primary Trainer when that aircraft had a 300-hp engine installed.

The **Seversky BT-8** in 1934 was the Army's first monoplane Basic Trainer. It was also the first new aircraft that satisfied the need for a Basic Trainer built as such rather than being a converted "O" plane or a souped up "PT" Trainer. The BT-8 was a low-wing two-seater with a 450-hp engine and a top speed of 175 mph. The landing gear was standard and the front wheels wore fancy pants. Thirty BT-8's rolled out of the factory for active duty.

Seversky BT-8, the Air Corps' first monoplane Basic Trainer. (USAF Photo 13350-AC)

The **North American BT-9** in 1936 was the forerunner of the famous AT-6 Texan yet to come for WWII. The BT-9 was a low-wing two-seater with a sliding canopy over the cockpit and a 400-hp engine giving it a top speed of 170 mph. One gun on the cowl and a swing gun in the rear seat distinguished the BT-9 as the closest thing to a combat plane yet designed as a Trainer. Production totaled 266 aircraft as result of a substantial increase in appropriations for expansion of the Air Corps. Some of the BT-9's were used as Trainers in World War II.

The **Fleetwings BT-12 Sophomore** in 1938 was the first stainless steel Army airplane ever built. It was a low-wing monoplane with single 450-hp engine and a top speed of 195 mph. Pilot and student sat in tandem with dual controls beneath a transparent hood. Only 25 BT-12's were delivered to the Air Corps and they were available for WWII service.

A grand total of approximately 1,300 Trainers of all types were supplied to the Army between WWI and WWII. No wonder the United States was unprepared at the time of the Japanese attack on Pearl Harbor.

TRAINERS OF WWII

When America entered World War II in December 1941, the Army Air Forces found itself caught up in a wild scramble to train hundreds of thousands of military aircrewmen for flying jobs. Those highly specialized duties included those of pilot, copilot, flight engineer, navigator, bombardier, glider pilot, radioman, radar operator, armorer, observer, photographer and crew chief. Trainer planes, schools and instructors were urgently needed. Aircraft factories all over America ceased civilian airplane production and joined in an all-out effort to supply war planes of all types. Authority was granted to commandeer private and commercial airplanes for active duty. The Faddis Act provided the Pentagon with a legal instrument to withhold any defense materials previously scheduled for shipment abroad.

During WWII the USAAF pilot training program averaged about 40 weeks in length and progressed through four major ten-week stages: Preflight training, Primary training, Basic training and Advanced training. Upon graduation from Advanced training, cadets were awarded their "wings" and commissioned Second Lieutenants in the USAAF. Postgraduate training involved two more stages of transition into operational aircraft before the trainees were certified as combat-ready pilots.

Fairchild PT-26 Cornell, the USAAF's first monoplane Primary Trainer to have an enclosed cockpit. (USAF Photo 24425)

Ryan PT-20 Primary Trainer, popular WWII monoplane. Snapshot taken July 3, 1940. (USAF Photo 173877)

Aspiring young cadets first reported to a Preflight School (earlier called Replacement Center) like the San Antonio Aviation Cadet Center that originated at Kelly Field, Texas. During the 10-week course they received instruction in customs of the service, academic subjects, visual target identification, and an introduction to aircraft technology—but no actual flying.

After successfully completing Preflight training, students were transferred to civilian-operated flight schools for ten weeks of Primary training in single-engine, low-powered, two-seated, dual-control trainers. At peak strength there were 56 such schools in operation; they stressed the fundamentals of flying and placed little emphasis on military discipline. The most popular Biplane Primary trainer was the Stearman-Boeing family of PT-13, PT-17 and PT-18 Kaydets, all tandem-seaters with single 220 horsepower engines and a top speed of 120 mph. The favorite Monoplane Primary trainer was the Fairchild series of PT-19, PT-23 and PT-26 Cornells having 175 to 220 horsepower engines and a top speed of 130 mph. Another popular Monoplane was the Ryan family of PT-20, PT-21 and PT-22 Recruits of low-wing, all-metal construction with tandem open seats, 125 to 160 horsepower engines, and a top speed of 130-140 mph. Other Biplanes to reach only limited production were the Waco YPT-14, St. Louis YPT-15 and Ryan YPT-25 with 185-225 horsepower engines and a top speed of 120-130 mph. The highest numbered Primary trainer was the Boeing PT-27 built entirely for export to Allied nations overseas.

Upon completion of Primary training at civilian fields, the cadets returned to a military flying school for ten weeks of Basic training in single-engine higher-powered Monoplanes with tandem seats and enclosed cockpits. The objective was to mold young minds and bodies for the purpose of functioning in a military team. Formation flying, instrument flying rules (IFR), navigation, acrobatics and night flying were stressed. During Basic training the students were evaluated to determine which ones would graduate into single-engine Advanced training to become Fighter pilots and which ones would go on to twin-engine Advanced school and become Bomber pilots. Basic training was almost completely dominated by the Vultee BT-13 and BT-15 Valiants having retractable landing gear, 450 horsepower engines and a top speed of 185 mph. Lesser Basic trainers were the

Stearman-Boeing PT-17 Kaydet, eminent WWII biplane Primary Trainer in which more than 60,000 cadets were taught to fly. (USAF Photo)

North American BT-9, Fleetwings BT-12 Sophomore and the North American BT-14 Yale, all having 400 to 450 horsepower and a top speed ranging from 170 to 195 mph. The Boeing XBT-17 received the highest designation number, but that contract was cancelled.

After receiving their diplomas from Basic training, the students selected for single-engine Advanced training moved on to USAAF schools for the final ten weeks of instruction in a more powerful aircraft similar to a combat plane. The objective was to turn out Day-Fighter pilots by continuing with more advanced techniques in all the phases taught during Basic training, while at the same time emphasizing combat maneuvering of the aircraft and developing instantaneous reflexes to cope with counter-moves by the enemy. Advanced single-engine training was done solely by means of the North American AT-6 Texan, an all-metal low-wing Monoplane with a standard tail section, retractable undercarriage and tandem dual-control seats under an enclosed canopy. The Pratt & Whitney R-1340 600-hp engine gave it a top speed of 210 mph on a normal range of 870 miles. Armament consisted of one to three guns including a forward-firing fuselage gun, a wing-mounted gun and/or a flexible gun installed in the back seat. Up to 400 pounds of practice bombs could be carried underneath. Fighter pilot students concentrated on fixed gunnery and precision dive-bombing practice. Upon graduation from Advanced training they received a set of pilot's wings in one hand and a pair of gold bars in the other. Advanced trainers built strictly for Lend-Lease were the Noorduyn AT-16 Harvard, a variant of the AT-6 Texan, and the Stinson AT-19 Reliant, which were not used in the USAAF training program.

But not by any stretch of the imagination were the ''shavetails'' ready for combat. Next came five weeks of Transition training in the Republic AT-12 Guardsman or Pursuits like the Bell P-39 Airacobra and Curtiss P-40 Warhawk, with stress on handling the speedier aircraft and fine-tuning their fixed gunnery techniques. Finally, they received orders assigning them to units equipped with Fighters or Attack planes for twelve weeks of operational training before going into combat. In reality, it took six stages and about 57 weeks to produce a combat-ready Fighter pilot.

Cadets selected for multi-engine Advanced training were sent to USAAF schools for their final ten weeks of instruction in twin-engine trainers similar to light cargo transports. The objective was to familiarize the students with twin-engine aircraft and mold them into a crew. Each member came to be dependent upon the others and, through cross-training, each man learned a little about the other's responsibilities. Flexible gunnery practice and lots of instrument time were stressed, but there were no acrobatics. Most produced of the Advanced twin-engine trainers was the Beechcraft AT-10 Wichita four-seated, all-wood, low-wing Monoplane with two Lycoming R-680 295-hp engines and a top speed of 200 mph. The family of Cessna AT-8, AT-17 and UC-78 Bobcats were 5-seaters with Jacobs R-755 245-hp engines and a top speed of 195 mph. Another twin-engined Advanced trainer was the Curtiss AT-9 Jeep 4-seater with 280-hp engines and a top speed of 200 mph. Canada's contribution to the American training program was the Federal AT-20 Avro-Anson, a 4-seated low-wing cabinplane with Jacobs 330-hp engines and a top speed of 180 mph. Later in WWII the North American AT-24 Mitchell, a conversion from the B-25 Medium Bomber, became the most popular transition and aircrew trainer. It was a six-seater with 1,700-hp engines and a top speed of 285 mph. The AT-24 turned out to be the highest number assigned to the Advanced trainers. The Martin AT-23 Marauder, a derivative from the B-26 Medium Bomber, was designed as an Advanced trainer but was used mostly as a target tug. Upon graduation from Advanced twin-engine schools the cadets were presented with their pilot's wings and began their careers as 2nd Lts. in the USAAF.

Next came ten weeks of Transition training into medium and heavy bombardment aircraft or night-fighters with accent on formation flying, navigation and flexible gunnery training. The final ten weeks of Operational training was taken after assignment to units equipped with medium or heavy bombers. Effectively, there were six stages covering approximately 60 weeks between the date of entry into the program and the date of being certified as a combat-ready Bomber pilot.

North American AT-6 Texan, prominent WWII single-engined Advanced Trainer. (USAF Photo C-23650-AC)

Vultee BT-13 Valiant, the most-produced of all monoplane Basic Trainers in WWII. (USAF Photo 21307)

The majority of navigator and bombardier trainees were selected from cadets who had begun but not completed pilot training for one reason or another. Other trainees were brought directly into the programs through recruitment and volunteering. Training included an elementary course in piloting, flexible gunnery practice, precision navigation, radar operation, and an introduction to wireless radio codes. Fifteen weeks of navigational training were completed in aircraft equipped with astrodomes, like the Beechcraft AT-7 Navigator and AT-11 Kansan, both members of the UC-45 Utility Cargo Transport family. The popular AT-7 was the first airplane supplied to the Air Corps solely for use as a navigator trainer. It was a low-wing monoplane with conventional 3-wheeled landing gear and a twin-finned tail section. Two Pratt & Whitney 450-hp engines gave it a top speed of 225 mph. Three student stations had charts, compass, and stabilized drift signals. A rotatable astrodome for taking sextant readings was shared by each student in turn. The Lockheed AT-18A Hudson was also a navigator trainer. It had two Wright R-1820 1,200-hp engines, a top speed of 250 mph and three student stations. Eighteen weeks of bombardier training was received in aircraft equipped with a bomb-bay and a bombardier position, such as the most popular Beechcraft AT-11 Kansan with Pratt & Whitney 450-hp engines and a top speed of 215 mph. Two guns and ten practice bombs were provided for training 3 or 4 students at a time. Many crews of the Boeing B-17 Flying Fortress were trained in the AT-11. Upon graduation from their respective schools, navigators and bombardiers were awarded their wings and commissioned 2nd Lieutenants in the USAAF.

Flexible aerial gunnery training was given in six weeks at one of seven specialized gunnery schools like the one at Harlingen Field, Texas. All bombcrew members except pilot and copilot were required to be proficient in operating swing guns from trainers with gun turrets and gunner positions, such as the Beechcraft AT-11 Kansan with two flex guns and room for 3 or 4 students. The Lockheed AT-18 Hudson had five flex guns and the

Fairchild AT-21 Gunner carried one flex gun in the plexiglass nose cone and two swing guns in a power-operated top turret. The AT-21 was a five-seated mid-wing monoplane with twin tailfins and retractable landing gear. Night flying equipment allowed for nocturnal training missions. Two Ranger 450-hp engines gave it a top speed of 210 mph.

Flight engineer training for the Boeing B-29 Superfortress was conducted in two stages. The 19-week basic course at Amarillo Field, Texas, included classes in maintenance, aircraft characteristics, flight procedures and operational performance factors. The 10-week advanced phase at Hondo Field, Texas, concentrated on fuel consumption, power settings, emergency procedures and cruise control parameters affecting power output. Upon graduation from the advanced course, flight engineers went through transition training in the course of which they learned to fly the B-29. Finally, they were assigned to an operational unit for full combat-crew training. The Consolidated AT-22 Liberator, a derivative of the B-24 Bomber, was adapted into a Flying Classroom for the specific purpose of training flight engineers. Four Pratt & Whitney R-1830 1,200-hp engines gave it a top speed of 305 mph. The crew of five men was supplemented by two instructors who gave inflight assistance and advice to eleven students applying in the air what they had learned on the ground.

Technical schools at Scott Field, Illinois, Keesler Field, Mississippi, Shepard Field, Texas, Chanute Field, Illinois, Lowry Field, Colorado, and Jefferson Barracks, Missouri, were supplemented by civilian schools in training radio operators, armorers, airplane mechanics and other support personnel.

The massive training program during WWII accomplished the feat of turning out approximately 1,214,800 personnel trained for direct participation in the job of getting

Cessna AT-17 Bobcat, popular WWII twin-engined Advanced Trainer, caught here in formation on a training exercise. (USAF Photo 40805-AC)

Fairchild AT-21 Gunner, WWII Advanced Trainer used to teach airborne gunnery techniques. Seen in this picture is the XAT-21 prototype. (USAF Photo 24427-AC)

aircraft off the ground and keeping them airborne. At peak performance the program was geared to produce about 200,000 aircrew members per year. On March 1, 1944, the USAAF reached an all-time peak strength of 2,411,294 men and women with 79,908 airplanes.

Round-figure values for each category of aircrew and support personnel trained during World War II are shown in the following table:

Pilots/Copilots	180,500
Navigators	36,500
Bombardiers	32,100
Gunners	157,200
Radio Operators	150,600
Armorers	95,200
Mechanics	341,100
Other Support	221,600
Total	1,214,800

The following tables show the numbers of Trainer aircraft produced during World War II:

PRIMARY TRAINERS

Designation	Produced
PT-13	691
YPT-14	15
YPT-15	14
PT-17	3,519
PT-18	150
PT-19	4,889
PT-20	30
PT-21	100
PT-22	1,023
PT-23	1,125
YPT-25	5
PT-26	1,056
Total	12,617

BASIC TRAINERS

Designation	Produced
BT-9	266
BT-12	25
BT-13	7,832
BT-14	251
BT-15	1,693
Total	10,067

ADVANCED TRAINERS

Designation	Produced	Training Role
AT-6	4,361	Pilot
AT-7	1,141	Navigator
AT-8	33	Transition
AT-9	790	Transition
AT-10	2,371	Transition
AT-11	1,606	Bombardier-Gunner
AT-12	50	Transition
AT-17	1,140	Transition
AT-18	300	Navigator-Gunner
AT-20	50	Transition
AT-21	175	Gunner
AT-22	5	Engineer
AT-23	558	Air-crew
AT-24	954	Air-crew
Total	13,534	

Final statistics of all Trainers produced in WWII:

Primary Trainers	12,617
Basic Trainers	10,067
Advanced Trainers	13,534
Grand Total	36,218

TRAINERS SINCE WWII

On September 18, 1947, the United States Air Force (USAF) became an autonomous branch of the mititary services, completely divorced from the Army. In 1948 the Air Training Command (ATC) Headquarters was located at Scott AFB, Illinois. The training program called for turning out 4,800 fully qualified aircrewmen annually. The main question faced by the Pentagon was, ''What type of aircraft do we plan and design for the training role?'' Conceive if you will an age of rapidly advancing aeronautical technology during the transition from prop-driven monoplanes to jet-powered aircraft. Represent to yourself if possible an Army Air Force not having the slightest idea what to do with all those aircraft left over from WWII. Imagine if you can the responsibility for maintaining a force that had to prepare for fighting a war mostly with antiquated WWII propeller aircraft, while at the same time looking ahead to an all-jet future. Picture the constant battle with Congress over decisions and funds for accomplishing such an uncertain and widely diversified mission. In retrospect, what type of training program would you have instituted—even if you had possessed a crystal ball to tell you that jet aircraft would not be suitable for all phases of the wars in Korea and Vietnam?

In 1948 basic pilot training schools were located at Goodfellow AFB, Perrin AFB, Connally AFB and Randolph AFB, all in Texas. Advanced single-engine pilot training schools were at Las Vegas AFB, Nevada, and Williams AFB, Arizona. Advanced twin-engine pilot training schools were at Vance AFB, Oklahoma, and Reese AFB, Texas. The navigator-bombardier school was at Mather AFB, California, the gunnery school was at Las Vegas AFB, Nevada, and the instrument school was at Tyndall AFB, Florida.

Training aircraft carried forward from the Army Air Forces were the T-6 Texan advanced single engine pilot trainer, The T-7 Navigator and TC-45 Expediter navigational trainers, the T-11 Kansan bombardier-gunner trainer, the T-19 Cornell monoplane primary trainer, the TB-25 Mitchell twin-engine pilot and aircrew trainer, and the TF-80C Shooting Star jet trainer. Already in the planning stage were the T-33 Silver Star (T-Bird) jet trainer, the T-28 Trojan prop-driven pilot trainer, and the T-29 Flying Classroom navigator-bombardier trainer.

In 1949 the decision was made to re-manufacture 2,068 of the North American T-6 Texans into T-6G's with modernized cockpit, better all-around visibility, full dual instrumentation, increased fuel capacity and a steerable tail wheel. With a 600-hp engine and a top speed of 210 mph, those aircraft formed the core of the USAF pilot training program. After completing training in the T-6G Texan, pilots moved up to advanced twin-engine training and/or took their jet training in the TF-80C, which was later redesignated the T-33. Gradual phasing out of the T-6G began in 1950 with the arrival of the T-28 Trojan; the phasing out was accelerated in 1953 when the Korean conflict ended.

North American T-28 Trojan, first all-new USAF post-WWII Trainer and first single-engine Trainer to have a tricycle landing gear. Photographed inflight with clear-view canopy and minus the turnover truss bar. (USAF Photo 47712-AC)

Meanwhile, the Beechcraft T-7, T-11 and TC-45 handled the training of navigators, bombardiers and gunners. All three aircraft had twin 450-hp engines with a maximum speed of 220 to 240 mph. Phasing out of the WWII aircrew trainers began in 1950 with the arrival of the T-29 Flying Classroom.

Consolidated-Vultee T-29 ''Flying Classroom,'' first plane designed especially for training navigators and bombardiers in groups. Seen during its initial flight on September 22, 1949. (USAF Photo 36552-AC)

The North American TB-25 (the old AT-24) carryover trainer conversion served as both a twin-engine advanced trainer and as an aircrew trainer right up until the last of them were retired at Reese AFB in 1959, being phased out in favor of an all-jet training program in the T-37 Tweety Bird. The TB-25 had two 1,700-hp engines with a top speed of 270 mph.

The Lockheed TF-80C, which became the T-33 on May 5, 1949, had a dual role: to act as a transition trainer for rated pilots changing over from propeller aircraft to jets, and to serve as an advanced trainer teaching jet propulsion to newly graduated pilots. The jet engine of the T-33 generated 4,600 pounds of thrust to produce a maximum speed of 545 mph. Phasing out of the T-33 began in 1961 with the arrival of the T-38 Talon. Some 27,000 pilots got their Jet training in the T-33 Silver Star (T-Bird).

The North American T-28 Trojan was planned and designed as a trainer that would serve in both the basic and advance pilot training phases. Beginning with deliveries in 1950 the T-28 was tried out in both those roles; however, the 800-hp engine and its 290 mph top speed proved to be a little too much for some of the young cadets to handle. So the Air Force went in search of a lower-powered trainer to use in the basic training phase, with plans for reverting back to the WWII philosophy of training pilots. Phasing out of the T-28 began in 1961; that aircraft gave way to the T-37 jet trainer.

The Consolidated-Vultee T-29 navigator-bombardier-radar trainer in 1950 rapidly replaced all the leftover WWII aircrew trainers. The T-29 was the largest aircraft of its type in history—a "Flying Classroom" in every sense of the phrase. It carried a crew of four and two instructors, and featured up to 14 fully equipped student stations. Twin 1,500-hp engines gave it a maximum speed of 300 mph. Twenty of the T-29's at the Air Force Academy in Colorado had two J44 turbojet boosters mounted near the wing ends for climb performance. Phasing out of the T-29 began in 1973 with the arrival of the jet-powered T-43 Aircrew Trainer.

The Beechcraft T-34 Mentor in 1953 represented a sort of reversal in the philosophy of pilot training programs. It was a realistic step backward to the WWII system of first

Lockheed T-33A Silver Star (T-Bird), first USAF all-jet Trainer to carry a "T" designation. Photo shows a formation of students on a cross-country mission. (USAF Photo 47286-AC)

Beechcraft T-34A Mentor, low-powered monoplane Trainer used in the primary phase of pilot training during the era between the Korean and Vietnam wars. (USAF Photo 46971-AC)

requiring young cadets to complete their introductory training in a lower-powered trainer. The T-34 had a 225-hp engine with a top speed of 190 mph. Cadets were required to complete 30 hours of familiarization flying in the T-34 before moving up to the T-28. The next step was to the T-33 jet trainer or, later, the T-37 jet trainer. The phasing out of the T-34 began in 1964 with the arrival of the T-41 Mescalero.

The Cessna T-37 Tweety Bird jet pilot trainer was delivered in 1955 and began to arrive in quantity at ATC training schools in 1957. Initially, T-37's were used as the next step up for cadets who had completed 30 hours of primary training in T-34 Mentors. In 1957 plans were formulated for an all-jet training program, and the first two groups used as guinea pigs were the November 1958 and March 1959 classes. In "Project All-Jet," as the exercise came to be known, cadets went straight into the T-37 without any previous training in prop-driven aircraft. Success with Project All-Jet resulted in T-37's becoming standard trainers beginning on April 1, 1961, in ATC schools at Moody AFB, Georgia, Craig AFB, Alabama, Vance AFB, Oklahoma, Webb AFB, Texas, Reese AFB, New Mexico, and Williams AFB, Arizona. Under that consolidated program, cadets received their primary and intermediate pilot training in the T-37 before moving up to the T-33 or T-38 to learn advanced techniques. The T-37 had twin turbojet engines generating 1,205 pounds of thrust each to provide a top speed of 425 mph. By July 1, 1978, a total of 1,287 T-37's had been built for the USAF, with production continuing at the rate of about a dozen planes per year.

The Northrop T-38 Talon advanced jet pilot trainer was delivered in 1961 as a replacement for the aging T-33. With two turbojets delivering 2,500 pounds of thrust each to give a top speed of 840 mph, the T-38 was the first supersonic trainer in Air Force history. Talons arrived on the scene just in time to receive some of the first cadets graduating from

North American T-39 Sabreliner, Utility Trainer eXperimental (UTX) aircraft shown on an evaluation flight over California in 1960. (USAF Photo 160608-AC)

training schools that provided "all-jet" training in T-37's under the brand new program. The T-38 achieved the most outstanding safety record of any supersonic aircraft in USAF history. When production ended in 1972 a total of 1,187 Talons had been delivered to the Air Force for use in pilot training.

The North American (later Rockwell) T-39 Sabreliner Transition jet trainer was delivered in 1960 to satisfy the need for a versatile aircraft to be used in the Utility-Trainer-eXperimental (UTX) program. Two turbojets with 3,000 pounds thrust each give it a top

Boeing T-43A Aircrew Trainer, all-jet successor in 1973 to the prop-drive T-29 "Flying Classroom." (USAF Photo 182972)

speed of 595 mph. Pilot and copilot/student have dual controls and the cabin can be arranged with four to six seats. Principle role of the T-39 is giving proficiency check-rides to Senior Officers. It can also be used as a classroom, staff transport or high speed communications aircraft. The T-39 was not a ''school'' trainer; it was used at USAF Headquarters, at the Strategic Air Command and by the Systems Command.

The Cessna T-41 Mescalero was delivered in 1964 as part of the accelerated program to turn out pilots for the war in Vietnam. It was also to serve as a replacement for the aging T-34 Talon primary trainer. With a 150-hp engine driving its two-bladed propeller at a top speed of 135 mph, the T-41 was used at the Air Force Academy and other schools to give cadets 30 hours of instruction, including full instrumentation, before moving up to the T-37 for jet training. By 1976 a total of 289 Mescaleros had been delivered to ATC schools.

Cessna T-41 Mescalero, low-powered successor to the T-34 primary Trainer in 1964. Pictured inflight over the Air Force Academy in Colorado. (USAF Photo 113971)

The Boeing T-43 spacious jet Flying Classroom began to arrive at Mather AFB in 1973 carrying a crew of four and three instructors, with sixteen navigator-bombardier-radar positions for proficiency students and cadets. Twin turbofans with 14,000 pounds thrust each give it a cruise speed of 580 mph. The equipment was the same or similar to that utilized in most USAF operational aircraft of the late 1970's, and has been periodically updated. Nineteen T-32's were successful in replacing 77 T-29's under a new under-graduate navigational training concept. In that program a Honeywell T-45 Ground-Based Electronic Simulator is used to fly imaginary missions in simulated environments before students go up in the T-43 to smooth out the wrinkles. Subsequent increased effectiveness and aircraft durability indicate a utilization factor more than double that of the predecessor T-29.

U—UTILITY: 1952 TO THE PRESENT

The "U" for Utility category was introduced in 1952 to designate small cabin planes that could haul substantial loads into and out of unprepared landing strips and serve in the roles of liaison, light cargo transport, air evacuation, observation, air-sea rescue, communications and VIP staff transport.

The first of the "U" planes was the DeHavilland U-1 Otter built strictly for the Army with deliveries beginning in March 1955. None of the U-1's were transferred to the Air Force.

The Lockheed U-2 Gray Ghost was proposed, designed, built, flown and accepted by the Air Force within an 8-month period in 1954-55. Urgently needed as a secret Photo-Recon plane for monitoring arms buildups in the Soviet Union, The U-2 was designated in the "Utility" category to hide its true identity; it should carry a "SR" Strategic Reconnaissance or "TR" Tactical Reconnaissance designation along with the Lockheed SR-71 or Lockheed TR-1. Also affectionately referred to as the "Shady Lady," the U-2 is virtually a high-powered sailplane capable of remaining aloft for eight to ten hours at a time. An astro compass takes automatic celestial fixes, a downward looking periscope magnifies the earth below, a sniffer samples the air for radioactivity, and belly cameras take detailed photographs. Other sophisticated equipment includes nose-mounted radar, inertial guidance system, autopilot, flap-type dive brakes, air intake scoops, slipper fuel tanks, ventral antenna, and a wet wing that holds extra fuel.

The U-2 is a mid-straight-wing monoplane with standard tail section and two 2-wheeled

Lockheed U-2 High Altitude Sampling Program (HASP) surveillance aircraft of the 4080th Strategic Reconnaissance Wing (SAC) shown on a flight over the Caribbean. (USAF Photo 164466-AC)

retractable landing units in tandem under the fuselage. Wheel struts under each wing provide balance while on the ground; they drop away during takeoff. Skids on the wingtips prevent tipovers when landing. A single Pratt & Whitney J57-P turbojet engine mounted in the fuselage supplies 11,200 pounds of thrust. The top speed is 600 mph, the range stretches to 3,800 miles with all tanks full, and the service ceiling is near 80,000 feet. The pilot sits in the cockpit at the front of the fuselage, but the U-2D had two tandem seats. Some of the U-2's bore no national markings but carried only a 3-digit tail number for identification, such as "449" on the one that made an emergency belly landing at Fujisawa, Japan, on September 24, 1959.

Flying a U-2 assigned to Incirlik AFB, Adana, Turkey, on May 1, 1960, Francis Gary Powers took off from Peshawa, Pakistan, on a surveillance flight. While flying at 68,000 feet he was forced to bail out near Sverdlovsk, a town in the USSR 1,400 miles inside Communist territory. It was never confirmed whether the aircraft exploded or was shot down by Russian missiles. That incident prompted Nikita Khrushchev to castigate the United States and President Dwight Eisenhower, thereby causing the Paris Summit conference to collapse at its opening session on May 16, 1960. Ike subsequently cancelled all spy-flights over the Soviet Union. On September 9, 1960, Gary Powers began serving a 10-year sentence for espionage. On February 10, 1962, he was released in exchange for the Russian spy, Abel: the exchange took place on a checkpoint bridge in East Berlin, Germany. For his ordeal Gary Powers was awarded the CIA Intelligence Star Medal. On August 1, 1977, while on a news reporting flight for the Los Angeles KNBC television station, Gary Powers was killed when his helicopter crashed in suburban Van Nuys.

When America lost its listening posts in Iran after the Shah was deposed in 1979, U-2 Strategic Reconnaissance planes were fitted with special listening devices for intercepting missile signals emitted by vehicles being launched out of Tyuratan, 650 miles north of the Iranian border. In 1979 the Lockheed TR-1 Tactical Reconnaissance plane was being developed as a possible successor to the U-2 sometime in the 1980's. Due to the classified nature of their business, accurate production figures on the U-2 planes are not available.

The Cessna U-3A Administrator in 1958 was a low-wing cabinplane exactly like the six-place L-27 Liaison plane delivered in 1957. The U-3B in 1960 had two Continental O-470 260-hp engines, a top speed of 240 mph and permanent wingtip tanks. It was an all-weather Utility plane with a range of 1,400 miles. Reporting for combat in Vietnam during 1964 the U-3B was used on a variety of missions, hauling personnel and light cargo packages while being constantly subjected to enemy fire. Production totaled 195 aircraft, of which 160 were U-3A's and 35 were U-3B's, the last being delivered in 1961. The U-3 Administrator was also nicknamed the "Blue Bird."

The Aero Design U-4B Commander in 1960 was a direct redesignation of all USAF L-26 Liaison planes delivered in 1956 with twin Lycoming O-540 350-hp engines, a top speed of 250 mph, a crew of two, five to nine passenger seats and a 500-pound cargo capacity. The U-4B has been used as a staff transport by both the Air University and the Military Airlift Command (MAC). A total of 18 L-26's were delivered to the Air Force and only the redesignated examples carried the U-4B designation. No new production contracts were issued.

The Helio-GAC (General Aircraft Corporation) U-5A Twin Courier in 1964 had an unusual looking power assembly with two large engine housings mounted above and forward of the wings, high off the ground to protect against propeller damage from flying debris when operating in and out of rugged countryside on Short Take-Off and Landing (STOL) missions. Normal fuel tanks were located in the wings, but optional tanks could be fitted in the nose and at the wingtips. The U-5A was a high-wing cabinplane with standard tail section and conventional landing gear having three fixed wheels. Two Lycoming O-540 250-hp engines gave it a top speed of 185 mph and the range was 2,400 miles when extra tanks were carried. Pilot and copilot sat in dual control seats up front and four passengers or a 550-pound payload could be carried in the cabin. Only seven U-5's were delivered to the Air Force for evaluation trials.

The Canadian DeHavilland U-6 Beaver in 1962 was a direct redesignation from the triphibian L-20 Liaison plane delivered in 1952 with a Pratt & Whitney R-985 450-hp engine, top speed of 160 mph and a combat range of 450 miles or a ferry range of 780 miles. Pilot and copilot sat up front, and six passengers or 1,000 pounds of cargo could be accommodated in the compact cabin. In 1964 U-6 Beavers joined the battle in Vietnam, operating as general Utility planes to gain the distinction of having been workhorses in two wars, Korea and Vietnam. A total of 212 L-20's were built for the Air Force and only the redesignated examples carried the U-6 designation. No new production contracts were submitted.

The Piper U-7 Super Cub in 1962 was a direct redesignation of all surviving L-21 Liaison planes that began their military career in 1951 for the Korean war. The U-7 was a high-wing cabinplane with standard tail section, three fixed wheels, a Lycoming O-290 125-hp engine, a top speed of 125 mph, a range of 750 miles, a two-man crew and a 50-pound luggage rack. U-7's were used as general Utility planes in Vietnam beginning in 1964. Production totaled 702 L-21's, mostly for the Army, and only the few that were redesignated carried the USAF U-7 designation. No new production contracts were negotiated.

The Helio-GAC U-10A Super Courier in 1961 was a direct redesignation of the L-28 Liaison plane delivered in 1958 for conducting Short Take-Off and Landing (STOL) experiments. The U-10B in 1964 was a long range version with a Lycoming O-480 295-hp engine, a top speed of 200 mph, a 120-gallon fuel tank, a ferry range of 1,200 miles, carrying a crew of two and four passengers or 1,000 pounds of cargo in the cabin. The U-10B entered combat in Vietnam during 1965 for use as an airborne troop carrier on COunter-INsurgency (COIN) operations in the jungles. A total of three L-28's were converted to U-10A's and 100 new U-10B's were produced, but many of these were transferred to Central and South American countries under the Military Assistance Program (MAP).

The Grumman HU-16 Albatross in 1962 was a direct redesignation from the SA-16 Search Amphibian that was first delivered to the USAF in 1949. The HU-16 has two Wright R-1820 1,425-hp engines, a top speed of 165 mph, an endurance of 23 hours and a range of 2,700 miles. The 6-man crew includes pilot, copilot, navigator, systems operator and two attendants. Up to 12 litters or 15 survivors can be accommodated in the cabin. HU-16's reported for combat in Vietnam during 1965, and gained fame on rescue missions, plucking aircrew members out of the water and saving ground troops from being overrun at forward fire bases. Production totaled 395 SA-16's for the Air Force, and only the redesignated aircraft carried the HU-16 designation. No new production contracts were drawn up.

The Cessna U-17 Skywagon was delivered to the Air Force in 1965 for channeling to friendly nations overseas as a general Utility plane under the Military Assistance Program.

The U-17 was a high-wing cabinplane with standard tail section, two front wheels and either a wheel or tailskid at the rear. Pilot and copilot sat up front and four passengers rode in the cabin. A detachable cargo pack carrying a capacity of 300 pounds could be hung under the fuselage. A Continental O-520 300-hp engine gave it a top speed of 180 mph. The Air Force kept for its own use only a few of the U-17's, most of which were destined for overseas shipment between 1965 and 1977.

The Fairchild-Hiller AU-23A Peacemaker Attack Utility plane of 1971 was developed in competition with the Helio-GAC AU-24 Stallion under a USAF program code-named "Credible Chase." Designed for tactical support with Short Take-Off and Landing (STOL) capability, the AU-23A can be adapted for many roles including COIN operations, armed photo-recon, forward air control, liaison, leaflet dropping, loudspeaker broadcasting, psywar, chemical spraying, staff transport and cargo hauling. Alternative armament configurations use a 20mm cannon firing 700 rounds per minute, two 7.62mm window guns firing 2,000 rpm, or two gunpods on underwing pylons firing 4,000 rpm. The 2,310-pound weapons load varies among bombs, rockets, napalm, flares and smoke grenades attached to four underwing pylons and one under-fuselage hardpoint. The AU-23A is a high-wing cabinplane with standard tail section and three fixed wheels. It can take off with a run of only 500 feet. A single Garrett-Airesearch TPE-331 665-hp engine gives it a top speed of 175 mph and the combat range is 560 miles. Pilot and copilot sit up front with dual controls and up to 8 fully equipped combat troops ride in the cabin. A payload of 2,000 pounds can be hauled by removing the passenger seats. A total of 15 AU-23A's were delivered to the Air Force. One was destroyed, thirteen were shipped to Thailand and one was retained by the USAF.

The Helio-GAC AU-24A Stallion Attack Utility plane was delivered in 1972 under the "Credible Chase" program for Short Take-Off and Landing (STOL) tactical support operations in a wide variety of combat roles. The AU-24A was a high-wing cabinplane with three fixed wheels and a standard tail section. All-metal flaps alowed it to leap into the air after a takeoff run of only 325 feet. A single Pratt & Whitney PT6A 680-hp engine gave it a top speed of 215 mph and a loaded range of 450 miles; the ferry range stretched out to 650 miles. Pilot and copilot-gunner sat side by side in the cockpit, and six to eight combat troops had individual seats with reclining backs and headrests. A 1,300-pound payload could be hauled when all of the seats were removed. Armament consisted of a 20mm rapid-fire cannon in the cabin. An assortment of rockets, flares, napalm, gunpods or bombs could be hung on four underwing pylons and a hardpoint under the fuselage; the capacity weapons load was 2,240 pounds. A total of 15 AU-24A's were delivered to the Air Force. One was retained for service trials and the other 14 went to Cambodia.

SPECIAL RESEARCH EXPERIMENTAL AIRCRAFT

X—SPECIAL RESEARCH: 1944 TO THE PRESENT

The Bell XS-1 Supersonic experimental Rocketship was the forefather of Special Research aircraft designed specifically to fly as fast as they could, as high as they could, and gather as much scientific data as possible. Work began on the XS-1 in 1944 and the first glide flight was made on January 19, 1946, the aircraft being carried aloft underneath a B-29 Superfortress Mothership and released over Pinecastle (later McCoy) Air Base, Orlando, Florida. The first powered flight was made on December 9, 1946 out of Muroc (later Edwards) Air Base, California. Official delivery was made in August 1947, just a month before the USAAF became the USAF. About that time the designation was changed from XS-1 to X-1. Powered by a Reaction Motors XLR-11 four-barrel rocket engine generating 6,000 pounds of thrust, the X-1 named "Glamorous Glennis" became the first aircraft in history to break through the sound barrier in level flight. Watched over by a P-51 Mustang and a P-80 Shooting Star on that historic flight, Chuck Yeager, the USAF pilot, became the first human being to fly an airplane faster than Mach-1. The term "Mach" is used to indicate the speed of sound, regardless of altitude or atmospheric

Bell X-1 "Rocketship," first rocket-engined Special Research aircraft to exceed a speed of Mach-2. Photo taken above Muroc AFB, California. (USAF Photo 34457-AC)

conditions. Mach-1 varies from 740 mph at sea level to about 650 mph in thin air at high altitudes. For that accomplishment Captain Yeager was awarded the 1947 Collier Trophy for outstanding achievement in aviation. The X-1 subsequently attained a top speed of 964 mph and soared to a maximum altitude of 70,120 feet.

The Bell X-1A was delivered to Edwards AFB, California, in November 1953 with a Reaction Motors XLR-11 8,000-pound thrust turbot-rocket engine. It had the same wingspan but was 4 feet 8 inches longer than the X-1. On December 12, 1953, Chuck Yeager became the first USAF pilot to break through Mach-2 when he set a record of 1,612 mph and flew at 76,000 feet. On August 21, 1954, Kit Murray of the USAF flew the X-1A to an altitude record of 98,400 feet. Supersonic means "up to five times the speed of sound."

Meanwhile, work began for the Navy in 1944 on the jet-powered Douglas D-558-I Skystreak, which flew for the first time in March 1947 and set a new world speed record of 650 mph in August of 1947. The Skystreak worked side by side with the X-1 on many of its missions.

The Bell X-2 Transonic swept-wing rocketship was conceived in 1946, delivered to the USAF in 1952 and flew in 1953 as the first American aircraft to have an escape capsule that floated down by parachute with the pilot inside. A Curtiss-Wright XLRM-25 rocket engine generated 15,000 pounds of thrust in short bursts. The undercarriage had a nose wheel at the front and skid-type gear instead of wheels underneath. On a test flight in May 1954 Skip Ziegler of Bell was killed when the X-2 was jettisoned from its B-50 Superfortress Mothership. On May 22, 1956, Frank "Pete" Everest (USAF) broke through Mach-2 and on July 23, 1956, he attained a speed of 1,900 mph. On September 7, 1956, Iven Kincheloe (USAF) pushed it up to 126,200 feet. On September 26, 1956, Mel Apt (USAF) became the first triplesonic human when he broke through Mach-3 and throttled the aircraft up to 2,148 mph. Tragically, Captain Apt was killed on that flight when the plane crashed while returning to home base; that brought an end to the X-2 project. Transonic means "approximately the speed of sound."

Concurrently, the Navy-Douglas D-558-II Skyrocket swept-wing aircraft first flew in 1949 with two power plants, a turbojet and a turborocket, but was soon converted to carry only the rocket engine. On that initial test hop the Skyrocket flew at 1,200 mph and reached 82,000 feet. On November 20, 1953, Scott Crossfield of Douglas became the first double-sonic human when he broke through Mach-2 and set a speed mark of 1,328 mph in the D-558-II.

The Douglas X-3 Stiletto was first flown at Edwards AFB, California, on October 20, 1952, by Bill Bridgeman of Douglas who asked, "Where are the wings?" Measuring a little over seven feet in length, each tough wing panel was milled from a solid sheet of aluminum alloy and had knife-sharp edges. A stabilator—a combination of stabilizer and elevator—was used to change vertical direction. Two Westinghouse J34-WE turbojet engines generated 3,000 pounds of thrust each to produce a top speed of 650 mph. The X-3 didn't set any speed or altitude records, but it gathered valuable scientific data about frictional effects, pressure and aerodynamic stresses on the airframe. The Stiletto was retired from service and subsequently turned over to the Air Force Museum at Wright-Patterson AFB, Dayton, Ohio.

The Northrop X-4 Bantam in 1948 was a one-seated twin-turbojet subsonic flying-wing designed for investigating stability and maneuverability of aircraft with no tail. Control of the low-swept-wing aircraft was established by the use of elevons, a combination of elevators and ailerons. Two Westinghouse J30-WE 2,900-pound thrust turbojets were mounted under the wings at the juncture of the fuselage. The tail unit had a vertical tailfin but no horizontal tailplane, and the landing gear was of the retractable tricycle type.

The Bell X-5 was the first American aircraft to have variable-sweep wings, meaning that the wings were mechanically adjustable to extend straight out on takeoff, sweep back at a sixty-degree angle while cruising, then return to the extended position for landing. Powered by an Allison J35-A 4,900-pound thrust turbojet engine, the first flight was made by Skip Ziegler of Bell on June 20, 1951, during which he pushed it up to a supersonic speed. On October 31, 1953, Ray Popson (USAF) was killed in a crash of the X-5. Knowledge gained from the X-5 led to development of the Rockwell B-1 Bomber and General Dynamics F-111 Fighter, both of which had variable geometry wings.

Unmanned (pilotless) vehicles to carry an ''X'' designation were: the Lockheed X-7 Ramjet that was launched from a B-29 or B-50 Superfortress Mothership and recovered by parachute; the Bell X-9 Shrike-Rascal (GAM-63) guided air missile launched from a B-47 Stratojet Mothership; the North American X-10 Navaho (SM-64) twin-jet, delta-wing, twin-fin vehicle for the Cruise Missile project; the Convair X-11 (SM-65) Inter-Continental Ballistics Missile (ICBM) with a thermo-nuclear warhead; the Lockheed X-17 Ramjet test vehicle during development of the Polaris submarine ballistics missile; the Boeing X-20 Dyna Soar space shuttle that was cancelled; and the Martin-Marietta X-23 PRIME wingless missile with a delta-shaped fuselage, twin vertical tailfins, and six nitrogen-gas jets for reentry into the earth's atmosphere.

The Ryan X-13 Vertijet was the first manned aircraft designed to take off from a mobile launching pad. It had one seat, a delta-wing, and a single Rolls-Royce turbojet engine generating 10,000 pounds of thrust. The initial trial flight was made on December 10, 1955, utilizing a conventional tricycle landing gear to take off and land. The first launched flight with the gear stripped was made on May 28, 1956. It was thrust into the air, flew forward horizontally, returned to a hover attitude with the nose straight up, and landed by sitting down on its tail. The pilot's seat pivoted to keep him in an upright position at all times.

The Bell X-14 was delivered in 1956 and made its first flight on February 19, 1957, as a Vertical Take-Off and Landing (VTOL) aircraft with two side-by-side turbojets mounted in the nose. Thrust-diverters deflected the exhaust downward, permitting a vertical takeoff even though the plane was in a horizontal attitude. Once airborne the diverters rotated to thrust the air backward for forward flight. Landing could be accomplished by assuming a hover position with the nose pointed up and then backing in. The X-14 was a small, one-seated, mid-wing monoplane with open cockpit, standard tail section and tricycle landing gear. It attained a forward speed of 160 mph.

The North American X-15 Hypersonic research aircraft made its initial captive flight on March 10, 1959, unreleased by its B-52 Stratofortress Mothership. It completed its first unpowered glide flight on June 8, 1959, and flew with power for the first time on September 17, 1959, breaking through Mach-2 and climbing up to 51,000 feet. Scott Crossfield of North American was at the controls on all three flights. The X-15 had a low-straight-wing, a four-finned tail assembly, a retractable double-tire nose wheel and two retractable steel landing skids under the tail section. A single Reaction Motors XLR-99 rocket engine produced 70,000 pounds of thrust in short bursts. In order to withstand the wide range of temperatures encountered, the airframe was constructed of stainless steel and titanium, coated with a nickel alloy steel. The nose cone was specially designed to meet the problems arising during exit and reentry at the upper atmospheric boundary. Hypersonic means "five or more times the speed of sound."

On March 25, 1960, flight testing of the aircraft was assumed by Major Bob White (USAF), Lt. Commander Forrest Petersen (USN) and Joe Walker (NASA). Almost every time the X-15 took to the air it established a new international performance record. On June 23, 1961, Bob White made the first mile-per-second flight when he gunned the aircraft to 3,603 mph. On November 28, 1961, President John Kennedy awarded the Harmon International Trophy for aviators to Scott Crossfield, Bob White and Joe Walker for their accomplishments in hypersonic research. On June 27, 1962, Joe Walker set the final record of 4,159 mph and on August 22, 1963, he climbed to 354,200 feet. On December 5, 1963, Bob Rushworth (USAF) reached Mach-6.06, the highest recording on the Mach register. All those marks were well beyond the original design expectations. Honorary "Astronauts Wings" were awarded to Bob White, Joe Walker and Bob Rushworth for having flown the X-15 above 50 miles (264,000 feet). Three X-15's were built; two B-52's served with the project that spanned four years—from 1959 to 1963—and covered some 160 flights.

North American X-15 Hypersonic research plane, world's fastest and highest flying Special Research airplane. Caught here being launched from its B-52 "Mothership" over Edwards AFB, California, November 3, 1965. (USAF Photo 177331)

The North American X-15A-2 first flew under power on June 28, 1964, as a rebuilt X-15 with the same span but two and a half feet longer. The Reaction Motors XLR-99 57,000-pound thrust rocket engine used 18,000 pounds of liquid oxygen and anhydrous ammonia as propellents. Bob Rushworth flew at 2,964 mph and climbed to 83,000 feet. On October 3, 1967, Pete Knight (USAF) set a new record of 4,534 mph and on October 18, 1967, earned his "Astronauts Wings" by reaching an altitude of 277,000 feet. The entire X-15 program ended with the X-15A-2 in 1968 after many performance records were set and much valuable scientific information had been gathered.

The Hiller X-18 Tilt-Wing Convertiplane first flew on November 24, 1959, at Edwards AFB, California, using a Chase XC-122 Cargo Transport fuselage as the framework. Designed for experiments in Vertical Take-Off and Landing (VTOL), the high-wing pivoted through ninety degrees to point straight up and take off like a helicopter or point forward for horizontal flight. The tail section was standard and the landing gear had three fixed wheels. Two Allison T40-A engines underwing drove 6-bladed contraprops to produce 5,850-hp each, giving it a forward speed of 250 mph. A single Westinghouse J34-WE turbojet engine inside the rear fuselage supplied 3,400 pounds of thrust for vertical liftoff and hover stability.

The Curtiss-Wright X-19 Tilt-Prop Radial-Lift Vertical/Short Take-Off and Landing (VSTOL) plane was delivered in 1963 with two Lycoming T55-L 2,650-hp engines, a top forward speed of 460 mph and a range of 520 miles. The payload was 1,200 pounds and six seats could be installed in the cabin area. When viewed from above the X-19 looked a lot like the letter "H" with four tilting props mounted in nacelles, one at each high-wingtip and one on each end of the equal-span high-tailplane. The prop nacelles rotated to point almost straight up on takeoff, then returned to the normal forward position for horizontal flight.

The Northrop X-21 Laminar-Flow aircraft was delivered in 1963 as a much-modified WB-66D Destroyer Bomber with a wider span. Behind the cockpit the fuselage ridged upward over the high-swept-wing. Pilot and copilot sat in the cockpit above the nose wheel and three other crewmen had positions in the fuselage. The landing gear was tricycle and retractable; the tail section had a tailplane mounted about one-third the way up the fin. Two General Electric J79-GE 1,400-pound thrust turbojets, mounted one on each side of the fuselage rear, produced a top speed of 530 mph. Two underwing compressors, in nacelles where the WB-66 engines used to be, sucked air into the wings and ducted it into a plenum chamber to produce a laminar flow, thereby reducing drag and increasing the thrust.

The Bell X-22 Tilt-Duct Vertical/Short Take-Off and Landing (VSTOL) aircraft first flew on March 17, 1966, and was delivered to the USAF in May 1969. Four propellers were housed in large circular ducts, two at the front where the wings normally are and two on a wing at the rear in the normal tailplane position. A tailfin was mounted on the rear fuselage. The retractable tricycle landing gear had four tires, two on the nose wheel and one each on the main units. Pilot and copilot sat side by side in the cockpit up front. Four General Electric T58-GE 1,250-hp engines produced a top speed of 345 mph and the range was 445 miles. The fan-type ducts tilted to point straight up and exhaust downward on takeoff. At flight altitude the ducts rotated to point forward for horizontal movement. Looking down at a top-view drawing, the X-22 vaguely resembled a distorted football goalpost (H) with the rear wing wider than the forward ducts—it actually looked as if it ought to be flying the other way.

The Martin-Marietta X-24A Lifting-Body was delivered to the USAF on July 11, 1967, and made its first powered flight on March 19, 1970, out of Edwards AFB, California, with Major Jerauld Gentry of the USAF at the controls. It had a wingless delta-shaped fuselage with one seat, three vertical tailfins on a long tailplane, and retractable tricycle landing gear. A single Thiokol XLR-24 rocket engine generated 8,000 pounds of thrust. The X-24A rode along under the wing of its B-52 Stratofortress Mothership until being released at around 45,000 feet. Then a blast from the rocket engine catapulted it forward to fly the assigned mission and glide home afterwards. The X-24A attained a top speed of 1,050 mph and a maximum altitude of 71,407 feet.

The Martin-Marietta X-24B Lifting-Body was a rebuilt X-24A with a span of 19'2", a length of 37'6", a standard triple-finned tail section, and a more efficient Thiokol XLR-11 8,000-pound thrust engine. The X-24B was a wingless delta-shaped orbiter for studies on soft reentry and landing; it helped pave the way for the Space Shuttle "Enterprise." The X-24B reached a peak of 1,188 mph and climbed to 90,000 feet before the project was completed.

The Bensen X-25 powerless Gyro-Glider was delivered to Wright-Patterson AFB, Dayton, Ohio, in 1978 for tests with folding rotors on descent vehicles. Studies have also been conducted on using it as an unpowered or powered rotor-kite during rescue missions. Having a 2-bladed overhead rotor, three fixed wheels, a single tailfin with rudder, but no tailplane, it can be towed behind an automobile. The USAF originally procured a single-seater for tests and a double-seater for training. Bensen makes a powered version, which appears to be the best bet for rescue work.

The Osprey X-28 Air Skimmer was procured by the Navy in July 1971 as a lightweight flying boat for coastal patrol duties in Vietnam. It had a Continental 90-hp engine, top speed of 155 mph, a range of 385 miles and a single open cockpit for the pilot. The X-28 was a tri-service aircraft, but was never employed by the USAF.

SPACE VEHICLES— 1974 TO THE PRESENT

NASA-905 SHUTTLE CARRIER AIRCRAFT (SCA)

Work began in August 1974 on a Boeing 747-123 civilian airliner to be developed into a piggy-back carrier for the OV-101 Orbital Vehicle "Enterprise." After completion of modifications at Edwards AFB, California, in 1976, the 747-123 was designated NASA-905 Shuttle Carrier Aircraft (SCA). Four Pratt & Whitney JT9D turbojet engines develop 46,950 pounds of thrust each to lift the combined weight of 584,000 pounds off the ground. Top speed while mated is 300 mph, but speed is irrelevant to the mission. Pilot, copilot and two flight engineers occupy the huge aircraft that was lightened by having all its passenger seats removed. The low-wing configuration has extra lifting surface; the tail section is standard except for two special fins for stability, one on each end of the tailplane. The under-carriage has two wheels in the nose and four wheels in each of four main bogie units under the wings and fuselage. The term "bogie" applies to four wheels arranged in a rectangle like those on an automobile. A special rig was constructed to hoist the Enterprise onto the dorsal three-frame, six-legged pylon mountings.

OV-101 ORBITAL VEHICLE—"ENTERPRISE"

The Rockwell International OV-101 Orbital Vehicle "Enterprise," delivered in 1976, was the pioneer in trials with manned Space Shuttles designed to be launched like a rocket, orbit like a spaceship and fly home like an airplane. The OV-101 has a low, stubby coke-bottle wing, retractable tricycle landing gear with double wheels in each unit, a single tailfin swept back at 45 degrees and no tailplane. The flight crew of aircraft commander and pilot sit side by side in the upper deck of the nose module. A middle module is available to carry the payload, but it remained empty on all mated trials of the "Enterprise." The tail module was designed to house five rocket engines, but during piggy-back operations the main engine section was sealed off by an aerodynamic tailcone cover. Ventral fittings were fastened to three two-legged braces on the NASA-905 by the use of explosive bolts, which could be detonated for synchronized release when ready to separate the two aircraft in flight. Fred Haise, Joe Engle, Gordon Fullerton and Richard Truly trained tirelessly for the demanding roles of primary and backup crews.

Mated taxi tests began in April 1976, and the first mated-unmanned flight took off from Edwards AFB on February 18, 1977. On that two-hour flight the aircraft climbed to 16,000 feet, flew at 290 mph and came in for a smooth landing. A total of eight such takeoffs and landings were completed between February and August of 1977. The two aircraft remained attached on all flights in that second phase of the program.

At 8:45AM on August 12, 1977, with Fred Haise and Gordon Fullerton at the controls, the Space Shuttle broke its umbilical cord with the Mothership at 23,000 feet above the Mojave Desert and glided away in powerless flight. During detachment proceedings the NASA-905 went into a shallow dive while Gordon Fullerton fired off the explosive bolts, causing the "Enterprise" to be catapulted abruptly upward like a cork popping off a champagne bottle. Quickly, both aircraft were rolled in opposite directions, and the separation was perfect. After five minutes and 23 seconds of free-fall soaring, the Enterprise darted low over a dry lake bed at Edwards, dropped its landing gear and came in at 200 mph, sending up clouds of turkey-tail dust until braking to a halt 5,000 feet from the touch-down point. Five Northrop T-38 Talons served as Chase aircraft to monitor the flight.

In April 1979 the "Enterprise" joined its sistership "Columbia" at the Kennedy Space Center, Cape Canavaral, Florida. While the "Columbia" was being readied for its first trip into outer space, the "Enterprise" was used as a preflight ground trainer on techniques of mounting a Space Shuttle on its launching pad.

OV-102 ORBITAL VEHICLE—"COLUMBIA"

The Rockwell International OV-102 Orbital Vehicle "Columbia" was flown piggy-back style atop the NASA-905 from Edwards AFB to the Kennedy Space Center for final preparations before being launched upon it first orbital flight. The mated aircraft landed at the Kennedy Center precisely at 11:00 AM on March 24, 1979, being escorted on the final leg by Deke Slayton in a Northrop T-38 Talon. The "Columbia's" job is to take off like a rocket, serve as a spaceship to deliver its 65,000 pounds of satellites into orbit, capture any wornout orbiting vehicles in need of repair or salvage, reenter the earth's atmosphere, and return home like a conventional aeroglider. Designed to be reused up to a hundred times, the "Columbia" is durably constructed to withstand all stresses encountered during exit and reentry.

The general appearance of the "Columbia" is like that of the "Enterprise." The triple-deck nose module has a cockpit in the upper deck for the flight crew of aircraft commander, pilot and two engineers. Three electronics experts are seated in the middle deck, and sophisticated equipment is housed in the lower deck. The fuselage midsection that holds the payload has upward swinging doors for launching satellites into orbit. The tail module houses three internal Rocketdyne liquid-fuel engines that develop 400,000 pounds of thrust each, and two pod-mounted liquid-fuel rocket engines that provide 6,000 pounds of thrust each. For liftoff, the three Rocketdyne engines are supported by an external belly tank 154'2" long that holds about one and a half million pounds of super-cooled liquid oxygen and liquid hydrogen fuel. Attached to the sides of the tank are two jettisonable Thiokol solid-propellent boosters that burn 10,000 pounds of fuel per second while producing over 2,600,000 pounds of thrust each. All five of the liftoff engines combine to generate more than six and a half million pounds of thrust and give the vehicle an upward speed of nearly 3,000 mph. The two reusable boosters burn for two minutes, then fall away at the top of the first stage—approximately 160,000 feet—and float down by parachute. Then the three Rocketdyne engines go it alone, still sucking on the nonreusable belly tank that is jettisoned just before entry into orbit. The two pod-mounted 6,000-pound thrust rocket engines are used to maneuver the vehicle into its orbital path that takes it out as far as 176 miles above earth traveling at its top orbit speed of 17,500 mph. For reentry and reaction control there are 44 small rocket engine-thrusters mounted at various locations around the fuselage.

At 8:45 EST on the morning of February 20, 1981, while anchored securely to a launch pad at the Kennedy Space Center, all three of the "Columbia's" internal Rocketdyne engines were "quietly" test-fired at full power for the first time. Observed by John Young and Bob Crippen in a NASA Trainer circling above the site, that perfectly successful 20-second dress rehearsal served as the curtain raiser for a new era in man's attempts to conquer outer space. The occasion marked the 19th anniversary of John Glenn's pioneer orbital flight and hurdled the last major barrier prior to a planned launch in April of 1981.

On Sunday morning, April 12, 1981, at 7:00 EST, the "Columbia" roared from its launch pad at the Kennedy Space Center on the first of four planned orbital test flights. With all systems operating smoothly the Space Shuttle was settled into its orbital path within fifteen minutes after blastoff. John Young, the 50-year old Flight Commander, was on his fifth trip into outer space—one of those trips was an Apollo flight to the moon. Robert "Crip" Crippen, the 43-year old Pilot and computer expert, was on his first space voyage—he had previously trained as part of the backup crews for "Skylab" and the "Apollo-Soyuz" hookup.

At 10:12 am PST (01:21 pm EST) on Tuesday, April 14, 1981, after a near perfect reentry and descent, the "Columbia" glided in for an exact landing where "X marked the spot" on a runway at Edwards AFB in Rogers Dry Lake, a part of the Mojave Desert, California. After 2 days 6 hours and 21 minutes, during which time the Shuttle covered 36 orbits around the earth and accomplished all of its objectives, the aircraft and its crew were safely home.

AMERICA HAD DONE IT AGAIN!

APPENDICES

	PAGE
Bibliography	316
A Chronology of High Level Command Organizations	318
World War II Command Organizations	319
Aircraft Supplement: U.S. Navy, U.S. Marine Corps, Allied, and Enemy Aircraft	320
German Fighters, Dirigibles, and Bombers of World War I	320
German and Japanese Prop-driven Aircraft of World War II	323
British and Soviet Piston-engined Combat Planes of World War II	329
U. S. Navy and U. S. Marine Corps Fighters of World War II	338
Jet Fighters and Bombers of World War II	343
Russian-built Jet Fighters of the Korean War and the Vietnam War	347
Birth of the Air Force—A Chronology	350
Air Force Bases	363
Historical Brief	371

BIBLIOGRAPHY

I am grateful to the authors of the following works, which I consulted to authenticate many of the facts in this book:

Aircraft of the World
William Green and Gerald Pollinger. Doubleday & Co., New York, 1965.

Air Facts and Feats
John W. R. Taylor. Sterling Publishing Co., Inc., New York, 1978.

Air Force Blue Book
Tom Compere and William Vogel. Bobbs-Merrill Co., Inc., New York, 1961.

Air Planes
Enzo Angelucci. McGraw-Hill Book Co., New York, 1971

Airwar
Edward Jablonski. Doubleday & Co., New York, 1971.

Allied Aces of WWII
W. N. Hess. Arco Publishing Co., New York, 1966.

American Aces of WWII and Korea
W. N. Hess. Arco Publishing Co., New York, 1968.

American Combat Planes
Ray Wagner. Doubleday & Co., New York, 1968.

Challenging Skies
C. R. Roseberry. Doubleday & Co., New York, 1966.

Chronology of World Aviation
Gene Gurney. Franklin Watts, Inc., New York, 1965.

Combat Aircraft of the World
John W. R. Taylor. G. P. Putnam's Sons, New York, 1969.

Fighter Aces
Raymond Toliver and Trevor Constable. The Macmillan Co., New York, 1965.

Figher Pilots of WWII
Robert Jackson. Arthur Barker Ltd., London, 1976.

Flying Army
W. E. Butterworth. Doubleday & Co., New York, 1971.

German Air Force Fighters
Martin Windrow. Doubleday & Co., New York, 1968.

Great Air Battles
Gene Gurney. Franklin Watts, Inc., New York, 1960.

Jane's All the World's Aircraft
John W. R. Taylor. McGraw-Hill Book Co., New York (all editions), 1941-1979.

Soviet Air Force in WWII
Leland Fetzer. Doubleday & Co., New York, 1973.

United States Military Aircraft Since 1909
F. G. Swanborough. G. P. Putnam, London-New York, 1963.

United States Navy and Marine Corps Fighters
William Green and Gordon Swanborough. Arco Publishing Co., New York, 1977.

U.S. Air Force Biographical Dictionary
Colonel Flint O. DuPre. Franklin Watts, Inc., New York, 1965.

Who Was Who in American History—*The Military*
Marquis Who's Who, Inc., Chicago, 1973.

World War II In the Air
James Sunderman. Franklin Watts Inc., New York, 1963.

Zeppelins
Edwin Hoyt. Lothrop, Lee & Shepard Co., New York, 1969.

A CHRONOLOGY OF HIGH-LEVEL COMMAND ORGANIZATIONS

August 1, 1907. The Aeronautical Division, Office of the Chief Signal Officer, Signal Corps, U. S. Army, was inaugurated as the first Army air unit, Captain Charles Chandler commanding.

July 18, 1914. The Aviation Section, Signal Corps, U.S. Army, replaced the Aeronautical Division, Lt. Colonel Sam Reber commanding.

June 2, 1917. The Airplane Division, Signal Corps, U. S. Army, replaced the Aviation Section, Major Ben Foulois commanding.

September 3, 1917. The Air Service, American Expeditionary Force, U.S. Army, was established to handle air operations overseas during WWI, Brigadier General Bill Kenly commanding.

May 20, 1918. The Division of Military Aeronautics, Major General Bill Kenly commanding, and the Bureau of Aircraft Production under Mr. John Ryan, replaced the Airplane Division, thereby divorcing military aeronautics from the Army Signal Corps.

June 4, 1920. The Army Air Service, U. S. Army, Major General Charles Menoher commanding, replaced both the Division of Military Aeonautics and the Bureau of Aircraft Production, returning control of military air operations to the U. S. Army.

July 2, 1926. The Army Air Corps, U. S. Army, replaced the Army Air Service, Major General Mason Patrick commanding.

June 20, 1941. The U. S. Army Air Forces (USAAF) was established under the command of Major General Hap Arnold. The Office of the Chief of Air Corps was the overall boss of air operations, and the Air Forces Combat Command was responsible for the defense of America at home.

March 2, 1942. The Office of the Chief of Air Corps and the Air Forces combat Command were abolished under a scheme that made the USAAF an autonomous branch of the Army on a parallel with the Army Ground Forces and Army Supply Channels. General Hap Arnold was appointed Commander of the USAAF, responsible directly to the Supreme Allied Commander and the President of the United States.

September 18, 1947. The United States Air Force (USAF), Department of the Air Force, became an autonomous arm of the Military Services, completely divorced from the Army. General Carl Spaatz was named Chief of Staff for the USAF.

WW II COMMAND ORGANIZATIONS

During World War II the USAAF operated with fifteen numbered Air Forces, each being composed of separate subordinate commands to manage Fighter, Bomber, Air Support and Air Service facilities. The 20th Bomber Command operated under direct control of General Hap Arnold for deployment of B-29 Bombers anywhere in the world. The 21st Bomber Command was organized in the Mariana Islands specifically for the purpose of bombing the Japanese mainland. Each flying command had surordinate Divisions, Wings, Groups, Squadrons (12-24 planes), Flights (6 planes), Elements (3 planes) and Sorties (one aircraft). Those Major Commands were spread around the globe bounded by a line from America through western Canada to Alaska, down to Australià, back across northern South America to Africa, over to India, through the Middle East and Mediterranean Sea, up through the British Isles into Iceland and catching the tip of Greenland on the way back to America by way of Labrador.

WWII CHART OF MAJOR COMMANDS

Command Number	Area of Responsibility	Location of Headquarters
1st Air Force	Northeast U. S.	New York
2nd Air Force	Northwest U. S.	Washington
3rd Air Force	Southeast U. S.	Florida
4th Air Force	Southwest U. S.	California
5th Air Force	Southwest Pacific	Australia
6th Air Force	Caribbean	Panama
7th Air Force	Central Pacific	Hawaii
8th Air Force	Europe (ETO)	England
9th Air Force	North Africa-Europe	Egypt-England
10th Air Force	Burma-India (CBI)	India
11th Air Force	Northern Pacific	Alaska
12th Air Force	North Africa	Algeria
13th Air Force	South Pacific	New Zealand
14th Air Force	China (CBI)	China
15th Air Force	Mediterranean-Italy	North Africa
20th Bomber Command	Worldwide	Pentagon-India-China
21st Bomber Command	Japan	Mariana Islands

AIRCRAFT SUPPLEMENT— U.S. NAVY, U.S. MARINE CORPS, ALLIED AND ENEMY AIRCRAFT

GERMAN FIGHTERS, DIRIGIBLES AND BOMBERS OF WWI

ALBATROS D.I., D.II, D.III, D.V BIPLANES

The German Albatros D.I Biplane Fighter made its first flight in August 1916, and the improved D.II reported at the Western Front in September 1916. The advanced D.III entered combat in March of 1917 and the model D.V joined the battle in July 1917. Many of Germany's top Aces flew the Albatros D.II, D.III and D.V Biplane Fighters in WWI.

The D.II was a one-seated Biplane with standard tail section, two fixed wheels up front and a landing skid at the rear. A single Mercedes engine generated 160-hp to produce a top speed of 110 mph within an operating range of 150 miles. Armament consisted of two Maxim 7.92 mm cowl guns manufactured at Spandau and synchronized to fire through the 2-bladed propeller.

Oswald Boelcke—the Father of German Fighters—was killed on October 28, 1916, when his Albatros D.II Fighter collided with a comrade during a dogfight against a British DH-2 pusher-type Fighter. At the time of his death Boelcke was credited with 40 air victories.

Freiherr (baron) Manfred von Richthofen shot down his first Allied plane, a British FE-2B Escort Fighter, while flying an Albatros D.II biplane at the Western Front on September 17, 1916. In December 1916 von Richthofen painted his Albatros D.II scarlet, thus leading to his sobriquet the "Red Baron." Flying Albatros Biplanes and Fokker Triplanes, the Red Baron scored 80 air victories against Allied flyers to rank as WWI's Ace of Aces for all nations. In April 1917 Baron von Richthofen was appointed Commandant of JG-1, the first Fighter Group in German military history; it was composed of four Jastas (squadrons). Because of the variety of bright colors used on their airplanes, these Jastas were referred to in the press as "Flying Circuses." Of the Red Baron's 80 victories, 59 were scored in Biplane Fighters.

April of 1917 was labeled "Bloody April" by the British because of heavy aircraft losses during that month, mostly to German Albatros D.III Fighters.

Paul Baumer, Germany's 8th ranking Ace of WWI with 44 air victories, scored most of his kills while flying an Albatros D.V with the "Green Tail" Jasta-5 Hunting Squadron.

Production totaled about 2,650 Albatros Fighters for WWI.

FOKKER DR.I TRIPLANE

The German Fokker V.4 Triplane prototype first flew in July 1917, and the production model FOK F-1 Fighter took to the air on August 8, 1917. The designation was changed to DR.I when deliveries were made to the German Aces Werner Voss and Manfred von Richthofen on August 28, 1917. The abbreviation "DR" stood for *Dreidecker*, meaning triple-winged.

The Fokker DR.I was a one-seated Triplane with standard tail assembly, two fixed wheels up front and a tailskid at the rear. Two fuselage-mounted Spandau-Maxim 7.92mm machine guns were synchronized to fire through the 2-bladed prop. A single LeRhone or Oberursel 110-hp engine gave it a top speed of 115mph, and enough fuel was carried for an hour and a half of powered flight.

Werner Voss was the first German pilot to score a victory in the DR.I Tripe, claiming his 28th victim on August 29, 1917. During the next 25 days he won 20 air duels before being shot down and killed on September 23, 1917, by a British SE-5A Scout Fighter. Werner Voss ranked 4th among the German Aces of WWI with 48 air victories.

Manfred von Richthofen won his 60th air battle on September 1, 1917, sending a British RE-8 Observation plane down in flames near Zonnebeke; that was the first of 21 victories to be scored in his new scarlet Fokker DR.I Triplane. His last victory came on April 20, 1918, when he shot down a British Sopwith Camel. During a dogfight along the Western Front on April 21, 1918, the Red Baron fell to his death on the Allied side of the Somme river, victim of the guns of a British Sopwith Camel flown by Canadian Ace Roy Brown. The greatest Ace of WWI died with 80 air victories to his credit.

Production totaled about 320 DR.I Triplanes ending in May 1918.

FOKKER D.III, D.VII BIPLANES AND D.VIII MONOPLANE

The German Fokker V.11 prototype first flew in December 1917, and the production model D.VII Biplane Fighter was delivered in April 1918, regarded by many as the best German dogfighter of WWI. The predecessor model D.III in 1916, Fokker's first combat-worthy Fighter, had been flown briefly by German Aces von Richthofen, Udet and Boelcke while awaiting the arrival of the superior Albatros D.II Fighter. The Fokker D.VIII "Flying Razor" in 1918 was a Monoplane Fighter, but it came too late to affect the outcome of the war.

The D.VII biplane had a single open cockpit, a conventional tail section, two wheels up front and a landing skid at the rear. A single Mercedes 180-hp or BMW 185-hp engine gave it a top speed of 125 mph. Twenty-four gallons of fuel and oil allowed a range of 180 miles. Twin Spandau 7.92mm machine guns on the cowling fired synchronously through the 2-bladed propeller.

Wilhelm Reinhard assumed command of Manfred von Richthofen's "Flying Circus," on April 22, 1918, one day after the Red Baron had met his death. The Reinhard JG-1

Fighter Group flew their Fokker D.VII biplanes into battle for the first time in May 1918 during the second Battle of the Aisne.

Erich Loewenhardt, interim Commandant of Reinhard's JG-1, was Germany's 3rd leading Ace of WWI. Loewenhardt died in an air collision with one of his comrade D.VII Fighters while chasing a British SE-5A Scout on August 10, 1918. His 53 victories at the time of the crash ranked second behind those of Manfred von Richthofen, but Loewenhardt's score was soon surpassed by Ernst Udet.

Hermann Goering became Commandant of JG-1 on July 14, 1918, succeeding Wilhelm Reinhard, who had been killed in a crash on July 3, 1918. Goering got his first kill with the unit on July 18, 1918, downing a French SPAD XIII Fighter; he finished the war with 22 air victories. Hermann Goering served in Adolf Hitler's Nazi party and became Reichsmarshall of the Luftwaffe Air Arm in WWII.

Ernst Udet, second ranking German Ace of WWI, scored most of his 62 victories in a Fokker D.VII Fighter. His final kill came on September 26, 1918, claiming an American DH-4 Liberty Plane while flying his Fokker D.VII out of Metz. Ernst Udet survived the war and joined the Nazis to serve as Hilter's Technical Advisor for Aeronautics during WWII.

Production totaled about 1,000 D.VII Fighters for WWI.

ZEPPELIN L-3 DIRIGIBLE BOMBER

The first really successful mating of an airship with an internal combustion engine was accomplished in 1898 by a Brazilian named Alberto Santos-Dumont. The pioneer airship built by Graf Ferdinand von Zeppelin was tested on July 2, 1900, and the first Zeppelin Dirigible passenger service began in 1910. The words ''Zeppelin'' and ''Dirigible'' were to become synonymous in referring to airships.

The Zeppelin L-3 Luftschiff Dirigible Bomber was delivered for trials in May 1914 as the first airship in German military history. When WWI began on August 1, 1914, the German Army and Navy possessed about a dozen Dirigibles manned by crews trained to conduct reconnaissance and bombing missions. Dirigibles were practically invisible at night and their silenced engines made them difficult to detect from the ground.

The L-3 Zeppelin Bomber was 518′5″ long with a diameter of 48′8″ and a four-finned tail section having two horizontal and two vertical stabilizers. Its 140,000-pound weight was powered by three Mayback 210-hp engines giving a forward speed of 53 mph. Its flight endurance was eleven hours and the service ceiling was limited to 9,000 feet. The defensive armament consisted of six machine guns fired from three gondolas underneath. The front gondola was the command car from which most of the 16-man crew controlled the airship. The two side cars housed engines, propellers and the remaining crew members. A normal bomb load of 1,130 pounds included eight 110-pound bombs and ten 25-pound incendiaries. On some of the Dirigibles an Observer was seated in a small retractable car suspended underneath the airship by a steel cable. He spotted targets and relayed information to the Commander by telephone. Larger Dirigibles like the L-30 Recon-Bomber in 1916 carried ten guns and 10,000 pounds of high explosives.

The first German Dirigible bombing raid of WWI was made on January 19, 1915, by L-3 under command of Hans Fritz and L-4 captained by Magnus von Platen. Flying a nocturnal mission out of their moorings at Fuhlsbuttel, the two Dirigible Bombers attacked targets at Norfolk, Sheringham and Snettisham, along the English coast. While conducting a spy mission to keep an eye on the British Navy off the coast of Norway on February 17, 1915, the L-3 was caught in a snowstorm and crashed at Fanoe, Denmark.

By the time of Ferdinand Zeppelin's death on March 8, 1917, Dirigible Bombers had become sitting-ducks for improved Allied aircraft. After that time airships were used mainly on surveillance missions. The last Zeppelin shot down over London was the LZ-48 on September 16, 1917, felled by a British DH-2 pusher biplane Fighter out of Oxfordness. The last Dirigible bombing raid over Britain was conducted on August 6, 1918, during which Peter Strasser, head of the German Naval Airship Divsion, was killed when his L-70 Zeppelin was shot down by Egbert Cadbury in an American DH-4 Attack-Bomber. The last Zeppelin to fall in WWI was the L-53 Recon-Dirigible that was shot down over Terschelling, The Netherlands, on August 10, 1918, by Stuart Culley in a British Sopwith Camel Fighter. The final German reconnaissance flight of WWI was conducted by Dirigibles out of Nordholz and Ahlhorn on October 12, 1918.

During WWI German Dirigibles made 201 raids on the British Isles, including 51 over London, and dropped some 380,000 pounds of high explosives on Allied targets.

Production totaled about 100 Zeppelins built for use in WWI.

Famous postwar Dirigibles included the Graf Zeppelin of 1928 and the Hindenburg of 1936. The Hindenburg, last and largest of the Zeppelin airships, was 804 feet long and had a diameter of 135 feet. Its 500,000 pounds of weight was powered by four Daimler diesel engines delivering 1,100-hp each to give it a top speed of 85 mph. Accommodations were provided for a crew of 40 to care for 70 passengers on commerical flights ranging out to 10,000 miles. On May 6, 1937, the Hindenburg crashed at Lakehurst, NJ, killing 35 of the 100 people on board. Then and there the 31-year lifespan of Zeppelin Diribles came to a tragic end.

GOTHA G-IV

The first German conventional mass-bombing raid over the British Isles in WWI was made on May 25, 1917, by Gothaer Waggonfabrik Gotha-IV long-range biplane Bombers led by Ernst Brandenburg out of Ghent, Belgium. Fourteen Gotha-IV Bombers attacked London for the first time on June 13, 1917, dropping 9,240 pounds of bombs and inflicting 594 casualties. Production totaled 230 G-IV Bombers.

GERMAN AND JAPANESE PROP-DRIVEN AIRCRAFT OF WWII

MESSERSCHMIDT BF-109 (ME-109)

The German Bayerische-Flugzeugwerke BF-109 prototype made its first flight in September 1935, just six months after Hermann Goering's Luftwaffe Air Arm was created on March 9, 1935. The BF-109A Fighter was delivered to the German Condor Legion for combat trials in December 1937. The ME-109E was the first mass-produced version after Willy Messerschmidt acquired control of the Company on July 11, 1938. Although Messerschmitt had been a member of the design team since 1927, subsequent ME-109 Fighters were more commonly referenced as BF-109's.

Variants of the BF/ME-109 family included a Fighter-Interceptor, Fighter-Bomber, Attack Fighter, Photo-Recon plane, Nightfighter, two-seat Trainer, shipboard Fighter and high-altitude Spyplane. The popular BF-109G was nicknamed "Gustav" by the early pilots who flew it. Because of their similarity in appearance, the BF-109 and the American P-51 Mustang were often mistaken for each other during the heat of combat.

The BF/ME-109 was a low-wing Monoplane with enclosed cockpit, a plus-type tail section and conventional 3-wheeled landing gear having retractable front legs; later models had a retractable tail wheel also. The single Daimler-Benz 1,475-hp engine drove a 3-bladed prop to produce a top combat speed of 387 mph within an operating range of 450 miles. Armament included two 13mm nose guns, one 20mm cowl cannon and two 20mm wing cannon, all carrying a total 990 rounds of ammo. Two underwing hardpoints could handle either 550 pounds of bombs, two gun gondolas, or two "stovepipe" tubes carrying one rocket each for launching as air-to-air missiles against Allied Bombers.

On November 11, 1937, a BF-109A set a new world speed mark of 379 mph. On April 26, 1939, a specially built non-combat variant flown by Fritz Wendel claimed a record of 469 mph, which was to stand until the age of the Jets.

The German Condor Legion was formed in August 1936 to support Francisco Franco's forces in the Spanish Civil War. Later commanded by Wolfram von Richthofen, cousin to the Red Baron of WWI fame, the Condor Legion exploited the Spanish "Proving Grounds" to test Hitler's latest aircraft and provide combat experience for the Luftwaffe's aircrew members.

Werner Molders, the top Ace of Germany's Condor Legion with 14 air victories, got his first kill in a BF-109C on July 15, 1938, shooting down a Russian I-16 Fighter at the Northern Front. His first air victory of WWII was in a BF-109 on September 20, 1939, when he downed a French Curtiss Hawk-75. In July 1941 Colonel Molders, Germany's leading Ace at the time, with 115 aerial victories, was appointed Supreme Commander of the German Luftwaffe Fighter Arm. He was killed on November 21, 1941, in the crash of a Heinkel HE-111 Bomber at Breslau, Poland. Molders was succeeded in command by Adolf Galland, an experienced BF-109 and FW-190 Fighter pilot.

Joachim Marseille was German's top Ace on the Western Front with 158 air victories, the majority of which were scored in Northeast Africa. Marseille got his first kill over France in 1940. On September 1, 1942, he shot down 17 Allied planes in a single day during fierce fighting in support of Rommel's Afrika Korps. His last victory came on September 26, 1942, when he shot down a British Spitfire Fighter near El Alamein, Egypt. Joachim Marseille was killed in a BF-109G on September 30, 1942, while returning to his base in Libya. He is famous as the leading RAF-killer of all times, shooting down 151 Allied planes in only 18 months of combat at the North African Front.

During the Battle of Britain in July-October 1940 a total of 610 BF-109 Fighters and 2,088 other planes were lost in combat. On October 12, 1940, Hilter cancelled plans for the invasion of Britain and ordered that a Night blitz by heavy Bombers begin. On June 13, 1944, the first of about 8,000 Jet-powered V-1 Buzz Bombs was launched by catapult from sites near the coast of the Continent. Approximately 2,500 V-2 Flying Bombs were launched between Autumn 1944 and March 1945, adding up to 10,500 warhead missiles fired at Britain during the first such blitz in the history of mankind.

Gerhard Barkhorn was the 2nd ranking German Ace of WWII, winning 301 air duels at the Russian Front. He got his first kill while flying a BF-109 in June 1941 and finished the war with a brief stint flying ME-262 Jet Fighters. Barkhorn survived the war to live in peace. Gunther Rall ranked 3rd in WWII with 275 air victories and Otto Kittel downed 267 Allied aircraft to hold down the 4th spot. Almost all of Rall's and Kittel's victories were scored in BF-109's at the Russian Front.

Erich Hartmannn, in just two and a half years of flying BF-109 Fighters, destroyed 352 Allied planes to rank as the world's Ace of Aces for all times. His first kill was a Russian Bomber on November 5, 1942, while flying a BF-109. His 300th victory was won against a Russian P-39 Airacobra Pursuit over Baranovka. His last victim was a Russian YAK Fighter over Brno, Czechoslovakia, on May 8, 1945, the last day of war in Europe. Convicted as a war criminal in a Russian court, Erich Hartmann spent ten years in a Soviet prison camp until obtaining his release in 1955.

Emil Lang holds the all-time record of air victories in a single day, downing 18 Russian aircraft on three missions at the Eastern Front. His total reached 173 victories before he met his death.

Erich Rudorffer claimed the record for most victories on a single mission when he shot down 13 Allied planes on November 6, 1943. He survived the war with 222 victories to rank 7th among Germany's WWII Aces.

Production totaled about 35,000 BF/ME-109 Fighters ending in 1961, the most-produced Fighter in German history.

JUNKERS JU-87 STUKA

The Junkers JU-87-1 prototype with a twin-finned tail section and two-bladed propeller first flew in November 1935, and the redesigned single-finned JU-87-2 prototype took to the air in March 1936. Deliveries of the JU-87A Stuka (Divebomber) to the Luftwaffe began in May 1937.

The JU-87B was a mid-bent-wing monoplane with standard tail section, three fixed wheels and a two-seated cockpit centered above the wings. A single Junkers JUMO 1,100-hp engine drove a 3-bladed prop to give it a top speed of 232 mph. The range was 370 miles with a full load and the service ceiling was 24,000 feet. Armament included two 7.9mm wing guns and a defensive flex gun operated by the rear-seat observer. A bomb load of 1,100 pounds was fitted on five racks underneath. The JU-87D carried 3,960 pounds of bombs, two 20mm cannon in place of the wing guns and added a second swing gun in the back seat; one package carried by the JU-87D housed 92 anti-personnel frag bombs. The JU-87G tank-buster was armed with two long-barreled 37mm cannon mounted underneath the wings.

The JU-87A joined German Condor Legion forces in December 1937 for combat trials in the Spanish Civil War. Designed as an adjunct to heavy field artillery, the JU-87A was effective in disrupting troop movements, bombing docks, and clogging harbors.

On September 1, 1939, JU-87B Stukas of the Wehrmacht struck the first blow of WWII, attacking Polish targets on the Vistula river. Against outmoded Allied aircraft during the early campaigns of WWII, the JU-87 was highly successful. However, in the opening days of the Battle of Britain that began on July 2, 1940, Ernst Udet's prized Divebombers were destroyed in great numbers by British Hurricane and Spitfire Fighters defending their homeland. Shortly thereafter the Junkers Stukas were pulled back and used in less hazardous areas.

Hans Rudel, Germany's leading ground-attack pilot of all times, flew 2,530 sorties in WWII and destroyed 519 motorized vehicles on strikes against the Russians in his JU-87 Stuka.

Production totaled about 5,750 JU-87 Stukas for WWII.

MESSERSCHMITT BF-110 (ME-110)

The German Bayerische-Flugzeugwerke BF-110A prototype made its initial flight on May 12, 1936. In July 1938 the Company was renamed Messerschmitt, and delivery of the first production model BF-110C Zerstorer (Destroyer) long-range Fighter began in January 1939. Although built by Messerschmitt as the ME-110, it was more commonly referenced by the "BF" prefix.

The BF/ME-110 was a low-wing monoplane with twin-finned tail section, tandem two-seated cockpit and conventional landing gear having retractable front wheels. Twin Daimler-Benz 1,475-hp engines drove 3-bladed props to give it a top speed of 343 mph. The range was 1,300 miles and its service ceiling reached 32,000 feet. Armament included two 30mm nose cannon, two 20mm nose cannon, two 30mm downward-firing belly cannon and two 7.9mm upward-firing flexible guns in the rear seat, all carrying a total 1,990 rounds of ammo. A normal bomb load of 1,100 pounds or two rocket-launching tubes could be carried on underwing hardpoints.

The BF-110E Fighter-Bomber carried 4,410 pounds of bombs. The BF-110F Nightfighter had new Lichtenstein radar, two air-to-air rockets launched from underwing "stovepipes," and carried a 3-man crew. The BF-110G Nightfighter substituted two upward-firing 20mm cannon for the swing guns in the back seat and added a 4th crew member.

The BF-110 entered WWII on September 1, 1939, being used as a ground-attack plane within a few hours after the first blow was struck against Poland.

Hermann Goering favored using the BF-110 as an Escort Fighter accompanying heavier bombers during the Battle of Britain in July-October 1940. Outclassed by British Spitfire and Hurricane Fighters, the BF-110 soon requested Escort help from more maneuverable BF-109 Fighters—Goering's vaunted "Destroyer" turned out to be an Escort that needed an Escort! More than 200 BF-110 Fighters were shot down during that 4-month sky war above the British Isles.

The BF-110 was used in all the major thrusts during the initial wide-ranging assault at the Russian Front on June 22, 1941. From that time on the BF-110 was employed mainly in the roles of Night Interceptor, ground-Attack plane and combat Trainer. As a Night Prowler the BF-110 was unparalleled, finishing WWII with the distinction of having shot down more enemy aircraft than any other Nightfighter in history.

One of the war's great air battles took place on March 30, 1944, when 200 BF-110 Nightfighters met 800 RAF Night Bombers over Nuremberg. When the fight was over RAF officials counted more than 90 Bombers that had failed to return home.

Heinz Schnaufer was history's leading Nightfighter Ace with 121 Night victories at the Western Front, most of which were scored in BF-110 Nightfighters. He survived the war but was killed in a 1950 automobile accident.

Helmut Lent was Germany's 2nd ranking Nightfighter Ace with 102 of his 110 victories scored in Nightfighters at the Western Front. He was shot down and killed while flying a BF-110 over Germany on October 5, 1944.

Production totaled about 6,150 BF-110 Fighters, ending in December 1944.

The Messerschmitt ME-210, ME-310 and ME-410 were twin-engined single-finned experiments in the field of long-range Fighter-Escorts, but of those three designs only the ME-410 met with any success in WWII.

FOCKE-WULF FW-190

The German Focke-Wulf FW-190 prototype, designed by Kurt Tank, made its first flight on June 1, 1939, and deliveries of the FW-190A Fighter to the Luftwaffe began in July 1941. The versatile FW-190 was manufactured in many variants, including Day Fighter, Nightfighter, Fighter-Bomber, Attack-Fighter, Torpedo Fighter, two-seated Trainer, long-nosed high-altitude Interceptor and rocket-armed tank-buster. The nickname of ''Wurger'' (Butcher Bird) was bestowed by the men who flew FW-190A's in the early days.

The FW-190 was a low-wing monoplane having a standard tail, conventional retractable landing gear and a canopy-covered cockpit. A single BMW 1,700-hp engine produced a top speed of 425 mph and the propeller had either two or three blades. The service ceiling of the Interceptor reached 37,500 feet and its range when loaded for combat was 520 miles. Armament included two 13mm fuselage guns and four 20mm wing cannon. The Fighter-Bomber variant could carry 12 underwing rockets or a maximum bomb load of 3,970 pounds.

Otto Behrens supervised operational trials of the FW-190 that began in February 1941. The first combat-ready unit to receive FW-190A Fighters at the Western Front in August 1941 was Fighter Group JG-26 commanded by Walter Schneider. On September 3, 1941, a flight of FW-190's from JG-26 scored their first victories, sending 3 British Spitfire Fighters down in flames over Dunkirk.

Adolf Galland was the first Luftwaffe pilot to fly an FW-190A into combat, leading a raid over England on August 2, 1941. His first kill of WWII had been achieved while flying a BF-109 Fighter on May 12, 1940, when he sent a British Hawker Hurricane Fighter down in flames over Liege, Belgium. On November 21, 1941, General Galland succeeded Werner Molders as Supreme Commandant of the Luftwaffe Fighter Arm, where he served until being dismissed by Hitler in January 1945. Galland counted 47 American P-51 Mustang Pursuits among his total of 103 victories in WWII.

The first unit to receive FW-190A Fighters at the Russian Front in August 1942 was JG-51 commanded by Heinrich Krafft. Its pilots scored their first victories on December 17, 1942, claiming two Russian PE-2 Divebombers.

On August 19, 1942, two FW-190 Fighters were shot down at about the same time to become the first air victims of American Fighter pilots in WWII. One was claimed by Frank Hill in a British Spitfire Fighter and the other was downed by Sam Junkin in support of the attempted Third Front at Dieppe, France. FW-190 Divebombers attacked Allied landing forces at Dieppe, and played a major role in repelling that invasion.

Josef Wurmheller claimed 102 air victories in WWII, of which 93 were scored in FW-190 Fighters along the Western Front. He was killed in a dogfight near the French border on July 22, 1944.

Robert Weiss commanded JG-54 in August 1944 flying FW-190D Dora-9 Interceptors to protect ME-262 Jet Fighters during takeoff and landing at Achmer. Major Weiss plunged to his death on December 29, 1944, victim of a British Spitfire Fighter.

The Tank TA-152 Fighter rolled out of Focke-Wulf factories in November 1944 as a redesignation of the much-modified FW-190D. Designed for use as a high altitude Interceptor, the TA-152 had a longer fuselage and a wider span to carry five cannon plus a variety of stores underneath. The TA-152H Recon-Fighter was the final version, having a span of 48′8″ and a new engine that gave it a top speed of 472 mph at 41,000 feet. But the TA-152H never officially broke the world's speed record of 469 mph held by a souped-up version of the BF-109 Fighter.

The first unit to receive TA-152H Interceptors in December 1944 was JG-301, Kommando Aufhammer, charged with the responsibility of protecting ME-262 Jets taking off and landing at Hesepe.

Production totaled about 20,200 FW-190 and TA-152 variants for WWII.

MITSUBISHI A6M ZERO

The Japanese Mitsubishi A6M prototype made its first flight on April 1, 1939, and the A6M-1 Zero-sen shipboard Fighter joined the Imperial Navy in September 1939. The A6M-2 Fighter entered combat on August 19, 1940, flying escort to Bombers on a raid at Chungking, China. Variants of the A6M included a shipboard Fighter and Fighter-Bomber, Nakajima "Rufe" Floatplane, Divebomber, "Zeke" Fighter with narrowed span, rocket-armed Attack plane, long range Kamikaze Bomber, land-based Interceptor and two-seated Trainer. Marine Fighter pilots who opposed the Zero at Guadalcanal nicknamed it "Meatball."

The A6M was a low-wing monoplane with standard tail assembly and conventional landing gear having retractable front wheels. A single Nakajima 1,210-hp engine gave it a top speed of 358 mph and a range of 1,100 miles. Armament included two 7.7mm cowl guns, two 13.2mm wing guns and two 20mm wing cannon. Racks underneath could carry 550 pounds of bombs or four rockets.

During WWII Japanese Fighters were classified into 3 types: the A-type was a lightweight shipboard or land-based Fighter; the B-type was a heavyweight land-based Interceptor; and the C-type was a Nightfighter. A6M Fighter variants were built as all three types, and were the most widely used Japanese combat aircraft of WWII.

On December 7, 1941, Japanese carrier-based A6M-2 Zero-sen Fighters took off from the decks of Vice Admiral Chuichi Nagumo's naval task force near the Hawaiian Islands. Flying the lead plane, a Nakajima B5N2 Bomber, General Mitsuo Fuchida issued the command to attack Pearl Harbor just before 8:00AM with the first wave of a two-wave attack. That raid brought America into WWII and caused Admiral Isoroku Yamamoto, who witnessed the battle, to utter the prophetic phrase, "We have awakened a sleeping Giant!"

Saburo Sakai, flying an A6M Zero out of Formosa on December 8, 1941, shot down the first American P-40 Pursuit to fall during the initial Japanese raid on Clark Field at Manilla in the Philippines. Sakai became the 3rd leading Japanese Ace of WWII with 62 air victories. He was wounded over Guadalcanal in August 1942, blinded in one eye, and spent most of the war thereafter as an instructor. His last combat mission was flown in a Mitsubishi J2M Raiden land-based Interceptor against an American B-29 Bomber over Japan on August 13, 1945. Sakai survived the war to live out his remaining years in peace.

Hiroyoshi Nishizawa was Japan's leading WWII Ace with 104 Allied aircraft to his credit while flying Japanese Zeros. Nicknamed "the Devil" by flyers who opposed him in aerial duels, Nishizawa was distinguished as the leading USAAF killer of all times with about 92 American planes on his list. He died in a crash on October 26, 1944, when an American Navy F6F Hellcat Fighter shot down the Nakajima L2D Transport plane he was flying on a shuttle hop to Clark Field in the Philippines.

Shoichi Sugita ranked 2nd on Japan's list of WWII Aces with 80 victories scored in Mitsubishi A6M Zeros and Kawanishi N1K2 Shiden Fighters. Sugita is generally credited with shooting down the P-38 Lightning Pursuit that carried America's 2nd leading Ace, Tommy McGuire, to his death on January 7, 1945. Sugita was killed on April 17, 1945, when his N1K2 Violet Lightning Fighter was shot down over Kanoya, Japan, and fell at the southern tip of Kyushu Island.

Equipped with A6M-5 Divebombers in October 1944, Japanese Air Group-201 "Shimpu" was the pioneer Kamikaze (suicide) unit operating out of the Philippines on attacks against Allied seagoing vessels supporting General MacArthur's fulfillment of his famous vow, "I shall return!" The Shimpu group got their first of 34 sinkings on October 25, 1944, when the American escort carrier *Saint Lo* was sent to the bottom.

Production totaled about 10,950 A6M variants, ending in August 1945.

NAKAJIMA B5N KATE—MITSUBISHI G4M BETTY

Two notable Japanese Bombers of WWII were the Nakajima B5N carrier-based single-engine torpedo-bomber of 1937 and the Mitsubishi G4M land-based twin-engine Medium Bomber that first flew in 1939. During the Pearl Harbor attack of December 7, 1941, 143 B5N2 Kates dumped over 200,000 pounds of explosives onto Hawaiian targets. On December 11, 1941, Mitsubishi G3M2 Nells and G4M1 Bettys out of Indo China sank the British cruiser *Repulse* and the battleship *Prince of Wales* that were standing guard in the South China Sea off the coast of Singapore, Malaya. Left defenseless, Singapore surrendered on February 15, 1942. Betty Bombers operated out of Rabaul, New Britain, under command of Sadayoshi Yamada and Admiral Isoroku Yamamoto. On April 18, 1943, Admiral Yamamoto was killed when his G4M Betty was shot down over Bougainville, New Georgia, by American P-38 Lightning Pursuits out of Guadalcanal in the Solomons.

BRITISH AND SOVIET PISTON-ENGINED COMBAT PLANES OF WWII

GLOSTER GLADIATOR

The British Gloster Gladiator Biplane Fighter, England's last two-winged combat plane, was first delivered to RAF units in February 1937. It was a standard biplane with fixed wheels, a Bristol Mercury 840-hp engine and a top speed of 257 mph. Armament consisted of two fuselage guns and two underwing guns.

On June 11, 1940, the day after Benito Mussolini declared war against the Allies, three Gloster Gladiators named "Faith," "Hope," and "Charity," rose from the island of Malta in the Mediterranean Sea to engage Italian Fighters escorting SM-79 Tri-motor Bombers out of Sicily on a raid over the airfield at Hal Far. On that day "Timber" Woods scored his first victory, downing an Italian CR-42 Biplane Fighter above the capital city of Valletta. George Burges got his first Italian SM-79 Bomber on June 22, 1940, while flying a Gladiator over Kalafrana on Malta. During October-December of 1940 Marmaduke Pattle flew Galdiators with Squadron-80 in Greece to score 24 air victories and help drive Italian ground forces back across the border into Albania.

Production totaled 768 Gladiators for the RAF.

HAWKER HURRICANE

The British Hawker F.36/34 protoype made its initial flight on November 6, 1935, as the first Monoplane Fighter in RAF history. The production model MK-I Hurricane was delivered in December 1937 to Fighter Squadron-111 at Northolt. One version of the Hurricane was launched by catapult from merchant ships to serve as a Cata-fighter convoy defender. The Sea Hurricane was a shipboard Fighter, the Hurribomber was a Fighter-Bomber, the Night Hurricane was equipped with radar and one variant was fitted with skis for operations in the Arctic. The MK-V Hurricane was the last version made in Britain. The MK-X, MK-XI and MK-XII Hurricanes were built in Canada, differing from the British models only in the substitution of Packard engines for the Rolls-Royce power plants.

The Hurricane was a low-wing monoplane with standard tail section, conventional 3-wheel retractable undercarriage and enclosed cockpit at the wing's trailing edge. A single Rolls-Royce Merlin 1,620-hp engine drove a 3-bladed airscrew to give it a top speed of 342 mph. The service ceiling reached 35,000 feet and the range was 700 miles. Armament consisted of 8 to 12 Browning .303-inch wing guns. The Nightfighter had four 20mm wing cannon, the Tropical Hurricane had a 40mm cannon underwing for tank-busting, and the Fighter-Bomber carried 1,000 pounds of bombs or 8 rockets on two underwing racks.

P.W. Mould scored the first RAF kill of WWII at the Western Front on October 30, 1939, flying his Hurricane Fighter to shoot down a German Dornier DO-17 Medium Bomber over Toul, France.

Marmaduke "Pat" Pattle, a South African by nationality, flying first in British Gladiators and later in Hurricanes, destroyed 41 enemy aircraft to rank 2nd among Great Britain's WWII Aces. His first kill was an Italian Fighter over Libya on July 24, 1940. His last victory came on April 20, 1941, downing a German BF-109 Fighter in a dogfight near Athens, Greece. During the same air battle he was shot down by a German BF-110 twin-engined Fighter and spiraled flaming into Eleusis Bay. Some historians claim that Pat Pattle should be considered the RAF's—and South Africa's—leading Ace of WWII.

During the Battle of Britain in July-October 1940, Hurricanes were used as Interceptors againt German Bombers and BF-110 Fighters, while the more agile Supermarine Spitfires engaged BF-109 Fighters.

James Nicholson became the only RAF Fighter Pilot to earn England's Victoria Cross, awarded for heroism over Southampton on August 16, 1940, against enemy Fighters in his flaming Hurricane.

Josef Frantisek was Czechoslovakia's leading Ace of WWII with 28 victories scored while serving under the flags of Czechoslovakia, Poland and Great Britain. Flying Hurricane Fighters during the Battle of Britain, Frantisek scored 17 victories to rank first among Allied Fighter Pilots who participated in that action. He was killed in a crash of his Hurricane on October 8, 1940, just a few days prior to Hitler's halting the raids over Britain, thereby conceding a major setback in his planned invasion of the British Isles.

Frank Carey was officially credited with 28 aerial victories while flying Hurricanes in Burma. Some historians claim that records proving about 20 additional kills were lost in the shuffle. Had those victories been confirmed, Carey could have been the Ace of Aces among American and British Fighter Pilots of World War II. He survived the war and served with the RAF until his retirement.

Stanislaw Skalski, Poland's top Fighter Ace of WWII with 18 air victories, led his Squadron of Hurricanes into action at Dieppe on August 19,1942. When the smoke cleared his volunteers had sent 15 German planes down in flames.

Production totaled about 15,866 Hurricanes, ending in 1944.

SUPERMARINE SPITFIRE

The British Supermarine F.37/34 prototype first flew on March 5, 1936, and the MK-1 Spitfire Fighter reported for duty with the RAF in August 1938. Subsequent Spitfires were built in 40 different versions with an ''A'' wing built to accommodate only machine guns, a ''B'' wing to carry only cannon, or a ''C'' wing that carried a mixture of guns and cannon. The Seafire was a shipboard Fighter or Floatplane, the Spitbomber was a Fighter-Bomber, the Nightspit had a special nose cone for radar, and the Spyspit was an unarmed surveillance plane. Other variants experimented with contraprops, retractable tail wheel, bubble-type canopy, 5-bladed prop and a wider span.

The Spitfire was a low-butterfly-wing monoplane having a standard tail assembly and conventional undercarriage with retractable front wheels. The pilot sat in an enclosed cockpit near the wing's trailing edge. A single Rolls-Royce Merlin 1,470-hp engine drove a three-bladed airscrew to give it a top speed of 370 mph. Service ceiling was 36,000 feet and the range was 470 miles. Armament included 8 30-cal wing guns carrying 300 rounds of ammo, or four guns and two 20mm wing cannon. The Spitbomber carried eight rockets underwing or 1,000 pounds of bombs on three racks underneath. One variant carried four 20mm wing cannon instead of the machine guns.

The first victories of WWII by Spitfires were recorded over the British Isles on October 16, 1939, when an RAF pilot from Fighter Squadron-602 shot down a German HE-111 Bomber at about the same time another Spit from Squadron-603 downed a Junkers JU-88 Bomber.

James ''Johnnie'' Johnson flew Spitfires to destroy 42 enemy aircraft and rank as the British Ace of Aces in WWII. His first kill was a German BF-109 Fighter during a sweep across the English Channel in June 1941. His last victim was a BF-109 on September 27, 1944, over the Rhine near Arnhem, Holland. Johnson survived the war to serve on with the RAF.

George ''Screwball'' Beurling, flying Spits with both the RAF and RCAF, was the top Canadian Ace of WWII. His first victim was a German FW-190 over England in May 1942. Most of his 31 victories were scored over the island of Malta in the Mediterranean Sea during the latter part of 1942. Beurling survived the war only to be killed in a plane crash near Rome, Italy, on May 20, 1948.

Brendan "Paddy" Finucane, flying Spitfires with the RAF, was Ireland's leading Ace of WWII with 32 air victories. He was hit by ground fire while crossing the French coast on July 15, 1942, and crashed into the English Channel.

Clive "Killer" Caldwell was the top ranked Australian Ace of WWII with 29 victories. His first kill was a German BF-109 Fighter while flying an American-built P-40 Tomahawk over Tobruk, Libya. His final air duel was won on June 30, 1943, when he sent a Japanese A6M Zero spinning down over Indonesia while flying an RAAF Spit on an Aussie mission out of Darwin.

Adolph "Sailor" Malan was South Africa's leading Spitfire Ace of WWII with 35 air victories. His first victory came while flying a Spitfire with Fighter Squadron-74 on May 29, 1940, when he downed a German HE-111 Bomber during the evacuation at Dunkirk. He won 16 air duels in the Battle of Britain and scored his last kill in August 1941, a German BF-109 Fighter while flying with the Biggin Hill Wing. Sailor Malan survived the war but died of natural causes in September 1963.

On October 5, 1944, a Spitfire from Squadron-401 was credited with the first RAF victory of WWII over a German Messerschmitt ME-262 Jet Fighter-Bomber.

Production totaled about 22,500 Spitfires, the most-produced Fighter in British history.

DEHAVILLAND MOSQUITO

The British DeHavilland B.1/40 prototype first flew on November 25, 1940, and delivery of the MK-I Mosquito Recon plane began in July 1941. Deliveries of the MK-II Mosquito Fighter began in April 1942, simultaneously with the MK-IV Mosquito Bomber; the MK-VI Mosquito Fighter-Bomber began to appear in February 1943. The MK-III Mosquito was a dual-control Trainer and the Canadian-built MK-XVI Mosquito Photo-Recon plane was adopted by the USAAF as the F-8 Mosquito ("F" designated Photography aircraft for the USAAF in WWII). Because the Mosquito was contructed of wood in an era of all-metal airplanes, it was often referred to as the "Wooden Wonder."

The MK-II Mosquito was a high-wing monoplane with standard tail section, side-by-side seats in the 2-man cockpit atop the nose cone, and a conventional 3-wheel retractable undercarriage. Two Rolls-Royce 1,390-hp engines drove 3-bladed airscrews to give it a top speed of 380 mph. The service ceiling reached 34,500 feet and the range with a normal load was 1,370 miles. Armament included four .303-inch nose guns carrying 500 rounds of ammo each and four 20mm chin cannon with 150 shells per gun—one variant carried a 57mm chin cannon. The Fighter-Bomber version carried 1,000 pounds of bombs in a belly-bay plus 1,000 pounds of bombs or eight 60-pound rockets on two underwing racks. The Bomber variant could carry a 4,000-pound Blockbuster bomb.

The first Mosquito to enter combat was the MK-I Photo-Recon version that flew a surveillance mission over Bordeaux, France, on September 19, 1941, easily evading German BF-109 chasers on the way back home.

The MK-II Nightfighter with AI/MK-V radar entered combat in April 1942 and shot down its first Flying Bomb with Squadron-151 in June 1942. Night-Mosquitos of Bomber Support Group-100 flew as Escorts over Europe to engage German BF-110 and FW-190 Nightfighters over their own territory. Crews of Group-100 destroyed 267 enemy planes and lost only 69 Mosquitos while flying Night Escort missions with British Wellingtons and Lancasters. Home-defense Squadrons of MK-II Fighters claimed 623 German V-1 Buzz Bombs shot down over the British Isles during WWII.

Svien Heglund, top Norwegian Ace of WWII with 16 air victories, got four of his kills while flying a MK-II Mosquito Fighter.

The MK-IV Mosquito Bomber was able to outperform many of the Luftwaffe's finest Fighters. Called upon to fly commando-like missions on quick-hitting strikes, the MK-IV could dart in, drop its bombs and head for home before the enemy knew what had hit him, relying for security on speed and agility in lieu of defensive armament and heavy protective armor. The MK-IV Bomber made its combat debut with Bomber Squadron-105 on May 31, 1942, flying in a low-level daylight raid on Cologne. On September 22, 1942, MK-IV's made a nuisance raid on the Gestapo Headquarters in Oslo, Norway, over 600 miles across the North Sea. That was the first of many bombings designed to destroy detailed records of underground resistance groups maintained by the Gestapo. On April 11, 1944, six Mosquitos of Squadron-613 led by Bob Bateson took off from Lasham and demolished the Gestapo Headquarters in The Hague, capital city of Holland. The first "Mosquito Bite" felt by Berlin came on January 30, 1943, during a celebration of Hitler's tenth anniversary in power. Three MK-IV's of Squadron-105 led by Bob Reynolds out of Marham airfield made a split-second arrival over the capital city, timed perfectly to bomb the Sportspalast where Hermann Goering was making a propaganda speech assuring the populace that they were safe from such bombings. The first Mosquito drop of a 4,000-pound Blockbuster bomb was made at Duisburg on November 29, 1944, and the last Mosquito raid of WWII was conducted at Kiel on May 2, 1945.

MK-VI Fighter-Bombers joined the battle at Malta in May 1943 and reported for duty with the 2nd Tactical Air Force, England, in October 1943, attacking Axis ships and targets over Europe. On February 18, 1944, Mosquitos led by Percy Pickard of Fighter-Bomber Wing-140 blasted a hole in the jailhouse walls at Amiens, freeing 258 members of the French underground to resume sabotage work against the Germans. In August 1944 a MK-VI Mosquito had the dubious distinction of being the first Allied aircraft attacked by a new German Messerschmitt ME-262 Jet Fighter-Bomber MK-VI's fitted with 57mm cannon sent ten German U-boats to the bottom in WWII. Their first sinking occurred on April 9, 1945, off the coast of Norway. MK-VI Mosquitos were instrumental in the final crushing of German forces holding out in the Po Valley of Northern Italy.

Production totaled about 7,781 Mosquitos ending in 1950.

HAWKER TEMPEST

The British Hawker F.10/41 prototype made its first flight on September 2, 1942, and deliveries of the MK-V Tempest Fighter-Bomber began in April 1944.

The Tempest was a low-wing monoplane with standard tail section, conventional retractable landing gear, chin-type radiator scoop and a one-man cockpit at the wing's trailing edge. A Napier-Sabre engine of 2,180-hp drove a 4-bladed prop to give it a top speed of 427 mph. Armament consisted of four 20mm cannon mounted in the wings and carrying 800 rounds of ammo. A bomb load of 2,000 pounds was slung on racks underneath.

MK-V Tempests entered combat with the home defense forces in May 1944 and proceeded to destroy 638 German V-1 and V-2 Buzz Bombs before Hitler's launching pads were knocked out of commission. The leading Buzz Bomb killer of WWII was Joe Berry, who was credited with shooting down 60 of them over England and the English

Channel. On June 8, 1944, Tempests scored their first victories over the Continent in support of Allied landings in Normandy on D-Day, shooting down three German BF-109G Gustav Fighters. When Messerschmitt ME-262 Jet Fighter-Bombers began to appear in quantity at the Western Front, Tempests met them head-on in combat. Between the time of their first Jet-kill in October 1944 and when the Jets ran out of fuel in April 1945, Tempest Fighters were credited with destroying 20 of the Luftwaffe's finest Jet Fighter design.

Pierre Closterman, flying first in Supermarine Spitfires and later in Hawker Tempests with the RAF, scored 33 air victories along the Western Front to become Free France's leading Ace of WWII..

Production totaled about 1,405 Hawker Tempests.

AVRO LANCASTER

The British Avro MK-I Lancaster was delivered in 1941 as the RAF's second Heavy Bomber design, following close on the heels of the Short MK-I Stirling 4-engined Bomber, giving England a weapon for the start of a night bombing blitz against German targets in Europe.

The Lancaster was a high-wing monoplane having a twin-finned tail section and conventional undercarriage with retractable front wheels. Four Merlin 1,640-hp engines gave it a top speed of 287 mph. Turrets in the nose, belly, tail and back were armed with 8 to 10 Browning .303 machine guns. A maximum bomb load of 14,000 pounds could be carried on a range of 1,660 miles, but a later version was designed to carry the 22,000 pound Grand Slam bomb.

During the night of May 30-31, 1942, Germany's 4th largest city, Cologne, was attacked by 1,046 British Bombers in the first thousand-Bomber raid of WWII. Heavy Stirlings were first over the target, and Lancasters brought up the rear, damaging over half of the city. On May 16-17, 1943, Lancasters of Bombing Squadron-617, led by Guy Gibson out of Scrampton, attacked three German Dams—Moehne, Eder and Sorpe—with Bouncing bombs, causing severe flooding in the surrounding area. Of nineteen Bombers on that mission, nine failed to return home. For valor on that raid Guy Gibson was awarded the Victoria Cross. On the night of July 24-25, 1943, 347 Lancasters joined the force of Bombers that hit Hamburg with bombs and incendiaries for ten days, causing firestorms that destroyed the city and killed 50,000 of its inhabitants. Radar-distorting slivers called "Window" (chaff) were used for the first time on that raid.

Production totaled about 7,366 British and Canadian-built Lancasters.

MIKOYAN-GUREVICH MIG-3—YAKOVLEV YAK FIGHTERS

The Russian Mikoyan-Gurevich MIG-3 was the best Soviet Fighter available at the outset of WWII. However, experience in combat along the Eastern Front after June 1941 soon revealed it to be vastly inferior to the opposing German combat planes. Flown by many of Russia's famous Aces while awaiting the arrival of modernized Fighters, the MIG-3 carried two machine guns and one cannon at a top speed of 365 mph. Later versions carried six rockets on two attachment points under the wings.

Alexander Pokryshkin got his first kill of WWII in a MIG-3 on June 23, 1941, downing a German BF-109 Fighter near Jassy (Iasi) along the Rumanian border.

Production totaled about 3,350 MIG-3 Fighters, ending in April 1942.

The Russian Yakovlev YAK-1 prototype first flew on January 1, 1940, and the production model YAK-1 Fighter went into battle on June 22, 1941, when Adolf Hitler ordered the massive blitzkrieg along the 2,000-mile Russian border from the Baltic Sea up north to the Black Sea at the south.

Stepan Suprun, American-born member of the Russian Air Force, was proclaimed Hero of the Soviet Union for his work during operational trials of the YAK-1 Fighter. Suprun was killed in action on July 4, 1941, while flying a MIG-3 Fighter out of Borisov. For bravery in that action he became the first two-time winner of the Hero award, albeit posthumously.

Mihail Baranov scored all 27 of his kills in a YAK-1 Fighter bearing the inscription, "Death to Fascists," on the fuselage.

Lilya Litvak, the Soviet—and the world's—female Ace of Aces for all times with 12 air victories against male German Fighter pilots, was shot down and killed in a YAK-1 Fighter on August 1, 1943, while flying a sortie with the 73rd Air Regiment.

The improved YAK-7 Fighter-Trainer was rushed into production and entered combat in January 1942 as a hurry-up replacement for obsolescent MIG-3 and YAK-1 Fighters. But in dogfights the YAK-7 was still no match for the German Messerschmitt BF-109 Fighter.

The YAK-7 was a low-wing monoplane with standard tail unit and retractable 3-wheeled landing gear. A single Klimov 1,210-hp engine produced a top speed of 370 mph operating within a range of 520 miles. Armament consisted of two 12.7mm wing guns and one 20mm nose cannon. The Fighter version had a one-seat cockpit and the Trainer variant had two tandem seats.

The all-woman 586th Fighter Air Regiment was equipped with YAK-7 Fighters in September 1942 and entered combat at the Battle of Stalingrad (Volgograd). These female Fighter pilots were credited with shooting down 38 German planes in defense of that vitally strategic city Premier Stalin had ordered "held at all cost." The Commanding General at Stalingrad proclaimed, "Not one more step backwards!"

Katia Budanova ranked 2nd among the Russian female Aces of WWII with eleven victories on 66 combat missions in YAK-7 Fighters during the savage blood-spilling at Stalingrad.

The advanced YAK-9 Fighter-Bomber first appeared in October 1942 to bolster Russian airpower at Stalingrad. The YAK-9 carried two 12.7mm machine guns, a 37mm cannon and 440 pounds of bombs at a top speed of 410 mph.

YAK-9 Fighters of the Red Army Air Force participated in strikes against the Japanese Kwantung Army along the border of the region known as Manchukuo (Manchuria) when Stalin declared war against Japan on August 8, 1945, just one week before the cessation of hostilities.

The lighter YAK-3 Interceptor delivered in February 1943 had a top speed of 440 mph and was the fastest Soviet Fighter in WWII.

After the fall of France in June 1940, Frenchmen were divided into two factions. The "Free French" supported Charles DeGaulle and aligned themselves with the Allies. The "Vichy French" accepted Philippe Pétain as their leader and adopted a policy of partial collaboration with the Axis forces.

In December 1942 a group of Free French legionaires formed the ''Normandy Squadron'' and united with Russian forces in May 1943 at the Battle of Smolensk. Flying mostly in YAK-3 Fighters riding the skyways to Berlin, the French flyers shot down 276 German combat planes to help crush Hitler's Third Reich.

Marcel Albert was the top French Ace with the Normandy Squadron, shooting down 23 enemy planes while flying YAK-3 Fighters on the way to being decorated as a Hero of the Soviet Union.

Major K.A. Yevstignev scored 56 air victories in WWII to rank in a 3rd place tie with Grigori Retchkalov on Russia's list of Aces. Yevstignev flew YAK-3 Fighter-Interceptors against the Luftwaffe during the counter-offensive in Rumania that began on August 20, 1944, three years and two months after German stormtroopers had swarmed across the border headed in the opposite direction.

Grigori Retchkalov scored 44 of his 56 victories in American-built Bell Airacobra Pursuits for his over-all 3rd place tie in the rankings. He was also ranked just behind Alexander Pokryshkin as the number two P-39 pilot of WWII.

Nikolai Gulayev was the 5th ranking Soviet Ace of WWII with 53 air victories. His 36 kills in American P-39 Pursuits made him the third best Airacobra pilot of the war.

When the Korean war began on June 25, 1950, Russian-built YAK-9 Fighters of the North Korean army attacked Kimpo airfield near the capital city of Seoul, South Korea. On June 26, 1950, a YAK-9 Fighter fell to the guns of an American F-82 Twin Mustang Fighter flown by Bill Hudson. That was the first air victory recorded by a USAF Fighter pilot since the Air Force became a separate arm of the American military services.

Production totaled about 30,000 Yakovlev YAK Fighters.

LAVOCHKIN LAGG-3, LA-5, LA-7, LA-9

The Russian Lavochkin-Gorbunov-Gudkov LAGG-1 prototype first flew on March 30, 1940, and the advanced LAGG-3 Fighter was delivered in March 1941. LAGG-3 Fighters went into combat during the initial Wehrmacht assault along the Eastern Front on June 22, 1941.

Konstantin Gruzdev flew a LAGG-3 Fighter to score his first kill against the Luftwaffe on June 23, 1941. Within four months he had recorded 19 air victories, an amazing feat considering the superiority of German planes at the outset of hostilities.

The much-improved Lavochkin LA-5 Fighter began to roll off production lines in July 1942, destined to become the most successful Soviet Fighter of WWII. The LA-5 made its combat debut when German forces started their initial assault at Stalingrad on August 22, 1942. Variants of the LA-5 included an Interceptor, Nightfighter, Fighter-Bomber and Attack-Bomber.

The LA-5 was a low-wing monoplane with enclosed cockpit, standard tail section and retractable 3-wheeled landing gear. A single Shvetsov 1,650-hp engine gave it a top speed of 390 mph with a range of 450 miles. Armament consisted of two 20mm cannon carrying 200 rounds of ammo each. A bomb load of 440 pounds or six rockets could be slung on four racks underneath.

Ivan Kozhedub, three-time winner of the Russian Hero award, was the Soviet's—and the Allies—leading Ace of WWII with 62 air victories. Flying an LA-5 Fighter on July 6, 1943, he shot down his first enemy aircraft, a German Junkers JU-87 Stuka Divebomber during the nightmare at Kursk. Kozhedub switched to the later model LA-7 Fighter in July 1944 and won his first duel with a Messerschmitt ME-262 Jet Fighter on February 24, 1945. His last kill came on April 17, 1945, when he sent a German Focke-Wulf FW-190 down to earth midst the pile of rubble that once was Berlin. Ivan Kozhedub survived the war and in 1950 at the age of 30 was promoted to the rank of Major General, becoming the youngest General Officer in Soviet Air Force history.

The later model LA-7 Fighter-Interceptor in May 1943 had a top speed of 420 mph and was armed with 3 fuselage cannon plus six rockets on two racks under the wings. The LA-7R in 1945 was a rocket-boosted experiment with a top speed of 460 mph. The LA-7 was carried over into the postwar Soviet Air Force and saw service against United Nations forces during the Korean conflict.

Alexander Pokryshkin swapped his American-built P-39 Airacobra Pursuit for the new LA-7 Fighter in August 1944. Pokryshkin had made his first kill of WWII in a MIG-3 on June 23, 1941, and later switched to Airacobras, in which he scored 48 kills to rank as the Soviet's leading P-39 pilot. His last few air victories were won in LA-7 Fighters, and his total of 59 kills ranked 2nd on Russia's list of WWII Aces. Alex Pokryshkin was the only Soviet Fighter pilot beside Ivan Kozhedub to be a 3-time recipient of Russia's highest award, Hero of the Soviet Union.

On June 27, 1950, just two days after the Korean conflict began, two North Korean LA-7 Fighters were shot down by American F-82 Twin Mustang Fighters over the Capital city of Seoul.

The larger LA-9 in 1945 carried four cannon and had a top combat speed of 430 mph. It was too late for WWII, but continued in service to be flown by the North Korean Air Force during the war in Korea.

Production totaled about 15,200 Lavochkin Fighters.

ILYUSHIN IL-2—PETLYAKOV PE-2

The Russian Ilyushin IL-2 prototype first flew in 1940 and the one-seated IL-2 Attack-Bomber was delivered in the spring of 1941, but was slow and defenseless against enemy Fighter attacks. In September 1942 the IL-2 was modified into a two-seater carrying a defensive gunner in the back seat to protect against attacks from the rear.

The IL-2 was a mid-wing monoplane with standard tail and conventional 3-wheeled landing gear having retractable front legs. The pilot sat under a sliding canopy in the front seat; the gunner rode in an open back seat. A single Mikulin 1,600-hp engine gave it a top speed of only 290 mph on a combat range of 475 miles. Armament included two 7.62mm wing guns, two 20mm wing cannon and a 12.7mm swing gun handled by the rear gunner. Racks underneath could carry 880 to 1,320 pounds of bombs, or up to 8 rockets.

The one-seated IL-2 entered combat in June 1941, striking at German stormtroopers immediately after the outbreak of hostilities along the Eastern Front. When the Battle of Moscow began on September 30, 1941, single-seated IL-2 Attack planes hit German motorized columns along the line, destroying tanks, trucks, armored carriers and gasoline cars.

Colonel P.S. Vinogradov was awarded the Order of Lenin for shooting down two German BF-109 Fighters along the Volga on September 7, 1942, while flying his one-seated IL-2 ''Black Death,'' a name bestowed by German infantrymen who dreaded to hear the drone of its engine.

The two-seated IL-2 Attack-Bomber flew its first combat mission on October 30, 1942, attacking trains and convoys behind the front line, and serving as a tank-buster up and down the battle zone.

Lt. G.M. Parshin was honored as Hero of the Soviet Union for heroism while leading IL-2 Attack-Bombers against German artillery emplacements during the Soviet offensive at Leningrad that began on January 14, 1945. Parshin was generally recognized as Russia's leading IL-2 pilot of WWII.

Production totaled about 36,150 Ilyushin Attack-Bombers, the most-produced combat plane in Soviet history.

The Russian Petlyakov PE-2 Attack-Divebomber-Recon plane of 1941 served alongside the IL-2 from the outset of WWII until the last shot was fired at Berlin. The PE-2 was a twin-engine, twin-tailfin monoplane manned by a crew of two to four men. It carried five machine guns and up to 2,600 pounds of bombs at a maximum speed of 335 mph.

Ivan Polbin, commander of the 150th Bomber Air Regiment, led his PE-2 Attack planes into battle at the outskirts of Stalingrad on July 28, 1942, destroying 90 German vehicles within four days. During the Battle of Kursk in October 1943 he shot down two German JU-87 Stukas in a rare dogfight between opposing Divebombers. Polbin was killed on February 11, 1945, during a raid in his PE-2 at the central prong of the final Russian counter-offensive on the road to Berlin. Ivan Polbin was praised as the Soviet's greatest PE-2 pilot of all times.

Vladimir Petlyakov was killed in the Autumn of 1942 while on a flight in one of the PE-2 Divebombers he had orginally designed.

Maria Dolina, one of the best female Divebomber pilots of the 125th Bomber Air Regiment, flew her PE-2 into the Battle of Borisov during June 1944, earning the respect of her comrades for tireless devotion to duty and country.

In January 1949, during the Soviet attempts to disrupt the Berlin Airlift, PE-2's harassed American Cargo Transports by conducting practice bombing runs along the air corridors leading into the airfield at Tempelhof, West Berlin.

Production totaled about 22,500 Petlyakov PE-2 variants for WWII.

UNITED STATES NAVY AND MARINE FIGHTERS OF WWII

GRUMMAN F4F (FM-2) WILDCAT

The American Grumman F4F-3 prototype first flew on February 12, 1939, and delivery of the firm-winged F4F-3 Wildcat Fighter began in December 1940 to the carriers *Wasp* and *Ranger*. The folding-wing F4F-4 Wildcat Fighter was delivered in December 1941 to become the first-line Navy and Marine Fighter in the early days of WWII until Grumman FM-2 Wildcats and F6F Hellcats began to arrive as replacements in 1943. Marine land-based F4F Wildcats were bolstered by F4U Corsairs in 1942.

The F4F-4 was a mid-folding-wing monoplane with standard tail section, conventional landing gear having retractable front wheels and a cockpit centered above the wings. A single Pratt & Whitney R-1830 1,200-hp engine drove a 3-bladed prop to give it a top speed of 318 mph. The service ceiling was 34,900 feet and the range reached 830 miles.

Armament consisted of six 50-cal wing guns carrying 1,440 rounds of ammo. A bomb load of 200 pounds could be carried on two pylons under the wings. The F4F-3 had only four guns and the later FM-2 Wildcat carried six rockets on underwing pods.

The F4F-3 entered combat at Wake Island on December 9, 1941, with Marine Fighting Squadron VMF-211 commanded by Paul Putnam. The men of VMF-211 held out courageously until their last Wildcat was destroyed on December 22, 1941. During the Battle of the Coral Sea on May 7-9, 1942, F4F-3 Wildcats flew from decks of the USS *Yorktown* and *Lexington* to participate in history's first Sea Battle fought entirely by airplanes. The *Lexington* was sunk during that Naval engagement.

Ed O'Hare became the Navy's first Ace of WWII on February 20, 1942, shooting down 5 Japanese G4M1 Bombers near New Britain while flying an F4F-3 Wildcat with Fighter Squadron VF-42 off the carrier *Lexington*. For that act of heroism, O'Hare was awarded the Congressional Medal of Honor. On November 26, 1943, "Butch" O'Hare, disappeared while flying a F6F-3 Hellcat on a night mission off the carrier *Essex*. His record stood at 12 victories.

The F4F-4 entered combat during the Battle of Midway on June 3-7, 1942, flying from the carriers *Yorktown, Enterprise* and *Hornet* to play an important part in the U.S. Navy's first convincing victory of WWII.

John Smith commanded VMF-223, the first Marine F4F-4 Fighter Squadron to land at Guadalcanal on August 20, 1942. During the following two months he flew with the "Cactus Air Force" out of Henderson field to shoot down 19 Japanese planes and rank 6th among WWII Marine Aces. For intrepidity during the air war over Guadalcanal, John Smith was awarded the Congressional Medal of Honor. He survived the war to live in peace.

Joe Foss flew F4F-4 Wildcats with Marine Fighter Squadron VMF-121 out of Henderson Field on Guadalcanal in the Solomon Islands. Foss scored 26 air victories to rank first among the Marine Wildcat Aces. His initial victory came on October 13, 1942, and during the next six weeks he sent 23 Japanese planes down in flames. For that display of heroism and devotion to duty he was awarded the Congressional Medal of Honor. Joe Foss survived the war to enter politics and become Governor of South Dakota.

The Grumman-Eastern Motors FM-1 prototype first flew on August 31, 1942, and the FM-2 Wildcat Fighter was delivered beginning in September 1943. The FM-2 used a modified F4F airframe and had a 1,350-hp engine. The range was 780 miles and the top speed reached 332 mph, fastest of the Wildcats. Armament consisted of four 50-cal machine guns in the wings carrying 1,720 rounds of ammunition. Six rockets could be slung on racks under the wings.

Ken Hippe became a Navy Ace when he scored all 5 of his victories on a single mission while slamming his FM-2 Fighter into a formation of Japanese Bombers over Leyte Gulf in the Philippines on October 24, 1944.

During WWII Marine Fighter pilots shot down about 2,350 Japanese planes and the Corps boasted 120 Aces.

Production totaled 7,815 F4F and FM-2 Wildcats ending in 1945.

CHANCE-VOUGHT F4U CORSAIR

The American Chance-Vought F4U-1 prototype made its initial flight on May 29, 1940, as the first United States combat plane to exceed 400 mph in level flight. The F4U-1 Corsair Fighter was first delivered to Marine Squadron VMF-124 on July 31, 1942, and initial deliveries to Navy Squadron VF-12 began on October 3, 1942. Early Corsairs carried only four wing guns and had upward-folding wings. The FG-1 firm-wing Corsair in February 1943 was built by Goodyear, and the F3A-1 hinged-wing Corsair was built by Brewster beginning in April 1943.

The F4U-1 was a low-bent-wing monoplane with standard tail section, conventional 3-wheeled retractable landing gear and a canopy-covered cockpit above the wing's trailing edge. The single Pratt & Whitney R-2800 2,000-hp engine drove a 3-bladed prop to give it a top speed of 390 mph. The service ceiling reached 36,200 feet and the range extended to 1,070 miles. Armament consisted of six wing-mounted 50-cal machine guns carrying 2,146 rounds of ammo. The F4U-1D could be loaded with 4,000 pounds of bombs under the center section or 8 rockets on underwing pylons. The F4U-1C substituted four 20mm wing cannon for the six guns.

The F4U-2 was a Nightfighter with AN/APS-6 radar equipment and the F4U-3 experimented with a 4-bladed propeller. The F4U-4 carried four wing cannon, eight underwing rockets and had a top speed of 425 mph. The F4U-5 delivered in December 1945 carried 4,000 pounds of bombs and was the fastest Corsair at 470 mph, but it was too late for WWII.

F4U-1 Fighters with Marine Squadron VMF-124 landed at Guadalcanal on February 12, 1943, and flew their first mission the following day, attacking targets at Bougainville. From that day forward Japanese troops referred to the Corsair as "Whistling Death."

Ken Walsh got his first kill on April 1, 1943, while flying his F4U-1 Fighter with VMF-124, downing a Japanese A6M Zeke over Guadalcanal. He became the first Corsair Ace of WWII, scoring for the fifth time on May 13, 1943—another Zeke over the Solomons. Walsh was awarded the Congressional Medal of Honor for shooting down four Japanese A6M Zeros while flying his Corsair as escort for American Liberator Bombers over Kahili on Bougainville. He ranked 4th on the Marine Corps' list of WWII Aces with 21 air victories. Ken Walsh survived the war to serve with the Marines until his retirement.

Greg Boyington was credited with 28 air victories to lead all the Marine Aces in WWII. Six of his kills were achieved as a pilot in American P-40 Pursuits with Claire Chennault's Flying Tigers in the China-Burma theater. His other 22 victories were scored while flying F4U-1 Corsairs up "the slot" between Guadalcanal and Bougainville. "Pappy" Boyington commanded Marine Fighter Squadron VMF-214, beginning with their first mission out of the Russell Islands in September 1943 until January 1944 after they moved up to Vella LaVella in the New Georgia group. Nicknamed the "Black Sheep," Boyington's bunch sent 127 Japanese aircraft down in flames, more than any other Marine Squadron in the Pacific. For action above and beyond the call of duty, Boyington was awarded the Congressional Medal of Honor. He was shot down and captured while leading a raid at Rabaul, New Britain, on January 3, 1944. Greg Boyington survived POW life and returned home after his release.

Bob Hanson was the Marine's 3rd leading Ace of WWII with 25 air victories while flying F4U-1 Fighters with the VMF-215 "Fighting Corsairs" unit in the Solomons. He was awarded the Congressional Medal of Honor and the Navy Cross for valor in action while shooting down 20 Japanese planes during January 14-30, 1944, from his base in the New Georgia Islands. On February 3, 1944, Bob Hanson was killed by enemy ground fire while on a mission out of Vella LaVella.

Don Aldrich flew Corsairs with Squadron VMF-215 in the Solomons to score 20 air victories and rank 5th among the Marine Aces of WWII. He survivied the war but was killed in a 1947 aircraft accident.

Navy F4U-1 Fighters with Squadron VF-17 from the *USS Bunker Hill* entered combat in the New Georgia Islands during September 1943. Pilots of VF-17 shot down an entire 18-plane force of Japanese Bombers attempting to sink the carriers *Essex* and *Bunker Hill* one day in November 1943.

F4U-2 Nightfighter pilots of VMF(N)-532 got their first kills on April 13, 1944, shooting down two Japanese G4M Bombers out of Kwajalein in the Marshall Islands.

John Bolt is distinguished as being the only Marine Corps double-war Ace. He scored six WWII victories in Corsairs with Sqadron VMF-214, then shot down six planes in Korea while serving as a Fighter pilot on strikes against the Communist forces.

During WWII F4U Fighters shot down some 2,140 Japanese planes while losing only 189 Corsairs in dogfights.

Guy Bordelon became the Navy's first Ace and the Navy's only Night Ace of the Korean conflict when he scored his 5th victory on July 17, 1953, while flying a F4U Night-Corsair in defense of Seoul, the capital city of South Korea.

Production totaled 12, 581 F4U, FG and F3A Corsairs, ending in December 1952.

GRUMMAN F6F HELLCAT

The American Grumman F6F-1 swivel-winged prototype first flew on June 27, 1942, and deliveries of the F6F-3 Hellcat Fighter began on January 16, 1943, to Squadron VF-9, replacing Grumman F4F Wilcats on the carrier *Essex* shortly thereafter.

The F6F-3 was a low-wing monoplane with standard tail section, conventional retractable landing gear and a cockpit above the center of the wings. A single Pratt & Whitney R-2800 2,000-hp engine drove a 3-bladed prop to give the aircraft a top speed of 373 mph. The range was 1,090 miles and the service ceiling reached 37,300 feet. Armament consisted of six 50-cal machine guns in the wings carrying 400 rounds of ammo per gun.

The F6F-3N Nightfighter had MIT AN/APS-6 radar equipment. The F6F-5 in April 1944 carried two 20mm cannon, four machine guns and a bomb load of 3,000 pounds on three racks underneath, or six underwing rockets and two rockets under the center section. The F6F-5N Nightfighter had new MIT AN/APS-6 radar with transmitter and detector in a radome that jutted out from the right wingtip. The F6F-6 in July 1944 was the fastest of the Hellcats at 417 mph.

The F6F-3 Fighter entered combat on August 31, 1943, off the aircraft carriers *Yorktown* and *Essex,* attacking targets on Marcus Island. During the "Marianas Turkey Shoot" of June 19, 1944, Hellcats of Task Force-58 shot down 220 Japanese planes; Air Group-15 from the *Essex* got 60 of them and Squadron VF-16 off the *Lexington* accounted for 44. Another 126 Japanese planes shot down by Naval guns added up to 346 destroyed during softening-up operations prior to the American invasion of Saipan.

Alex Vraciu, the Navy's 4th leading Ace of WWII with 19 victories, shot down six Japanese Judy Divebombers on a single mission with Fighting Squadron VF-16 in the Mariana Islands on June 19, 1944.

David McCampbell flew F6F Hellcats as Commander of Air Group-15 operating from the deck of the *USS Essex* to score 34 air victories and lead the Navy Aces in WWII. His first kill was a Japanese A6M Zeke over Tinian in the Mariana Islands on June 11, 1944. McCampbell also holds the all-service record for most victories on a single mission, downing 9 Japanese planes on October 24, 1944, while flying his Hellcat in an air battle off Luzon in the Philippines. For that display of valor he was awarded the Congressional Medal of Honor. David McCampbell survived the war and continued on active duty with the Navy until his retirement.

Cecil Harris was the 2nd ranking Navy Ace of WWII with 24 air victories while flying Grumman Fighters in both the Atlantic and Pacific oceans. He entered combat on November 8, 1942, flying a F4F Wildcat Fighter in support of Operation "Torch," the Allied landings in Northwest Africa. During the Philippines campaign of late 1944, Harris flew F6F Hellcats with Fighting Squadron VF-18. He returned to civilian life when WWII ended, but after being recalled to active duty for the Korean conflict he decided to make the Navy a career.

Gene Valencia ranked 3rd on the Navy's list of WWII Aces with 23 victories. Operating over enemy territory from Okinawa to Tokyo, his 4-plane combat team of Hellcats with Squadron VF-9 shot down 43 Japanese planes during the offensive actions of early 1945.

F6F-3N Night-Hellcats entered combat with Fighting Squadron VF(N)-76 off the *USS Yorktown* and got their first kills on February 16, 1944.

Bill Henry was the Navy's leading Night Ace of WWII, scoring 9 of his 12 victories in a F6F carrier-based Nightfighter.

Bob Baird became the first Marine Night Ace of WWII when he scored 5 kills in a land-based F6F Hellcat Nightfighter.

Tom Reidy scored the last Navy Hellcat victory of WWII on August 15, 1945, shooting down a Japanese Recon-Bomber over Tokyo Bay for his tenth kill of the war.

During WWII carrier-based and land-based F6F Fighters were credited with destroying 5,156 Japanese aircraft for the loss of only 270 Hellcats in dogfights.

F6F-5K Flying Bombs were launched from the carrier Boxer in August 1952 during the Korean war to attack coastal and inland targets north of the 38th Parallel—the Hellcat's career ended with a great big bang.

Production totaled 12,275 F6F Fighters, ending in December 1945.

The Grumman F7F Tigercat was a twin-engine Fighter-Bomber and the F8F Bearcat was an improved single-engine Fighter, but both of those parallel designs were too later for WWII.

JET FIGHTERS AND BOMBERS OF WWII

GLOSTER/WHITTLE E.28/39—GLOSTER METEOR

The British Gloster/Whittle E.28/39 experimental protoype, piloted by Gerry Sayer, made its initial flight on May 15, 1941, as Britain's first Jet-powered aircraft. It was given the name of "Pioneer," but was often referred to affectionately by members of the design team as "Squirt."

The E.28/39 was a low-straight-wing monoplane with standard tail, retractable tricycle landing gear, and a small cockpit above the wing's leading edge. A single Whittle 860-pounds thrust turbojet engine was installed in the fuselage with intake on the nose and an exhaust pipe at the rear. Top speed attained was 466 mph with a more powerful later model engine; the highest altitude reached was 42,000 feet. Design specs called for an 81-gallon fuel tank and four guns carrying 2,000 rounds of ammunition, but the Squirt was never fitted with arms.

Only two E.28/39 prototypes were built. One exploded and the other was used in extensive trials before being preserved at the Science Museum in London, England.

The British Gloster F.9/40 twin-turbojet prototype made its first flight on March 5, 1943, as a progressive design of the single-engine E.28/39 Pioneer Jet aircraft. Delivery of 16 Gloster MK-I Meteor Fighters to Squadron-616 near London began in June 1944. The original nickname of "Thunderbolt" was changed to "Meteor," thus avoiding conflict with the prop-driven American P-47 Thunderbolt Pursuit of WWII fame.

The Meteor was a low-straight-wing monoplane with plus-type tail section, retractable tricycle landing gear and an enclosed cockpit above the wing's leading edge. Two Rolls-Royce 2,000-pounds thrust turbojet engines were embedded in the wings. Top speed was 490 mph, the service ceiling reached 44,000 feet, and the range stretched to 1,340 miles. Firepower was provided by four 20mm cannon mounted two on each side of the nose cone.

MK-I Meteor Interceptors entered combat on July 27, 1944, against German unmanned Flying Bombs that were terrorizing the London area. The first RAF downing of a V-1 Buzz Bomb by a MK-I Meteor was accomplished on August 4, 1944, with T.D. Dean as the pilot.

Late model MK-III Meteors stationed at Brussels, Belgium, departed on April 16, 1945, for their first combat mission to engage German ME-262 Swallow Jet Fighters. But the confrontation failed to materialize, and WWII ended without the Meteor and the Swallow meeting head-to-head in a dogfight.

On November 7, 1945, a special-built Meteor flown by H.J. Wilson set a world speed record of 606 mph, and on September 7, 1946, Captain E.M. Donaldson upped the mark to 616 mph.

Australian Meteors of RAAF Squadron-77 entered combat in Korea, but were outclassed in dogfights by Russian-built, Chinese-owned, North Korean-driven MIG-15 Jet Fighters. Therefore, those Meteors were used mainly in the ground-attack role.

Production totaled 304 Meteors for WWII, but more were built later.

HEINKEL HE-176—MESSERSCHMITT ME-163, ME-263

The German Heinkel HE-176 prototype made its initial flight on June 20, 1939, with Erich Warsitz serving as test pilot for the world's first rocket-powered aircraft. Based upon studies in rocket propulsion by Werner von Braun, the Heinkel Company developed a Walter rocket motor that generated 1,320 pounds of thrust, with which the HE-176 attained a speed of 170 mph. A later, more powerful engine gave it a top speed of 502 mph.

Two Heinkel HE-176 prototypes were proposed, but only one of them was ever built. Following many successful flights, the rocket program was abandoned and the HE-176 never got past the experimental prototype stage.

The German Messerschmitt ME-163A prototype was first flown by Heini Dittmar in August 1941, and the ME-163B Komet Fighter-Interceptor took to the air for trials in August 1943. The Komet became operational in March 1944 with JG-400 commanded by Wolfgang Spate at Wittmundhafen. The ME-163B was the first and only rocket-propelled Interceptor of WWII. The ME-163C would have been a little longer with a wider span, and the ME-163D experimented with a retractable tricycle landing gear.

The ME-163 was a short, stubby, tailless flying-wing having a large tailfin but no tailplane. The two jettisonable front wheels were dropped after takeoff, leaving a single skid under the center of the fuselage for glide-in landings. The enclosed one-seated cockpit was located in the fuselage at the wing's leading edge. A single Walter rocket motor mounted inside the fuselage generated 3,750 pounds of thrust to permit a climb from runway to 30,000 feet in less than three minutes. Top speed attained was 596 mph, the service ceiling reached 39,500 feet, and the powered endurance lasted only 8 to 12 minutes. Armament consisted of two 30mm wing cannon carrying 60 rounds of ammo per gun.

Hanna Reitsch, a pioneer German woman flyer, was one of the test pilots involved in development of the ME-163 during the unpowered glide-phase of the project. Later assigned to trials of V-1 Buzz Bombs, she was badly injured in a crash while gliding a flying missile back in for a landing.

Takeoffs in an ME-163 were made under rocket power, and when airborne the front wheels were jettisoned. After a quick climb to altitude it made intercept passes at enemy aircraft until the fuel supply was exhausted. Then the pilot made a couple of unpowered passes and glided home for a touchdown using the landing skid for a belly flop. Spontaneous explosions caused by fumes in the empty tank destroyed some of the Komets while skidding down the runway.

The ME-163B first appeared in combat on July 28, 1944, with JG-400 out of Leipzig, and recorded its first kill on August 16, 1944, downing an American B-17 Flying Fortress Bomber. A single Komet once destroyed three B-17's on three quick passes through a formation over Politz. Debut of the ME-163B in the skies over Germany created serious problems in strategy for Allied Bomber crews. Fortunately, most of the Komets delivered were temperamental and had frequent malfunctions. During their short tour of combat in WWII, ME-163 Interceptors just about broke even with their opponents, losing 14 Komets while shooting down twelve Allied aircraft in defense of vital production plants near the fatherland. The last combat mission by an ME-163B Komet was flown on May 7, 1945, out of the Base at Husum.

The ME-263A Interceptor was a redesignation of the ME-163D with increased fuel load, sophisticated improvements and a 15-minute powered flight duration. Factories geared up for production, but Germany's supply of petrol ran out before the ME-263 rocket-propelled Fighter could join the war.

Production totaled about 400 ME-163 and ME-263 variants.

HEINKEL HE-280—MESSERSCHMITT ME-262

The German Heinkel HE-280 prototype took to the air on April 5, 1941, as the world's first twin-engined Jet-powered aircraft. Distinguished as the first Jet airplane designed specifically for use as a combat Fighter, the HE-280 also pioneered an ejection seat for the pilot.

The HE-280 was a low straight-wing monoplane with retractable tricycle landing gear, twin-finned tail section and a cockpit positioned above the wing's leading edge. Two Heinkel turbojet engines, mounted underneath the wings, generated 1,650 pounds of thrust each to produce a top speed of 510 mph. Design specifications called for armament to include cannon, rockets and bombs, but the HE-280 was never fitted with weapons.

The HE-280 did not reach the production stage because Hitler was convinced that he could win the war with prop-driven combat planes. When in 1944 that philosophy proved false, the HE-280 was not revived because the Luftwaffe favored Messerschmitt's ME-262 Jet Fighter-Bomber. Therefore, only one HE-280 prototype was built.

The German Messerschmitt ME-262 Jet prototype first flew on July 18, 1942, but production was slowed by high-level disagreements on policy. The operational ME262A-2A Sturmvogel (Stormbird)—insisted upon by Hitler—joined the Luftwaffe for trials during the spring of 1944 as Germany's first production model Jet Fighter-Bomber. The ME-262A-1A Schwalbe (Swallow)—favored by Goering and Galland—was the Fighter-Interceptor version, which didn't begin to show up in quantity with the Luftwaffe Fighter Arm until October 1944.

The ME-262 was a twin-jet low-straight-wing monoplane with retractable tricycle landing gear and a tailplane mounted about one-third the way up the tailfin. Two Junkers JUMO 1,980-pounds thrust turbojet engines were slung under the wings near the fuselage. The pilot sat in an enclosed bubble-type cockpit on the fuselage near the wing's trailing edge. The maximum speed was 542 mph, the operational ceiling reached 36,000 feet and the combat range was 650 miles. Armament consisted of four 30mm nose cannon carrying a total of 370 shells. One Fighter-Bomber carried a 2,200-pound bomb or two 1,100-pounders on hardpoints under the fuselage forward of the wings. Another version could carry 24 rockets on two underwing racks.

One ME-262 Fighter variant carried six cannon; a Recon-Fighter had two nose cannon, two nose cameras and three fuel tanks underneath; one version was a two-seated Trainer; a two-seated Nightfighter was four feet longer; and a rocket-boosted Interceptor—named "Heimatschutzer" or "Home Protector"—could climb to 36,000 feet in less than five minutes.

Walter Nowotny, Austria-Germany's 5th ranked Ace of WWII with 255 kills, commanded the first ME-262 Fighter-Bomber Group with units at Achmer and Hesepe. Stormbirds entered combat on August 11, 1944, conducting strikes against General Patton's American Tank Corps. Nowotny's   Stormbirds   got their first air kill on October

4, 1944, a Boeing B-17 Bomber on a daylight raid. The first ME-262 that fell to American guns was downed on August 28, 1944, by Joe Myers in a P-51 Mustang Pursuit of the 78th Fighter Group, Nowotny had scored his initial victory in a BF-109 Fighter on July 19, 1941, over Estonia at the Russian Front. He was shot down and killed in an ME-262 Fighter-Bomber on November 8, 1944, by American P-51 Mustang Pursuits accompanying daylight Bombers over Bramsche. The German KG-51 Jet Fighter Group, Kommando Edelweis, was grounded in April 1945 because of exhausted fuel supplies. Surviving members were transferred to Salzburg-Maxglam where they surrendered on May 3, 1945.

Adolf Galland, Supreme Commander of the Luftwaffe Fighter Arm since November 1941, was demoted in January 1945 to organize Fighter Squadron JV-44 equipped with ME-262 Fighter-Interceptors. In addition to his 96 air victories in piston-engined Fighters, Galland also became a Jet Ace by destroying 7 Allied planes in ME-262 Swallows. His last kill came on April 26, 1945, an American B-26 Marauder Medium Bomber over the Danube. Galland survived the war and surrendered his JV-44 unit to American ground troops at Salzburg-Maxglam on May 3, 1945.

Heinz Bar was the top Jet Ace of WWII with 16 kills in ME-262 Fighters of Galland's JV-44 Squadron. Bar's 124 victories at the Western Front ranked 2nd behind Joachim Marseille in that arena, and his total of 220 kills on all fronts placed him 8th over-all among the German Aces.

Franz Schall was the 2nd ranking Jet Ace of WWII with 14 kills in his ME-262 Swallow. Erich Rudorffer and Hermann Buchner tied for 3rd place with 12 victories apiece while flying ME-262 Fighters with the JG-7 Group.

On April 10, 1945, ME-262's defending Berlin shot down ten American daylight Bombers, the largest number of kills scored by German Jets on any single day in WWII. One Swallow claimed 58 Allied kills.

Production totaled about 1,433 ME-262 Jets for the war, but only a few hundred of them were used in combat. Approximately 120 ME-262's were destroyed in actual combat by Allied aircraft.

ARADO AR-234—NAKAJIMA J8N1

The German Arado AR-234A prototype flew initially on June 15, 1943, as the world's first twin-turbojet Bomber. The AR-234B Blitz Bomber appeared in combat for the first time in December 1944 during the Allied crossing of the Rhine at Remagen.

The AR-234 was a high-wing monoplane with standard tail section, retractable tricycle landing gear and a one-man plexiglass-covered cockpit in the nose. Two JUMO turbojets, housed in nacelles under the wings, generated 1,960 pounds of thrust each to give it a top speed of 475 mph. Armament consisted of two 20mm cannon. A bomb load of 3,300 pounds was carried in twin belly-bays and on racks underneath the engine nacelles. The AR-234C in October 1944 had four jet engines and a top speed of 595 mph.

Production totaled about 212 Arado AR-234 Bombers, ending in 1945.

The Japanese Nakajima J8N1 Kikka twin-engined Jet prototype made its first flight on August 7, 1945, just one week before Emperor Hirohito's surrender brought an end to hostilities in WWII. The J8N1, Japan's pioneer entry into the age of the Jets, resembled the German Messerschmitt ME-262 Jet Fighter in appearance. The Kikka was powered by two Nakajima 1,050-pounds thrust turbojets mounted underneath the wings to give it a top speed of 425 mph.

Only two J8N1 prototypes were built, and one of them never got off the ground.

HEINKEL HE-178, HE-162

The German Heinkel HE-178 prototype took to the air on August 27, 1939, as the world's first Jet-powered aircraft. The test pilot on that pioneer flight was Erich Warsitz, who became famous as the first man in history to fly both a Rocket-ship (HE-176) and a Jet airplane (HE-178).

The HE-178 was a high-straight-wing monoplane with conventional 3-wheel retractable landing gear and a tail section having a horizontal stabilizer on each side of the fuselage. The single Heinkel 1,300-pounds thrust turbojet engine, designed by Hans von Ohain, was mounted inside the fuselage with a nose-fitted intake and an exhaust tube at the rear. The loaded weight was only 4,400 pounds and the top speed attained was 435 mph.

The HE-178 served as a pioneer in Jet aircraft research, but it never evolved into a production model. Only two HE-178 prototypes were built, and one of them was used as an engine test bed.

The German Heinkel HE-162 prototype made its first flight on December 6, 1944, and the HE-162A Volksjager-Salamander Jet Fighter-Interceptor reported for duty with the phantom JG-84 Fighter Group in March 1945. The Heinkel Company proved they were geared for production by delivering the first HE-162A within six months after the original order was received.

The HE-162 was a high-straight-wing monoplane with down-turned wingtips, enclosed cockpit atop the fuselage front, retractable tricycle landing gear and a twin-finned tail assembly. The single VMW 1,760-pounds thrust turbojet engine was mounted in a tube-like housing on top of the fuselage and exhausted over the tail section between the tailfins. Top speed was 522 mph, the service ceiling reached 39,300 feet, and the range was limited to 410 miles. Firepower was provided by two 20mm cannon, carrying 120 shells per gun, fitted on the front fuselage belly.

One HE-162 prototype experiemented with rocket-assisted takeoffs and another version was fitted with twin ramjets for propulsion. Neither of them succeeded—but only because time ran out.

Organized on paper and Commanded by Heinrich Dahne, Fighter Group JG-1 became operational when the first HE-162A Volksjagers were received on April 14, 1945. Dahne was killed ten days later when his HE-162 crashed at Warnemunde. The HE-162 saw only brief action in combat for a few days before the German storage dumps ran out of fuel.

Production totaled about 116 HE--162 Salamander Fighters before the factories were all closed in 1945 with the approach of the Allied Armies from every direction.

RUSSIAN-BUILT JET FIGHTERS OF THE KOREAN WAR AND THE VIETNAM WAR

YAKOVLEV YAK-15—MIKOYAN—GUREVICH MIG-15

The Russian Yakovlev YAK-15 Fighter made its initial flight on April 24, 1946, as the first Soviet Jet-powered aircraft in history. It was a progressive development, mating WWII vintage YAK-3 Interceptor airframes with captured German JUMO turbojet engines intended for use with Messerschmitt ME-262 Jet Fighters.

The YAK-15 was a mid-straight-wing monoplane with standard tail section, conventional retractable landing gear and a one-seat cockpit above the wing's trailing edge. A single German JUMO 1,875-pounds thrust turbojet engine gave it a top speed of 488 mph. Armament consisted of two 23mm cannon in the nose.

Olga Yamschikova became one of history's first female pilots to fly a Jet Fighter when she took a YAK-15 up for a test hop in April 1947.

Only a few YAK-15 conversions were made for use in technological experiments with other Yakovlev developments.

The Russian Mikoyan-Gurevich MIG-15 prototype made its initial flight on December 30, 1947, and the production model MIG-15 Jet Fighter was delivered to the Soviet Air Force in 1948. Designed as the first swept-wing Jet Fighter in Russian history, the one-seated Fighter version was referred to by NATO officials as "Fagot" and the two-seated Trainer variant was code-named "Midget."

The MIG-15 was a mid-swept monoplane with a plus-type tail section, retractable tricycle landing gear and enclosed cockpit forward of the wing's leading edge. A single Klimov 5,950-pounds thrust turbojet engine was installed in the fuselage with intake through the nose and a tailpipe at the rear. Top speed was 650mph, the service ceiling reached 49,000 feet, and the range was 560 miles. Armament consisted of one 37mm cannon and two 23mm cannon in a triangle under the nose, all carrying a total of 200 shells. A 2,000-pound bomb load or an assortment of rockets and missiles were carried on two racks under the wings.

Developed as a high-altitude Interceptor, the swept-wing MIG-15 was superior to straight-wing American F-80 and F-84 Jet Fighters in the months immediately following the outbreak of war in Korea on June 25, 1950. With arrival of the American F-86 Sabre Jet swept-wing Fighter in "MIG Alley" on the Yalu river during November 1950, the scales of air superiority began to tip in favor of the United Nations forces.

The first MIG-15 destoyed in the air by an American USAF Jet Fighter was downed on November 8, 1950, by Russ Brown in an F-80 Shooting Star over Sinuiju at the Chinese border.

The first MIG-15 destroyed in the air over Korea by an American Navy Fighter was downed on November 9, 1950, by the pilot of a Grumman F9F Panther Fighter in MIG Alley.

Allied Fighters destroyed approximately 815 MIG-15 Fighters during the Korean conflict.

Production totaled about 15,000 MIG-15 Fighters.

The MIG-17 in 1953 was too late for Korea, and the MIG-19 in 1954 was the first Soviet Jet aircraft to exceed the speed of sound in level flight, coinciding almost exactly with the arrival of the American F-100 Super Sabre supersonic Fighter.

MIKOYAN-GUREVICH MIG-21

The Russian Mikoyan-Gurevich MIG-21 prototype made its initial flight in 1955, and appeared in public for the first time during a 1956 Soviet air parade in Moscow. The production model MIG-21 Fighter was delivered to USSR Air Force units in 1959 as a double-sonic Interceptor. The MIG-21 one-seated Fighter-Interceptor is code-named

"Fishbed" by NATO forces, and the two-seated Trainer version is referred to as "Mongol."

The MIG-21 is a mid-delta-wing monoplane with a large tailfin and a horizontal stabilizer on each side of the fuselage rear. The landing gear is tricycle retractable and the cockpit is located in the fuselage above the wing's leading edge. A single Tumansky 9,900-pounds thrust turbojet engine is mounted in the fuselage with intake on the nose and exhaust at the rear. Top speed is 1,320 to 1,380 mph, the service ceiling reaches 57,000 feet, and the range of action is 750 miles; a 130-gallon droptank gives it an extended range of 1,250 miles. Armament includes two 30mm cannon under the fuselage nose and two air-to-air missiles or 38 rockets underwing. A bomb load of 1,100 pounds can be carried on two hardpoints underneath. Rocket boosters assist takeoffs from short airstrips and a drag-chute provides braking action when landing.

In October 1959 a specially built MIG-21 set a new world speed record of 1,484 mph, and in April 1961 raised the altitude mark to 113,892 feet.

On April 26, 1966, a North Vietnamese MIG-21 was shot down by an American F-4 Phantom two-seated Jet Fighter during the first all-supersonic Jet dogfight in history.

Production of the MIG-21 continues today.

BIRTH OF THE AIR FORCE— A CHRONOLOGY

1903
December 17. Orville and Wilbur Wright successfully flew the world's first flying machine (Aeroplane) at Kitty Hawk, N.C.

1907
August 1. The Aeronautical Division, Army Signal Corps, was formed as the Army's first Air unit. B/Gen James Allen was the Chief Signal Officer and Capt. Charles Chandler was appointed Officer in Charge, Aero Division. Cpl Ed Ward and Pfc Joe (AWOL) Barrett completed the three-man Staff. Equipment included no Balloons, no Dirigibles and no Aeroplanes.
November. B/Gen James Allen ordered a powered, lighter-than-air Airship (Dirigible) from Thomas Baldwin.
December 5. The Board of Ordnance and Fortification, War Department, authorized the Signal Corps to advertise for a heavier-than-air Flying Machine.

1908
February 10. An Army contract was signed with the Wright brothers for construction of a Wright Model-A pusher-type heavier-than-air Flying Machine.
April 11. Lt Frank Lahm relieved Capt Charles Chandler as Officer in Charge of the Aeronautical Division.
May 19. Lt Tom Selfridge became the first unofficial Army pilot when he flew Alexander Graham Bell's Aeroplane, the "White Wing."
August 12. The Aero Division received its first Baldwin lighter-than-air powered Airship (Dirigible). Lts Tom Selfridge, Frank Lahm and Ben Foulois were the first Airship pilot trainees.
August 27. The Wright Model-A Flying Machine was delivered to Fort Myer, VA., for tests.
September 3. The Wright Model-A Flying Machine was demonstrated to Army and War Department officials at Fort Myer, VA.
September 7. Trials of the Wright Model-A Flying Machine began at Fort Myer, VA.
September 9. Lt Frank Lahm rode with Orville Wright as the first Army Officer to fly in the Wright Model-A Flying Machine.
September 17. Lt Tom Selfridge was killed while taking a check ride with Orville Wright in the Wright Model-A Flying Machine at Fort Myer, VA. Lt Selfridge was the

Army's—and the world's—first aeroplane fatality. The Model-A was badly damaged and Orville Wright was seriously injured in the accident.

October. While Orville Wright was in the hospital recuperating from his injuries, Wilbur Wright began construction of a new Wright Model-B Flying Machine for the Aero Division.

1909

July 20. The Wright Model-B Flying Machine was delivered for trials to Fort Myer, VA.

July 27-30. Lts Frank Lahm and Ben Foulois made test flights with Orville Wright in the Wright Model-B Flying Machine.

August 2. The Wright Model-B biplane Trainer was accepted by the Army Signal Corps at Fort Myer, VA. It was the Aeronautical Division's first Aeroplane.

September. Ground was cleared for the Army's first Airfield at College Park, MD.

October 26. Lt Fred Humphreys was distinguished as the Army's first official pilot when he flew solo in the Wright Model-B Trainer at College Park, MD. Later that same day, Lt Frank Lahm soloed in the Model-B to become the second Army pilot.

November 3. Navy Lt George Sweet rode as passenger in the Wright Model-B, thus becoming the U.S. Navy's first airborne Officer.

1910

January. Lt Fred Humphreys left the Aero Division and returned to line duty, disappearing from the Aviation scene.

January. Lt Frank Lahm left the Aero Division and went back to line duty; later he would return for flying duties.

February. An Army winter Flying Field began operation at Fort Sam Houston, San Antonio, Texas, with the Wright Model-B as its only Aeroplane and Lt Ben Foulois as its only Officer; he had not yet learned to fly.

March 2. Lt Ben Foulois taught himself to fly in the Wright Model-B Trainer at Fort Sam Houston, Texas, to become the Army's 3rd pilot—at that time he was the Aero Division's only pilot.

March 19. A Wright civilian Flying School began operations at Montgomery, Ala. (later Maxwell Field).

1911

January 21. A Curtiss civilian Flying School opened at North Island, San Diego, Calif. Later it became an Army Pilot Training School.

January 27. Navy Lt Ted Ellyson flew solo in a Curtiss Trainer to become the Navy's first pilot.

February. Lts Paul Beck, John Walker and George Kelly entered the Curtiss civilian Flying School at San Diego, Calif. for training.

February. A Wright Model-8 aeroplane was lent to the Signal Corps temporarily by the Robert Collier Publishing Company.

March 3. Congress authorized $125,000 for expansion of the Aero Division. Chief Signal Officer B/Gen James Allen immediately ordered 3 Wright Model-B and 2 Curtiss Model-D Trainers, all pusher-type biplanes.

March 17. A Curtiss Model-D Trainer was delivered to the Signal Corps as the Aero Division's 2nd aeroplane.

March 25. A new Wright Model-B Trainer was delivered to the Aero Division at Fort Sam Houston, Texas.

April. Lts Paul Beck, John Walker and George Kelly were graduated from the Curtiss Flying School, San Diego, Calif., and reported for duty to Lt Ben Foulois at Fort Sam Houston, Texas.

April. Lts Hap Arnold and Tom Milling entered the Wright civilian Flying School at Simms Station, Dayton, Ohio.

April 11. A Curtiss Model-D Trainer was delivered to Fort Sam Houston, Texas.

April 11. The Army's first Pilot Training School was inaugurated at College Park, Md. Capt Charles Chandler returned to the Aero Division as School Commandant.

April 15. The first of two new Wright Model-B Trainers was delivered to College Park, Md.

May. Lt Roy Kirtland began training at the Wright civilian Flying School, Dayton, Ohio.

May 4. The original Wright Model-B Trainer was retired to the Smithsonian Institute.

May 10. Lt George Kelly was killed in a Curtiss Model-D Trainer at Fort Sam Houston, Texas. He was the first official Army pilot to die in an aeroplane crash.

May. Capt Charles Chandler soloed at College Park, Md.

May 13. Lts Hap Arnold and Tom Milling were graduated from the Wright Flying School at Dayton, Ohio, and reported for duty to Capt Charles Chandler at College Park, Md.

July. Lt Roy Kirtland reported for duty to Capt Charles Chandler at College Park, Md.

July 19. The Navy received its first aeroplane, a Wright Trainer.

September. Col George Scriven became assistant Chief Signal Officer under B/Gen James Allen.

November. An Army Pilot Training School opened at Augusta, Ga., with Lt Hap Arnold as Instructor.

December. At the end of 1911 there were 26 aeroplane pilots in all of America: eight of them were in the Army and three in the Navy. The strength of the Aero Division included 3 gasbag Balloons, 3 Wright Model-B Trainers, two Curtiss Model-D Trainers and 8 pilots: Capt Charles Chandler and Lts Ben Foulois, Paul Beck, John Walker, Hap Arnold, Tom Milling, Roy Kirtland and Frank Lahm.

December. The Navy had a Flying Field at Greenbury Point, Md., with three biplane Trainers and 3 pilots: Lts Ted Ellyson, John Towers and John Rodgers.

1912

January. During 1912 the Signal Corps began expansion of the Aero Division. New pusher biplanes procured included the Wright Model-C, Wright Model-F, Curtiss Model-E and Curtiss Model-F. Also delivered was the Burgess Model-H tractor biplane, the Army's first puller-type Trainer.

February. Lt Frank Lahm organized an Aero Detachment in the Philippine Islands. The strength was one Wright Model-C Trainer, two Officers and 4 enlisted men.

March 11. Lt Frank Lahm inaugurated the Pilot Training School in the Philippines. Lt Moss Love and Cpl Vernon Burge were his first students.

May 22. The U.S. Marines received authorization to form an Aviation unit.

May 27. The rating "Military Aviator" was introduced for pilots of the Aeronautical Division.

May 30. Wilbur Wright died—he was 45 years old.

June 7. Capt Charles Chandler fired a hand-held Lewís machine gun from the right seat of a Wright Model-B Trainer flown alternately by Lts Tom Milling and Roy Kirtland at College Park, Md. That was the Army's first armed aeroplane.

June 14. Cpl Vernon Burge flew solo in the Philippines to become the Army's first enlisted pilot.

July 5. Capt Charles Chandler, Lt Hap Arnold and Lt Tom Milling were the Army's first pilots to be awarded their "Flying Bald Eagles" as "Military Aviators."

September 28. Cpl Frank Scott, mechanic, was killed in the crash of a Wright Model-B Trainer at College Park, Md. He was the first Army enlisted man to die in an aeroplane.

November 5. Lt Hap Arnold demonstrated at Fort Riley, Kan., that the aeroplane could be used as an Observation craft to direct Field Artillery fire.

November 27. A Curtiss Model-F Amphibian was delivered to the Aero Division. It was the Army's first Floatplane.

December. By the end of 1912 the Aeronautical Division had expanded to a strength of 12 aeroplanes, 14 pilots and 39 enlisted men.

1913

January. During 1913 the Army's new aeroplane designs included the Wright Model-HS pusher biplane, Curtiss Model-G tractor biplane and Burgess Model-I Seaplane.

February. A winter flying field was opened at Texas City, Texas.

February 13. Chief Signal Officer B/Gen James Allen retired from the Army.

March. The Army's first combat unit, the 1st Aero Squadron, was formed at North Island, San Diego, Calif. Its strength was 9 airplanes, 9 Officers and 51 enlisted men.

June. An Army Pilot Training School was opened in Hawaii.

July 1. At the beginning of the fiscal year the Aero Division had 15 planes, 23 Officers and 91 enlisted men.

September. B/Gen George Scriven was appointed Chief Signal Officer.

October 20. Lt/Col Sam Reber was named Commander of the Aero Division.

December. The Pilot Training School at College Park, Md., was closed.

1914

January. New planes acquired by the Army in 1914 included the Curtiss Models-J/N tractor biplanes and the Martin Model-T tractor Amphibian.

January. The Army Pilot Training School was opened at North Island, San Diego, Calif., on the site of the original Curtiss civilian Flying School. Oscar Brindley was the first civilian instructor.

January. A Naval Training Station was established at Pensacola, Fla.

February 24. At a meeting of high Army officials, pusher-type aeroplanes were condemned as outmoded, unsafe and worthless for military purposes.

June 24. The Pilot Training School at San Diego, Calif., received its first Curtiss Model-J tractor biplane Trainer. The Model-J was the forerunner of the JN-4D Jenny.

July 18. The Aviation Section, Army Signal Corps, replaced the Aeronautical Division. B/Gen George Scriven was Chief Signal Officer and Lt/Col Sam Reber continued to command the force of 28 planes, most of which had mechanical problems.

July 23. Ben Foulois was promoted to Captain.

August 1. World War I was declared in Europe. The Army Aviation Section had 20 serviceable Trainers and no combat planes.

September 14. Capt Ben Foulois assumed command of the 1st Aero Squadron, San Diego, Calif.. The strength of the Squadron was 8 airplanes, 16 Officers, and 75 enlisted men.

1915

January. New Army planes delivered in 1915 included the Curtiss Model-R tractor biplane, Martin Model-TT tractor triplane Amphibian, Wright-Martin Model-R tractor Amphibian, Lowe-Willard-Fowler Model-V tractor biplane, and Sturtevant Model-S tractor biplane.

October. Lt Carl "Tooey" Spaatz entered the Army Pilot Training School at San Diego, Calif.

October. Capt Billy Mitchell became Deputy Commander of the Aviation Section under Lt/Col Sam Reber. B/Gen George Scriven remained Chief Signal Officer.

November. The 1st Aero Squadron, Capt Ben Foulois commanding, was transferred to Fort Sam Houston, Texas.

1916

January. During 1916 new Army planes included the Curtiss JN-4 Jenny, Curtiss Model-L triplane Amphibian, Curtiss Model-S armed triplane and Sloan-Standard Model-H tractor biplane.

January. The 2nd Aero Squadron moved into the Philippine Islands, thus becoming the first Aviation unit of the Army to serve overseas.

Februrary. Capt Billy Mitchell earned his "Flying Bald Eagle" insignia, paying his own way while learning to fly at the civilian Curtiss Flying School, Newport News, Va.

March 9. Pancho Villa crossed the border from Mexico into the United States and raided Columbus, N.M., killing 17 American civilians.

March 15. Capt Ben Foulois and Lt Carl Spaatz led the 1st Aero Squadron to Columbus, N.M., in support of General John J. "Black/Jack" Pershing's ground forces sent to drive Pancho Villa back across the Mexican border. That was the first warlike action participated in by the Aviation Section. The 1st Aero Squadron had a strength of 11 Officers, 82 enlisted men and eight airplanes, including the Curtiss Model-R, Curtiss Model-JN, Sloan Model-H and Sturtevant Model-S designs.

April 1. Frank Lahm was promoted to Captain.

May. Hap Arnold was promoted to Captain.

May 20. Lt/Col George Squier became Commander of the Aviation Section, succeeding Lt/Col Sam Reber. Capt Billy Mitchell remained as Deputy Commander.

June. Billy Mitchell was promoted to Major and named Commander of the Aviation Section, succeeding Lt/Col George Squier. B/Gen George Scriven remained Chief Signal Officer.

July 1. Sam Reber was promoted to Colonel.

July 1. During the fiscal year new Flying Schools were opened at Chicago, Il., Essington, Pa., Memphis, Tenn. and Mineola, N.Y.

1917

February. Capt Hap Arnold organized and Commanded the 7th Aero Squadron in the Panama Canal Zone.

February 13. B/Gen George Scriven retired from the Army.

February 14. George Squier was promoted to Brigadier General and succeeded George Scriven as Chief Signal Officer, from which position he took an active interest in, and supervised the operations of the Aviation Section.

March 17. Maj Billy Mitchell sailed for Spain as a technical representative to observe military aviation.

March 18. Capt Ben Foulois reported to the Aviation Section as aeronautical consultant to B/Gen George Squier.

April 6. America entered WWI. The Aviation Section had a strength of 55 airplanes, 35 pilots and 1,088 enlisted men spread thinly throughout 7 Aero Squadrons: the 1st, 3rd, 4th and 5th Squadrons were in the States, the 2nd was in the Philippines, the 6th in Hawaii and the 7th in Panama.

April 10. Maj Billy Mitchell arrived in Paris, France, and began making arrangements for an Aviation Section in the war zone.

April 24. Maj Billy Mitchell flew over enemy lines as an observer in a French Reconnaissance plane, thus becoming the first member of the Aviation Section to enter air combat in WWI.

May. Billy Mitchell was promoted to Lieutenant Colonel.

May. Carl Spaatz was promoted to Captain.

May 28. Sgt Eddie Rickenbacker was shipped to France. Shortly thereafter he became a staff car driver for General John J. Pershing's aides.

June 2. The Airplane Division, Army Signal Corps, replaced the Aviation Section. B/Gen George Squier was Chief Signal Officer, and a few days later Maj Ben Foulois was named Commander of the Airplane Division.

June 13. General John J. Pershing arrived in Paris, France, as Commanding General of the American Expeditionary Force (AEF).

June 17. Hap Arnold was promoted to Major.

June 20. Lt/Col Billy Mitchell succeeded Maj Townsend Dodd as Aviation Advisor to General Pershing.

June 26. Army Ground Forces of the 1st Division arrived in France as the first combat unit in the AEF.

July 24. Congress authorized $640 million for expansion of the Airplane Division, Army Signal Corps.

July 25. Ben Foulois was jump-promoted from Major to Brigadier General (temporary).

August. The Curtiss JN-4D Jenny, Standard Model-SJ and Thomas-Morse Model-S tractor biplane Trainers were ordered into mass production. Work was begun on the British-American DeHavilland DH-4 Attack-Bomber.

August. Billy Mitchell, Ray Bolling and Hap Arnold were promoted to Colonel.

August 5. Charles Chandler was promoted to Colonel as Commander of Balloons for the AEF in France.

August 8. Col Billy Mitchell was appointed Commander of Air combat operations for the AEF, and Col Ray Bolling was named Commander of non-combat operations, which included Administration, Supply, Communications and Training.

August. Sgt Eddie Rickenbacker entered the French Flying School at Tours, France, to become an airplane pilot.

August 13. The 1st Aero Squadron, Commanded by Maj Ralph Royce, arrived in France as America's first Air combat Observation unit in WWI.

September 3. Air Service, AEF, was formed with Headquarters at Chaumont, France. B/Gen Bill Kenly, a non-flyer, was appointed Chief, Air Service, AEF. Col Billy Mitchell was designated Commander of Air Combat Operations, Col Ray Bolling headed Supply and Communications, and Maj Tom Milling was placed in charge of Training.

October. Cadet Jimmy Doolittle entered the Pilot Training School at Rockwell Field, Calif.

October 6. Chief Signal Officer George Squier was promoted to Major General.

October 10. Eddie Rickenbacker was graduated from the French flying school at Tours with the rank of 1st Lieutenant.

October 10. An Army Engineering Laboratory was established at McCook Field, Dayton, OH.

November. B/Gen Ben Foulois succeeded Bill Kenly as Chief, Air Service, AEF, Col Billy Mitchell remained Commander of Air Combat Operations.

1918

January 20. Col Billy Mitchell became head of Air Combat Operations for Air Service, 1st Army Corps, with Headquarters at Neufchateau. The 1st Army Corps was Commanded by General Hunter Liggett.

February. Frank Lahm was promoted to Major.

March. Lt. Eddie Rickenbacker reported for air duty with the 94th Aero Squadron, Commanded by Maj Raoul Lufbery.

March 11. Lt Jimmy Doolittle was graudated from the Pilot Training School at Rockwell Field, Calif.

March 19. American Aero Squadron Fighters entered combat for the first time on the Western Front.

March 26. The French General Ferdinand Foch was named Commander in Chief of all Allied forces on the Western Front.

April 14. Lts Doug Campbell and Alan Winslow of the 94th Aero Squadron scored the first Army air victories in Nieuport 28C-1 Fighters at the Western Front.

April 15. The first U.S. Marines Air unit was formed at the Naval Air Station, Miami, Fa.

April 24. The Military Aeronautics Division, Army Signal Corps, briefly replaced the Airplane Division. M/Gen George Squier was Chief Signal Officer and B/Gen Bill Kenly was appointed Commander, Military Aeronautics Division.

April 29. Bill Kenly was promoted to Major General.

May 20. M/Gen George Squier retired from the Army.

May 20. For the first time since the Aero Divsion was born on August 1, 1907, control of Army Air activities was removed from the Signal Corps when President Woodrow Wilson and Congress authorized two new Bureaus: the Bureau of Aircraft Production, directed by Mr. John Ryan, was responsible for procurement of aircraft and parts; the Division of Military Aeronautics, Directed by M/Gen Bill Kenly, was charged with all active duty Army military functions.

May 24. The Bureau of Aircraft Production and the Division of Military Aeronautics were combined under the Air Service by Secretary of War Newton Baker.

May 29. B/Gen Mason Patrick replaced B/Gen Ben Foulois as Chief, Air Service, AEF. B/Gen Ben Foulois was named Commander, Air Service, First Army (yet to be formed). Col Billy Mitchell remained Commander, Air Service, 1st Army Corps, responsible directly to B/Gen Foulois.

June 20. The 1st Air Brigade was formed to conduct Air Combat operations at Chateau-Thierry. Col Billy Mitchell became Commander, Air Service, 1st Air Brigade, with Headquarters at Haute Feuille. Col Tom Milling Commanded the 1st Pursuit Wing at Toul.

July 27. Col Billy Mitchell replaced B/Gen Foulois as Commander, Air Service, First Army, with Headquarters at La Ferte sous Jouarre, choosing Col Tom Milling as his Chief of Staff. Col Mitchell was later to commend Tom Milling as the best Deputy he ever had.

August. Carl Spaatz was promoted to Major.

August 27. Mr. John Ryan was appointed Director of Air Service from his new post as Air Assistant to the Secretary of War. Thus a civilian was officially placed at the head of all Army air activities.

September 12. Col Billy Mitchell directed the start of the Allied air offensive at St. Mihiel that involved 1,481 American, French and British combat aircraft.

September 25. Lt Eddie Rickenbacker was appointed Commander of the 94th Aero Squadron on the same date that he earned the Congressional Medal of Honor.

September 29. Frank Luke was killed. Subsequently he became the first Army pilot to be awarded the Congressional Medal of Honor for actions at St. Mihiel.

October 1. Billy Mitchell was promoted to Brigadier General.

October 14. B/Gen Billy Mitchell was moved up to Chief of Air Service, Group of Armies, AEF. Col Tom Milling was named to succeed B/Gen Mitchell as Commander, Air Service, First Army, AEF.

October 21. B/Gen Mitchell and Col Milling assumed the above two Commands; it took a week for the General Order to arrive.

October 28. Eddie Rickenbacker was promoted to Captain.

November 11. World War I ended. Peak strength of all the Army Air Services reached 8,403 airplanes and 195,023 personnel, including 20,708 Officers.

November 14. B/Gen Billy Mitchell was appointed Chief of Air Service, 3rd Army of Occupation, with Office at Coblenz, Germany.

December 23. M/Gen Charles Menoher was appointed Director of Air Service, succeeding Mr. John Ryan . M/Gen Bill Kenly remained as Director of the Division of Military Aeronautics.

1919

February B/Gen Billy Mitchell came home to a hero's welcome as America's only "Flying General."

March 10. B/Gen Billy Mitchell succeeded M/Gen Bill Kenly as Director of the Division of Military Aeronautics under M/Gen Charles Menoher, Director of Air Service. M/Gen Menoher openly opposed a separate Air Arm for the military services, the very thing Billy Mitchell staunchly advocated.

June 28. The Treaty of Versailles was signed, officially ending World War One.

June 29. M/Gen Mason Patrick came home; the Air Service of the AEF ceased to exist.

July. Ben Foulois reverted to the rank of Major and was assigned as Assistant to the Director of Air Service.

October 31. M/Gen Bill Kenly retired from the Army.

November 30. Col Sam Reber retired from the Army.

1920

May.Maj Ben Foulois returned to Europe on Air Attache duty.

June 1. Strength of the Air Service had dwindled to 9,050 personnel,including only 969 Officers.

June 4. The Army Air Service was reorganized to replace the Air Service. M/Gen Charles Menoher became Chief, Army Air Service. Col Oscar Westover was the Executive Officer. Control of the Air Service was returned to the Army.

June 4. The rating "Airplane Pilot" replaced "Military Aviator," which had been introduced on May 27, 1912.

June. Hap Arnold reverted from Colonel to his permanent rank of Captain.

July 1. Col Charles Chandler was assigned to Army Air Service as Technical Advisor.

July 2. Frank Lahm was promoted to Lieutenant Colonel.

July. Hap Arnold was promoted to Major the second time around.

October 18. Col Charles Chandler retired from the Army.

1921

April 27. B/Gen Billy Mitchell was appointed Deputy Commander of the Army Air Service under M/Gen Charles Menoher.

July 21. Army Martin MB-2 Bombers led by B/Gen Billy Mitchell out of Langley Field, Va., sank the German WWI battleship *Ostfriesland* in a bombing demonstration off the coast of Virginia, thus proving that airplanes could sing large Naval vessels.

August. M/Gen Menoher attemped to discharge B/Gen Mitchell as his Deputy, but the move was rejected by the General Staff.

October 4. M/Gen Charles Menoher resigned as Chief, Army Air Service.

October 4. M/Gen Mason Patrick succeeded M/Gen Menoher as Chief, Army Air Service.

1923

February. Ben Foulois was promoted to Lieutenant Colonel.

1925

March 26. Billy Mitchell reverted to Colonel and was relieved as Deputy Chief of Army Air Service.

March 27. B/Gen James Fechet succeeded Billy Mitchell as Deputy Chief, Army Air Service.

October. M/Gen Mason Patrick was reaffirmed as Chief, Army Air Service.

October 28. Col Billy Mitchell's general court martial began in Washington, DC.

December 17. Col Billy Mitchell was found guilty of "conduct that would bring discredit to the Army" by the general court martial board of officers.

1926

February 1. Billy Mitchell resigned his commission.

March 20. M/Gen Charles Menoher retired from the Army.

July 2. The Army Air Corps replaced the Army Air Service. M/Gen Mason Patrick was appointed Chief, Army Air Corps, and retained B/Gen James Fechet as his Deputy. Frank Lahm was promoted to Brigadier General (temporary) to serve 4 years as head of Air Corps training.

1927

December 12. M/Gen James Fechet was appointed Chief, Army Air Corps, replacing M/Gen Mason Patrick.

December 12. Ben Foulois was jump-promoted to Brigadier General from Lt/Col and appointed Deputy Chief of Army Air Corps.

December 13. M/Gen Mason Patrick retired from the Army.

1929

June. B/Gen Ben Foulois became Chief of Material Division, Wright Field, Ohio.

1930

February 15. Lt Jimmy Doolitte resigned his commission, but remained in the Air Corps Reserve.

July. Frank Lahm reverted from B/Gen to his permanent rank of Lt/Col and was assigned as Commander of the Air Corps Training Center at Kelly Field, Texas.

1931

February. Hap Arnold was promoted to Lieutenant Colonel.

October 7. Frank Lahm received his permanent promotion to Colonel.

December 19. Ben Foulois was promoted to M/Gen and succeeded M/Gen James Fechet as Chief, Army Air Corps.

December 19. M/Gen James Fechet retired from the Army.

1932

July. B/Gen Oscar Westover became Deputy Chief of the Army Air Corps under M/Gen Ben Foulois.

1935

February. Hap Arnold was jump-promoted to B/Gen from Lt/Col.

March 1. General Headquarters (GHQ) Air Force was created parallel to the Army Air Corps for the purpose of controlling Tactical Air operations.

December 22. Oscar Westover was promoted to M/Gen and appointed Chief, Army Air Corps, succeeding M/Gen Ben Foulois.

December 31. M/Gen Ben Foulois retired from the Army.

1936

January. B/Gen Hap Arnold became Deputy Chief, Army Air Corps, under M/Gen Oscar Westover.

February 19. Billy Mitchell died.

1938

September 21. M/Gen Oscar Westover was killed in an airplane crash at Burbank, Calif.
September 29. Hap Arnold was promoted to Major General and appointed Chief, Army Air Corps.

1939

April 3. Congress authorized $300 million for expansion of the Army Air Corps.
June. The GHQ Air Force became a part of the Army Air Corps.
September 1. World War II was declared in Europe.
November. Carl Spaatz was promoted to Colonel.

1940

July 1. Maj Jimmy Doolittle was recalled to active duty with the Army Air Corps. He subsequently rose to the rank of Lt/Gen.
October. Carl Spaatz was promoted to Brigadier General and appointed Deputy Chief, Army Air Corps, under M/Gen Hap Arnold.
December 19. The GHQ Air Force was taken from the Air Corps and placed under control of Field Force Commanders.

1941

June 20. The United States Army Air Forces (USAAF) was established, M/Gen Hap Arnold Commanding. The USAAF was composed of two paralled Commands: the Office of the Chief of Air Corps, and the Air Forces Combat Command.
November 30. B/Gen Frank Lahm retired from the Army.
December. Hap Arnold was promoted to Lieutenant General and remained Commanding General of the USAAF.
December 7. America entered WWII in the Pacific.
December 11. America entered WWII in Europe.

1942

January. Carl Spaatz was promoted to Major General.
January 29. M/Gen Mason Patrick died.
February 22. The USAAF Bomber Command was formed in England, B/Gen Ira Eaker Commanding.
March 2. The USAAF, commanded by Lt/Gen Hap Arnold, became a parallel Command to the Army Ground Forces (AGF) and the Army Service Forces (ASF); the umbilical cord holding the USAAF under control of Ground Forces was broken.
June 6. The War Department authorized light Liaison airplanes for Field Artillery units of the Army Ground Forces—thus a new Air Arm was born under direct control of the Ground Forces. Pilots were not to be supplied by the USAAF, but were to be recruited from within the Ground units.

1943

March. Hap Arnold was promoted to General (4-star).
March. Carl Spaatz was promoted to Lieutenant General.

1944
December 21. Hap Arnold was promoted to General of the Army (5-star).

1945
March 11. Carl Spaatz was promoted to General (4-star).
May 8. WWII ended in Europe.
August 15. Japan surrendered.
September 3. World War II ended.

1946
January. General Carl Spaatz began the transition to replace Hap Arnold as head of the USAAF.
June. General Carl Spaatz assumed full responsibility as Commanding General of the USAAF, succeeding Hap Arnold.
June 30. General of the Army Henry H. ''Hap'' Arnold retired from the USAAF after 39 years of active duty. His career spanned WWI and WWII, during which time he had a hand in practically every major happening that nurtured the air arm of the Army from its infancy to the mightiest air armada in the history of mankind.

1947
January 30. Orville Wright died at the age of 76.
July 18. Congress awarded Billy Mitchell the Congressional Medal of Honor (posthumously) and promoted him to Major General, retroactive to the date of his death—February 19, 1936.
September 18. The United States Air Force (USAF) became an autonomous arm of the military services, completely divorced from the Army.
September 25. General Carl Spaatz was named the first Chief of Staff, USAF.

1948
April. General Hoyt Vandenberg succeeded General Carl Spaatz as Chief of Staff, USAF.

1950
January 15. General of the Army Hap Arnold died in retirement at Sonoma, Calif. Although he was not on active duty in September 1947, he lived to see his lifelong dream come true when the USAF became an independent branch of the military services, parallel to the Army and Navy. America owed a debt of gratitude to pleasantly smiling Hap Arnold.

1953
July. General Nathan Twining succeeded General Hoyt Vandenberg as Chief of Staff, USAF. Subsequently, General Twining became the first USAF Chairman of the Joint Chiefs of Staff (JCS).

1957
July. General Tom White succeeded General Nathan Twining as Chief of Staff, USAF.

1961

July. General Curtis LeMay succeeded General Tom White as Chief of Staff, USAF.

1965-1978

General LeMay was succeeded by General John McConnell in 1965, followed in order by Generals John Ryan in 1969, George Brown in 1973, David Jones in 1974 and Lew Allen in 1978.

AIR FORCE BASES

This section provides an alphabetical listing of many Air Force Bases (AFB) and brief notices of the men in whose honor they were named.

Albrook AFB, Panama. Lt. Frank Albrook died on September 17, 1924, of injuries received in an aircraft accident five weeks earlier at Chanute Field, IL.

Andersen AFB, Guam. B/Gen James Andersen was killed in February 1945 when his aircraft crashed at Kwajalein, Marshall Islands.

Andrews AFB, MD. Lt/Gen Frank Andrews, advocate of the Boeing B-17 Bomber, was killed in a plane crash at Keflavik, Iceland on May 3, 1943.

Bakalar AFB, IN. Lt John Bakalar was killed in action when his Fighter was shot down over France on September 1, 1944.

Barksdale AFB, LA. 2nd Lt Gene Barksdale was killed on August 11, 1926, in the crash of an Observation plane on a test flight out of Wright Field, OH.

Bergstrom AFB, TX. Capt John Bergstrom, non-flying Admin Officer, was killed in action on December 8, 1941, when the Japanese forces bombed Clark Field, Philippines.

Biggs AFB, TX. Lt James Biggs was killed in action on October 27, 1918, when his DH-4 Attack plane crashed at Belrain, France.

Bolling AFB, DC. Col Ray Bolling, Deputy to Billy Mitchell in WWI, was killed in action by sniper fire near the Western Front on March 26, 1918.

Brookley AFB, AL. Capt Wendell Brookley was killed on February 28, 1934, in the crash of a BT-2 Basic Trainer at Bolling Field, DC.

Brooks AFB, TX. Cadet Sydney Brooks was killed on November 13, 1917, when his Trainer crashed at Kelly Field, TX.

Cannon AFB, NM. General John Cannon, WWII combat Commander in the Mediterranean theater, died in retirement at Arcadia, CA, on January 12, 1955.

Carswell AFB, TX. Maj Horace Carswell, Congressional Medal of Honor recipient, was killed in action on October 26, 1944, while conducting a night B-24 Bomber mission against Japanese ships in the South China Sea.

Castle AFB, CA. B/Gen Fred Castle, Congressional Medal of Honor winner, was killed in action on December 24, 1944, during a B-17 Bomber raid at Liege, Belgium.

Chennault AFB, LA. M/Gen Claire Chennault, leader of the Flying Tigers in China, died in retirement on July 27, 1958, at New Orleans, LA.

Clark AFB, Philippines. Maj Harold Clark was killed in a plane crash in the Panama Canal Zone on May 21, 1919.

Connally AFB, TX. Col James Connally was killed in action on May 29, 1945, while flying a B-29 Bomber over Yokohama, Japan.

Davis-Monthan AFB, AZ. 2nd Lt Sam Davis was killed in a crash at Arcadia, FL, on December 28, 1921. 2nd Lt Oscar Monthan was killed in the crash of a Martin Bomber at Oahu, Hawaii, on March 27, 1924.

Dow AFB, ME. 2nd Lt James Dow was killed in a mid-air collision above Queensboro, NY, on June 17, 1940.

Dyess AFB, TX. Lt/Col Bill Dyess survived the Bataan Death March and escaped, only to lose his life in the crash of a P-38 Pursuit at Burbank, CA, on December 23, 1943.

Edwards AFB, CA. Capt Glen Edwards was killed in the crash of a YB-49 Flying Wing Bomber at Muroc Field, CA, on June 5, 1948. Muroc was renamed Edwards on December 8, 1949.

Eglin AFB, FL. Lt/Col Fred Eglin was killed when his plane crashed on January 1, 1937, during a flight out of Headquarters, GHQ Air Force, Langley Field, VA.

Eielson AFB, AK. Col Carl Ben Eielson was killed in the crash of a civilian airliner on November 9, 1929, near Teller, Alaska, while on a rescue mission to the icebound boat, *Nanuk*.

Ellington AFB, TX. 2nd Lt Eric Ellington was killed in the crash of a Trainer at the Curtiss Flying School, San Diego, CA, on November 24, 1913.

Ellsworth AFB, SD. B/Gen Richard Ellsworth, often-decorated WWII combat pilot in the CBI theater, was killed in a crash at Newfoundland on March 18, 1953.

Elmendorf AFB, AK. Capt Hugh Elmendorf was killed in the crash of a YP-25 Pursuit out of Patterson Field, OH, on January 13, 1933.

England AFB, LA. Lt/Col John England was killed on November 17, 1954, when his F-86 Fighter crashed in France.

Ent AFB, CO. M/Gen Uzal Ent, mastermind of the WWII B-24 Bomber raid at Ploesti, Rumania, died in retirement on March 5, 1948, at Fitzsimmons Hospital, CO.

Fairchild AFB, CA. General Muir Fairchild died of natural causes on March 17, 1950, while serving as Deputy Chief of Staff, USAF.

Forbes AFB, KS. Maj Dan Forbes was killed in the crash of a YB-49 Flying Wing Bomber on June 5, 1948, at Muroc Field, CA. That was the same crash in which Capt Glen Edwards was killed.

George AFB, CA. B/Gen Harold H. George, WWI Fighter Ace, was killed in a crash on April 29, 1942, at Darwin, Australia.

Goodfellow AFB, TX. 2nd Lt John Goodfellow was killed in action on September 17, 1918, when his Fighter was shot down near Metz, France.

Griffiss AFB, NY. Lt/Col Townsend Griffiss disappeared during a flight from Russia to Britain on February 15, 1942.

Hamilton AFB, CA. Lt Lloyd Hamilton, WWI Fighter Ace, was killed in action when his Fighter was shot down over Belgium on August 24, 1918.

Harmon AFB, Newfoundland. Capt Ernest Harmon was killed in the crash of an O-25 Observation plane at Stamford, CT, on August 27, 1933.

Hickham AFB, HI. Lt/Col Horace Hickam was killed on November 5, 1934, when his A-12 Attack plane crashed at Fort Crockett, TX.

Hill AFB, UT. Maj Pete Hill was killed in the crash of a XB-17 Bomber prototype at Wright Field, OH, on October 30, 1935.

Holloman AFB, NM. Col George Holloman was killed in the crash of a B-17 Bomber at Formosa on March 19, 1946.

Howard AFB, Panama. Maj Charles Howard was killed in an airplane crash at Bryans Mill, TX, on October 25, 1936.

Hunter AFB, GA. M/Gen Frank Hunter, WWI Fighter Ace and native of Savannah, GA, was one of the few men to have a Base named after him while still alive.

Johnson AFB, Japan. Lt/Col Gerald Johnson, WWII Fighter Ace, was killed in a postwar crash at Tokyo, Japan, on October 7, 1945.

Keesler AFB, MS. 2nd Lt Sam Keesler was killed in action while flying as Observer in a reconnaissance plane at the Western Front on October 9, 1918.

Kelly AFB, TX. 2nd Lt George Kelly, the first Army pilot to die in an aeroplane crash, was killed in a Curtiss Model-D Trainer at Fort Sam Houston, TX, on May 10, 1911.

Kincheloe AFB, MI. Capt Iven Kincheloe, Korean Fighter Ace, was killed in the postwar crash of a F-104 Fighter out of Edwards AFB, CA, on July 26, 1958.

Kindley AFB, Bermuda. Capt Field Kindley was killed on February 1, 1920, in the crash of a British SE-5 Scout while on a training flight at Kelly Field, TX.

Kingsley AFB, OR. 2nd Lt Dave Kingsley, Bombardier, was killed on June 23, 1944, in a B-17 Bomber during a raid at Ploesti, Rumania.

Kirtland AFB, NM. Col Roy Kirtland, pioneer Signal Corps aviator, died of natural causes on May 2, 1941, at Moffett Field, CA.

Lackland AFB, TX. B/Gen Frank Lackland died in retirement at Walter Reed Hospital, DC, on April 27, 1943.

Larson, AFB, WA. Maj Don Larson, WWII Fighter Ace, was killed on August 4, 1944, when his P-47 Pursuit was shot down over Germany.

Laughlin AFB, TX. Lt Jack Laughlin was killed in action on January 29, 1942, while flying a B-17 Bomber over Java.

Loring AFB, ME. Maj Charles Loring, recipient of the Congressional Medal of Honor, was killed in action on November 22, 1952, while flying a F-80 Fighter-Bomber over North Korea.

Lowry AFB, CO. Lt Frank Lowry, Observer, was killed in action on September 26, 1918, on a reconnaissance sortie at the Western Front.

Luke AFB, AZ. 2nd Lt Frank Luke, WWI Fighter Ace and first Military Aviator to be awarded the Congressional Medal of Honor, was killed in action on September 29, 1918, while flying a SPAD-XIII Fighter near St. Mihiel, France.

McChord AFB, WA. Col Bill McChord was killed in the crash of an A-17 Attack plane near Richmond, VA, on August 18, 1937.

McClellan AFB, CA. Maj Hez McClellan was killed in the crash of a PB-2 Pursuit-Biplace on May 25, 1936, at Centerville, OH.

McConnell AFB, KS. Named for two brothers: Capt Fred McConnell was killed on October 25, 1945, when his private biplane crashed at Garden Plain, KS, and 2nd Lt Tom McConnell was killed in action on July 10, 1943, while flying a B-24 Bomber over Bougainville in the Solomon Islands.

McCoy AFB, FL. Col Mike McCoy was killed on October 9, 1957, when his B-47 Jet Bomber exploded over Orlando, FL.

MacDill AFB, FL. Col Les MacDill, WWI Flying School Commandant, was killed in a crash near Washington, DC, on November 8, 1938.

McGuire AFB, NJ. Maj Tommy Mickey McGuire, WWII Fighter Ace and recipient of the Congressional Medal of Honor, was killed in action on Januuy 7, 1945, during a dogfight in his P-38 Pursuit over Negros, Philippines.

Malmstrom AFB, MT. Col Einar Malmstrom was killed on August 21, 1954, in the crash of a T-33 Jet Trainer at Great Falls, MT.

March AFB, CA. 2nd Lt Peyton March died on February 18, 1918, at Fort Worth, TX, of injuries received in a plane crash at Kelly Field, TX.

Mather AFB, CA. 2nd Lt Carl Mather was killed in the crash of a Trainer at Ellington Field, TX, on January 30, 1918.

Maxwell AFB, AL. 2nd Lt Bill Maxwell was killed in a crash near Manilla, Philippines, on August 12, 1920.

Mitchel AFB, NY. Cadet John Mitchel, former Mayor of New York City, was killed on July 6, 1918, when he fell out of a Trainer at Lake Charles, LA.

Moody AFB, GA. Maj George Moody was killed in the crash of an AT-10 Advanced Trainer at Wichita, KS, on May 5, 1941.

Nellis AFB, NV. Lt Bill Nellis was killed in action on December 27, 1944, when his P-47 Pursuit was shot down over Luxembourg.

Norton AFB, CA. Capt Leland Norton was killed in action on May 27, 1944, when his A-20 Attack plane was hit by anti-aircraft fire over France.

Offutt AFB, NE. Lt Jarvis Offutt was killed while ferrying a British Fighter to the Western Front on August 13, 1918.

Olmstead AFB, PA. 2nd Lt Bob Olmstead was killed in a lighter-than-air gasbag over Holland on September 23, 1923, during an International Balloon competition.

Otis AFB, MA. Lt Frank Otis, Air National Guard, was killed on January 11, 1937, in an O-46 Observation plane at Hennepin, IL.

Paine AFB, WA. 2nd Lt Top Paine died in Salt Lake City, UT, on May 1, 1922, from accidental gunshot wounds.

Patrick AFB, FL. M/Gen Mason Patrick, first Chief of the Army Air Corps, died in retirement on January 29, 1942, at Walter Reed Hospital, DC.

Patterson AFB, OH, later Wright-Patterson AFB. Lt Frank Patterson was killed on June 19, 1918, while on a test hop in a DH-4 Attack plane out of McCook Field, OH.

Pease AFB, NH. Capt Harl Pease, winner of the Congressional Medal of Honor, was killed in action on August 7, 1942, when his B-17 Bomber was shot down over New Britain.

Perrin AFB, TX. Lt/Col Elmer Perrin was killed in the crash of a Martin B-26 Bomber on June 21, 1941, at Baltimore, MD.

Pope AFB, NC. Lt Harley Pope was killed in the crash of a JN-4 Jenny Trainer at Fayetteville, NC, on January 7, 1919.

Randolph AFB, TX. Capt Bill Randolph was killed in the crash of an AT-4 Advanced trainer at Gorman, TX, on February 17, 1928.

Reese AFB, TX. Lt Gus Reese was killed in action while flying a Fighter mission over Sardinia on May 14, 1943.

Richards-Gebaur AFB, MO. Lt John Richards was killed in action when his French Fighter was shot down on September 27, 1918, during the Battle of the Argonne. Lt/Col Art Gebaur was killed in action while flying a F-84 Fighter-Bomber during a mission over North Korea on August 29, 1952.

Robins AFB, GA. B/Gen Warner Robins, developer of the Air Corps Supply cataloging system, died on active duty at Randolph Field, TX, on June 16, 1940.

Schilling AFB, KS. Col Dave Schilling, much-decorated WWII Fighter Ace, was killed in an automobile accident in England on August 14, 1956.

Scott AFB, IL. Cpl Frank Scott, first Army enlisted man killed in an aeroplane, died in the crash of a Wright Model-B Trainer at College Park, MD, on September 28, 1912.

Selfridge AFB, MI. Lt Tom Selfridge, first Army Officer to die in an aeroplane crash, was killed in a Wright Model-A Flying Machine on September 17, 1908, at Fort Myer, VA.

Sewart AFB, TN. Maj Allan Sewart was killed in action on November 18, 1942, when his Bomber was shot down over the Solomon Islands.

Shaw AFB, SC. 2nd Lt Ervin Shaw was killed in action on July 9, 1918, during a mission over France in a British Bristol Recon-Fighter.

Stead AFB, NV. 2nd Lt Croston Stead was killed in the crash of his F-51 prop-driven Fighter at Reno, NV, on December 11, 1949.

Tinker AFB, OK. M/Gen Clarence Tinker was killed in action on June 7, 1942, when his B-24 Bomber went down while attacking Japanese naval vessels near Wake Island.

Travis AFB, CA. B/Gen Bob Travis was killed in the crash of a B-29 Bomber at Fairfield-Suisun AFB, CA, on August 6, 1950. Fairfield-Suisun was renamed in his honor on October 1, 1950.

Truax AFB, WI. Lt Tom Truax was killed in the crash of a P-40 Pursuit on November 2, 1941, near San Francisco, CA.

Turner AFB, GA. 2nd Lt Sullins Turner was killed in the crash of his P-36 Pursuit at Langley Field, VA, on May 23, 1940.

Tyndall AFB, FL. Lt Frank Tyndall was killed on July 15, 1930, when his P-1 Pursuit crashed near Mooresville, NC.

Vance AFB, OK. Lt/Col Leon Vance disappeared in an airborne Ambulance on July 26, 1944, after takeoff from Iceland to be evacuated back to the States.

Vandenberg AFB, CA. General Hoyt Vandenberg, former USAF Chief of Staff, died of illness on April 2, 1954, at Washington DC.

Walker AFB, NM. B/Gen Ken Walker, Congressional Medal of Honor recipient, was killed in action on January 5, 1942, when his Bomber was shot down over New Britain.

Webb AFB, TX. Lt Jim Webb was killed in the crash of an F-51 prop-driven Fighter at Hokkaido, Japan, on June 16, 1949.

Westover AFB, MA. M/Gen Oscar Westover, Chief of the Army Air Corps, was killed in a crash at Burbank, CA, on September 21, 1938.

Wheeler AFB, HI. Maj Sheldon Wheeler was killed when his Observation plane crashed in the Hawaiian Islands on July 13, 1921.

Whiteman AFB, MO. 2nd Lt George Whiteman, first pilot to die in WWII, was killed when his Pursuit was destroyed during takeoff from Wheeler Field, Hawaii, on December 7, 1941.

Williams AFB, AZ. Lt Charles Williams was killed in the crash of a Bomber in the Hawaiian Islands on July 6, 1927.

Wright AFB, OH; later became Wright-Patterson AFB. Named in honor of the Wright Brothers, inventors of the aeroplane. Wilbur died at the age of 45 on May 30, 1912, and Orville died at the age of 76 on January 30, 1948. Both expired in Dayton, Ohio.

Wurtsmith AFB, MI. M/Gen Paul Wurtsmith was killed in the crash of a B-25 Bomber on September 13, 1943, at Asheville, NC.

HISTORICAL BRIEF

This section includes all the remaining aircraft not listed anywhere else in this book. The format and order are the same as those used in the Historical Index earlier in this Work.

NUMBER	NAME-VARIANT	DESIGNER	SPAN	LENGTH	YEAR

THE EARLY PLANES

Wright	Kitty Hawk	Wright Brothers	40′4″	21′1″	'03
Pusher Biplane—Tailfirst—Single Wright 50-hp engine					
Model-A	Flying Machine	Wright Brothers	36′4″	28′	'08
Pusher Biplane—Tailfirst—Single Wright 50-hp engine—One built					
Model-F	(Wright-B)	Burgess	38′	30′	'11
Pusher biplane—Single Wright 60-hp engine—One built					
Model-C		Wright	38′	30′	'12
Pusher biplane—Single Wright 60-hp engine—7 built					
Model-E		Curtiss	41′	29′	'12
Pusher biplane—Single Curtiss 60-hp engine—3 built					
Model-F		Curtiss	43′10″	27′10″	'12
Pusher biplane—Seaplane—Curtiss 60-hp engine—3 built					
Model-F		Wright	34′	26′6″	'12
Pusher biplane—Single Daimler 90-hp engine—One built					
Model-J	(Wright-C)	Burgess	38′	30′	'12
Pusher biplane—Single Wright 60-hp engine—One built					
Model-HS		Wright	32′	26′6″	'13
Pusher biplane—Single Wright 60-hp engine—One built					
Model-I		Burgess	39′8″	31′6″	'13
Pusher biplane—Seaplane—Sturtevant 60-hp engine—One built					

NUMBER	NAME-VARIANTS	DESIGNER	SPAN	LENGTH	YEAR
Model-K	(Dunne)	Burgess	46'6"	23'	'13
Pusher biplane—Flying Wings—Canton 200-hp engine—One built					
Model-J		Curtiss	40'2"	26'4"	'14
Puller (Tractor) biplane—Curtiss-OX 80-hp engine—Two built					
Model-N	(JN-3)	Curtiss	53'4"	32'7"	'14
Tractor biplane—Single Curtiss-OX 90-hp engine—19 built					
Model-T		Martin			'14
Tractor biplane—Single Curtiss-OX 120-hp engine—13 built					
Model-D		Thomas	52'9"	29'9"	'15
Tractor biplane—Single Thomas 135-hp engine—Two built					
JN-2	(Model-J)	Curtiss	40'2"	26'4"	'15
Biplane—Single Curtiss-OX 80-hp engine—Ten built					
Model-O	(Model-N)	Curtiss	53'4"	32'7"	'15
Biplane—Single Curtiss-OX 90-hp engine—One built					
Model-R		Curtiss	48'4"	29'	'15
Biplane—Single Curtiss 200-hp engine—65 built					
Model-R		Wright-Martin			'15
Biplane—Single Wright 150-hp engine—14 built					
Model-S		Martin			'15
Biplane—Single Hall-Scott 125-hp engine—14 built					
Model-S		Sturtevant	48'8"	29'	'15
Biplane—Single Sturtevant 150-hp engine—Eleven built					
Model-TT		Martin			'15
Tractor triplane amphibian—Sturtevant 135-hp engine—4 built					
Model-V		LWF			'15
Tractor biplane—Single Sturtevant 200-hp engine—135 built					
Model-D		Gallaudet			'16
Pusher biplane—Twin Dusenberg 150-hp engines—4 built					
Model-H		Sloan-Standard	40'	27'	'16
Tractor biplane—Single Hall-Scott 125-hp engine—12 built					

NUMBER	NAME-VARIANTS	DESIGNER	SPAN	LENGTH	YEAR
JN-5	Twin Jenny	Curtiss	52'9"	29'4"	'16

Tractor biplane—Twin Curtiss—OX 90-hp engines—8 built

Model-L		Curtiss			'16

Tractor triplane—Seaplane—Single engine—4 built

Model-S		Curtiss			'16

Tractor triplane—Single Curtiss-OX 100-hp engine—4 built

Model-F		LWF			'17

Biplane—Single Liberty 400-hp engine—One built

JN-6	Jenny (JN-4H)	Curtiss	43'7"	27'4"	'17

Biplane—Single Wright 150-hp engine—807 built

ATTACK PLANES: POST-WWI

Model-D		Engineering Division	45'11"	30'3"	'18

Biplane—Single Liberty-12 400-hp engine—12 built

Model-G		LWF	41'7"	29'	'18

Biplane—Single Liberty-12 435-hp engine—Two built

BVL-12		Pomilio	48'3"	31'7"	'19

Biplane—Single Liberty-12 400-hp engine—6 built

ATTACK

A-2	(O-2)	Douglas	39'8"	29'6"	'25

Biplane—Single Liberty-12 420-hp engine—One built

A-4	Falcon (A-3)	Curtiss	38'	28'4"	'27

Biplane—Single P & W R-1340 410-hp engine—One built

A-5	(A-3)	Curtiss			—

Biplane—Single engine—Cancelled

A-6	(A-3)	Curtiss			—

Biplane—Single engine—Cancelled

A-7		Fokker-Atlantic	46'9"	31'	'31

Monoplane—Single Curtiss V-1570 600-hp engine—One built

NUMBER	NAME-VARIANTS	DESIGNER	SPAN	LENGTH	YEAR
A-9	(P-24)	Detroit-Lockheed			—

Monoplane—Single engine—Cancelled

| A-10 | Shrike (A-8) | Curtiss | 44′3″ | 32′6″ | '32 |

Monoplane—Single P&W R-1690 625-hp engine—One built

| A-13 | Gamma (A-16) | Northrop | 48′ | 29′2″ | '34 |

Monoplane—Single Wright R-1820 710-hp engine—One built

| A-14 | (A-18) | Curtiss | 59′6″ | 40′4″ | '34 |

Monoplane—Twin Curtiss R-1670 775-hp engines—One built

| A-15 | (B-10) | Martin | | | — |

Monoplane—Twin engines—Cancelled

| A-16 | Gamma (A-13) | Northrop | 48′ | 29′8″ | '35 |

Monoplane—Single P&W R-1830 950-hp engine—One built

| A-21 | | Stearman | 65′ | 53′ | '40 |

Monoplane—Twin P&W R-2180 1,400-hp engines—One built

| A-22 | Maryland (A-23) | Martin | 61′4″ | 46′8″ | '39 |

Monoplane—Twin P&W R-1830 1,050-hp engines—One built

| A-23 | (A-30) | Martin | | | — |

Monoplane—Twin engines—Cancelled

| A-32 | | Brewster | | | '43 |

Monoplane Divebomber—Single P&W R-2800 2,150-hp engine—Two built

| A-34 | (Navy SB2A) | Brewster | | | — |

Single engine—Cancelled

| A-37 | | Hughes | | | — |

Twin engines—Cancelled

| A-38 | Destroyer | Beech | | | '44 |

Twin Wright R-3350 2,200-hp engines—Two built

| A-39 | | Fleetwings | | | — |

Single engine—Cancelled

| A-40 | (Navy SB3C) | Curtiss | | | — |

Single engine—Cancelled

NUMBER	NAME-VARIANTS	DESIGNER	SPAN	LENGTH	YEAR
A-41		Vultee	54'	48'8"	'44
Single P&W R-4360 3,400-hp engine—One built					
A-42	(B-42, B-43)	Douglas	70'6"	53'7"	'44
Twin Allison V-1710 1,800-hp engines—One built					
A-43	Blackhawk (P-87, F-87)	Curtiss	60'	62'	'45
Single Westinghouse J34-WE turbojet engine—Two built					
A-44	(B-53)	Convair			—
Three turbojets—Cancelled					
A-45	(B-51)	Martin	53'1"	85'1"	'46
Three General Electric J47-GE turbojet engines—Two built					

NIGHT BOMBERS

NBS-2		LWF			—
Biplane—Twin engines—Cancelled					
NBS-3		Elias	77'5"	48'5"	'22
Biplane—Twin Liberty—12 425-hp engines—One built					
NBS-4	(NBS-1)	Curtiss	90'2"	46'5"	'23
Biplane—Twin Liberty-12 435-hp engines—Two built					
NBL-2		Martin	98'	48'	'24
Monoplane—Twin Engineering Division W-2779 700-hp engines—1 built					

HEAVY BOMBERS

HB-2		Atlantic			—
Monoplane—Twin engines—Cancelled					
HB-3		Huff-Daland			—
Monoplane—Twin engines—Cancelled					

LIGHT BOMBERS

LB-2		Atlantic-Fokker	72'10"	51'5"	'26
Monoplane—Twin P&W R-1340 410-hp engines—One built					
LB-3	(LB-1)	Huff-Daland-Keystone	67'	45'	'27
Biplane—Twin P&W R-1340 410-hp engines—One built					

NUMBER	NAME-VARIANTS	DESIGNER	SPAN	LENGTH	YEAR
LB-4		Martin			—
All-metal biplane—Twin engines—Cancelled					
LB-7		Keystone	74′8″	48′10″	'28
Biplane—Twin P&W R-1690 525-hp engines—18 built					
LB-8	(LB-7)	Keystone	75′	42′6″	'29
Biplane—Two P&W R-1860 550-hp engines—One built					
LB-9	(LB-7)	Keystone	75′	43′5″	'29
Biplane—Two Wright R-1750 525-hp engines—One built					
LB-10	(LB-6, B-3)	Keystone	74′8″	49′3″	'30
Biplane—Two Wright R-1750 525-hp engines—Became B-3—64 built					
LB-11	(LB-6)	Keystone	75′	43′5″	'29
Biplane—Twin Wright R-1750 525-hp engines—Two built					
LB-12	(LB-7)	Keystone	74′8″	48′10″	'28
Biplane—Twin P&W R-1860 575-hp engines—One built					
LB-13	(B-4, B-6)	Keystone	74′8″	48′10″	'30
Biplane—Twin P&W R-1860 575-hp engines—7 built					
LB-14	(B-5)	Keystone	74′8″	48′10″	'30
Biplane—Twin Wright R-1750 525-hp engines—3 built					

BOMBERS

B-1		Keystone	85′	62′	'27
Biplane—Twin Curtiss V-1570 600-hp engines—One built					
B-4	(LB-13)	Keystone	74′8″	48′10″	'32
Biplane—Twin P&W R-1860 575-hp engines—Redesignation					
B-5	(LB-14)	Keystone	74′8″	48′10″	'32
Biplane—Twin Wright R-1750 525-hp engines—Redesignation					
B-6	(LB-13, B-3)	Keystone	74′8″	48′10″	'32
Biplane—Twin Wright R-1750 525-hp engines—Redesignation					
B-8	(O-27)	Fokker	64′	47′	'30
Monoplane—Twin Curtiss V-1570 600-hp engines—One built					

NUMBER	NAME-VARIANTS	DESIGNER	SPAN	LENGTH	YEAR
B-11	(O-44, OA-5)	Douglas	95'	69'10"	'31

Amphibian—Twin P&W R-1690 550-hp engines—One built

| B-13 | (B-10) | Martin | | | — |

Monoplane—Twin engines—Cancelled

| B-14 | (B-12) | Martin | 70'6" | 45'3" | '33 |

Monoplane—Twin P&W R-1830 950-hp engines—One built

| B-16 | | Martin | | | — |

Monoplane—Six engines—Cancelled

| B-20 | | Boeing | | | — |

Four engines—Cancelled

| B-21 | | North American | 95' | 61'9" | '38 |

Twin P&W R-2180 1,200-hp engines—One built

| B-22 | (B-18A) | Douglas | | | — |

Twin engines—Cancelled

| B-27 | | Martin | | | — |

Twin engines—Cancelled

| B-28 | | North American | 72'6" | 56'6" | '42 |

Twin P&W R-2800 2,000-hp engines—Two built

| B-30 | | Lockheed | | | — |

Four engines—Cancelled

| B-31 | | Douglas | | | — |

Four engines—Cancelled

| B-33 | | Martin | | | — |

Two or four engines—Cancelled

| B-38 | (B-17E) | Boeing | 103'9" | 74' | '41 |

Four Allison V-1710 1,425-hp engines—One built

| B-39 | (B-29) | Boeing | 141'3" | 98'2" | '41 |

Four Allison V-3420 3,000-hp engines—One built

| B-40 | (B-17F) | Boeing | 103'9" | 74'9" | '41 |

Four Wright R-1820 1,200-hp engines—15 built

NUMBER	NAME-VARIANTS	DESIGNER	SPAN	LENGTH	YEAR
B-41	(B-24D)	Consolidated	110'	66'4"	'42

Four P&W R-1830 1,200-hp engines—One built

| B-44 | (B-29) | Boeing | 141'3" | 99' | '42 |

Four P&W R-4360 3,000-hp engines—One built

| B-54 | (B-50) | Boeing | | | — |

Four engines—Cancelled

| B-59 | | Boeing | | | — |

Four turbojets—Cancelled

B-71	No record
B-73	No record
B-74	No record
B-79	No record
B-81	No record
B-82	No record
B-84	No record
B-85	No record
B-86	No record

TRANSPORT

| T-3 | | LWF | | | '22 |

Biplane—Liberty-12 400-hp engine—One built

AMBULANCE

| A-1 | | Cox-Klemin | | | '23 |

Biplane—Liberty-12 400-hp engine—Two built

CARGO TRANSPORT

| C-3 | Trimotor | Ford | 73'11" | 50' | '28 |

Monoplane—Three Wright R-790 235-hp engines—8 built

| C-5 | Trimotor | Fokker | | | '29 |

Monoplane—Three Wright R-975 300-hp engines—One built

| C-7 | Trimotor (C-2) | Fokker | 74'2" | 48'4" | '29 |

Monoplane—Three Wright R-975 330-hp engines—7 built

NUMBER	NAME-VARIANTS	DESIGNER	SPAN	LENGTH	YEAR
C-9	Trimotor (C-3)	Ford	73'11"	50'	'30
Monoplane—Three Wright R-975 300-hp engines—7 converted					
C-10	Robin	Curtiss			'29
Single R-420 engine—One built					
C-11	Fleetster (C-22)	Consolidated			'31
Pratt & Whitney R-1860 500-hp engine—One built					
C-12	Vega (C-101)	Detroit-Lockheed	41'	27'8"	'31
Monoplane—Single P&W R-1340 450-hp engine—One built					
C-13	No record				
C-15	Ambulance (C-14)	Fokker	59'	43'3"	'31
Monoplane—Single Wright R-1820 575-hp engine—One built					
C-16		Fokker			—
Amphibian—Cancelled					
C-17	Speed Vega (C-12)	Lockheed	41'	27'8"	'31
Monoplane—Souped-up engine—One built					
C-18	Monomail	Boeing			—
Monoplane—Single engine—Cancelled					
C-19	Alpha	Northrop			'31
Single P&W R-1340 450-hp engine—3 built					
C-20		Fokker-General			—
Monoplane—Cancelled					
C-22	(C-11)	Consolidated			'31
Single Wright R-1820 525-hp engine—3 built					
C-23	Altair	Lockheed	42'9"	—	'32
Monoplane—Single engine—Two seats—One built					
C-24	Pilgrim-YIC	Fairchild	57'	39'2"	'32
Monoplane—Single Wright R-1820 575-hp engine—4 built					
C-25	(C-23)	Lockheed	42'9"	—	'32
Monoplane—Single engine—Two seats—One built					

NUMBER	NAME-VARIANTS	DESIGNER	SPAN	LENGTH	YEAR
C-28		Sikorsky			'32
P&W R-985 450-hp engine—One built					
C-31		Kreider-Reisner			'34
Monoplane—Single Wright R-1820 575-hp engine—One built					
C-41	(C-39)	Douglas	85'	61'6"	'39
Twin P&W R-1830 1,200-hp engines—Two built					
C-42	(C-39)	Douglas	85'	61'6"	'39
Twin Wright R-1820 1,200-hp engines—One built					
C-44		BFW-Messerschmitt	34'5"	27'2"	'39
For Air Attache in Berlin—Single engine—One built					
C-55	(C-46)	Curtiss	108'	76'4"	'41
Twin Wright R-2600 1,700-hp engines—One built					
C-62		Waco			—
Monoplane—Twin engines—Cancelled					
C-65	Skycar (C-107)	Stout			—
Commercial Transport—Cancelled					
C-76	Caravan	Curtiss			'43
Twin P&W R-1830 1,200-hp engines—16 built					
C-77		Cessna			'42
Commercial Transport—Eleven drafted					
C-79	(JU-52)	Junkers			'42
German Twin-engine transport—One procured					
C-80		Harlow			'42
Commercial Transport—4 drafted					
C-85	Orion-I	Lockheed			'42
Commercial Transport—One drafted					
C-88		Fairchild			'42
Commercial Transport—Two drafted					
C-89		Hamilton			'42
Commercial Transport—One drafted					

NUMBER	NAME-VARIANTS	DESIGNER	SPAN	LENGTH	YEAR
C-90		Luscombe			'42
Commercial Transport—Two drafted					
C-91	Trimotor	Stinson			'42
Commercial Transport—Three engines—One drafted					
C-92		Akron-Funk			'42
Commercial Transport—One drafted					
C-93	Canestoga	Budd			—
All-steel monoplane—Cancelled					
C-94		Cessna			'42
Commercial Transport—3 drafted					
C-98		Boeing			'42
Amphibian transport—4 built					
C-102	Speedster	Rearwin			'42
Commercial Transport—3 drafted					
C-103		Grumman			'42
Biplane—Single seat—2 drafted					
C-104		Lockheed			—
Twin engine—Cancelled					
C-106		Cessna			'43
Twin P&W R-1830 1,200-hp engines—Two built					
C-107	Skycar (C-65)	Stout			'43
Commercial Transport—One drafted					
C-110	(DC-5)	Douglas			'44
Commercial Transport—3 built					
C-112	(C-54)	Douglas	117'6"	100'7"	'46
Original DC-6 prototype—Four engines—One built					
C-113	(C-46G)	Curtiss	108'	76'4"	'45
Two mixed piston and turboprop engines—one built					
C-114	(C-54)	Douglas	117'6"	93'10"	'46
Four Allison V-1710 1,620-hp engines—One built					

NUMBER	NAME-VARIANTS	DESIGNER	SPAN	LENGTH	YEAR
C-115	(C-114)	Douglas			—
Four engines—Cancelled					
C-120	Packplane (C-119)	Fairchild	110′	87′	'52
Detachable pod—Twin P&W R-4360 3,500-hp engines—Two built					
C-127		Boeing			—
Twin turboprops—Cancelled					
C-128	(C-119)	Fairchild			—
Twin engines—Cancelled					
C-129	(C-47F, R4D-8)	Douglas	95′6″	63′9″	'50
Twin P&W R-1830 1,200-hp engines—One built					
C-132		Douglas			—
Four engines—Cancelled					
C-136	No record				
C-138	No record				
C-139	No record				

CARGO GLIDERS

CG-1		Frankfort			—
Cancelled					
CG-2		Frankfort			—
Cancelled					
CG-3		Waco	73′	43′4″	'42
9-seat—Glider Trainer—101 built					
CG-5	(CG-3)	St Louis			'41
9-seat—Glider Trainer—One built					
CG-6		St Louis			—
15-seat—Cancelled					
CG-7	(CG-3)	Bowlus			'41
9-seat—One built					
CG-8	(CG-4)	Bowlus			'41
15-seat—One built					

NUMBER	NAME-VARIANTS	DESIGNER	SPAN	LENGTH	YEAR
CG-9 30-seat—Cancelled		AGA Aviation			—
CG-10 42-seat—Ten built	(G-10)	Laister-Kauffman			'45
CG-11 30-seat—Cancelled		Snead			—
CG-12 30-seat—Cancelled	(CG-11)	Read-York			—
CG-14 32-seat—3 built	(CG-18, G-14)	Chase	95'8″	61'8″	'44
CG-16 42-seat—One built		General			'44
CG-17 27-seat version of C-47—One converted	(C-47)	Douglas	95'6″	63'9″	'44
CG-18 32-seat—All-metal—Became C-122—7 built	(CG-14, G-18)	Chase	95'8″	61'8″	'46
CG-19 30-seat—Cancelled		Douglas			—

TRAINER GLIDERS

NUMBER	NAME-VARIANTS	DESIGNER	SPAN	LENGTH	YEAR
TG-7 Polish sailplane—One bought		Orlick			'42
TG-9 One-seat sailplane—3 built		Briegleb			'42
TG-10 Two-seat—4 built		Wichita			'43
TG-11 German sailplane—One bought		Schemp			'42
TG-12 Two-seat—4 built		Bowlus			'42
TG-13 Two-seat—3 built		Briegleb			'42

NUMBER	NAME-VARIANTS	DESIGNER	SPAN	LENGTH	YEAR
TG-14 One-seat—One bought		Steiglemaier			'42
TG-15 Two-seat—8 built		Franklin			'42
TG-16 One-seat—Two built		ABC			'42
TG-17 One-seat—One built		Franklin			'42
TG-18 One-seat—3 built		Midwest			'42
TG-19 Two-seat—One built		Schweyer			'42
TG-20 One-seat—4 built		Laister-Kauffman			'42
TG-21 One-seat—One built		Notre Dame			'42
TG-22 One-seat—One built		Melrose			'42
TG-23 One-seat—One built		Harper- Corcoran			'42
TG-24 One-seat—One built		Bowlus-Dupont			'42
TG-25 One-seat—One built		Plover			'42
TG-26 One-seat—One built		Universal			'42
TG-27 One-seat—One built		Grunau			'42
TG-28 One-seat—One built		Haller			'42

NUMBER	NAME-VARIANTS	DESIGNER	SPAN	LENGTH	YEAR
TG-29 One-seat—One built		Vollmer-Jensen			'42
TG-30 Two-seat—One built		Smith			'42
TG-31 One-seat—One built		Aero			'42

GYROPLANE

G-2 Single Wright R-975 300-hp engine—One built		Pitcairn			'35

ROTATING WING

R-1 Twin rotors—Single P&W R-985 300-hp engine—Two built		Platte-LePage			'41
R-3 Single Jacobs R-775 225-hp engine—One converted	(G-1)	Kellett	40'	28'10"	'41
R-7 Single engine—Cancelled	(R-6)	Sikorsky			—
R-8 Twin rotors—Single Franklin O-405 240-hp engine—Two built	(R-10)	Kellett			'43
R-9 Single Lycoming O-290 125-hp engine—One built		Firestone			'46
R-10 Twin P&W R-985 450-hp engines—Two built	(R-8)	Kellett			'45
R-11 Continental A-100 220-hp engine—One built		Rotocraft			'46
R-14 Single engine—Cancelled		Firestone			—

CORPS OBSERVATION

CO-1 Monoplane—Liberty-12 400-hp engine—3 built		Engineering Division— Gallaudet			'21

NUMBER	NAME-VARIANTS	DESIGNER	SPAN	LENGTH	YEAR
CO-2		Engineering Division			'22
Biplane—Liberty-12 400-hp engine—One built					
CO-3	(CO-1)	Engineering Division			—
Monoplane—Cancelled					
CO-5	(TP-1)	Engineering Division			'24
Biplane—Two seats—Liberty-12 400-hp engine—One built					
CO-6		Engineering Division			'23
Liberty V-1410 420-hp engine—Two built					
CO-7	(DH-4)	Boeing	42'6"	29'11"	'24
Biplane—Liberty-12 420-hp engine—3 built					
CO-8	(DH-4)	Atlantic	45'	29'11"	'23
Biplane—Liberty-12 420-hp engine—One built					

OBSERVATION

NUMBER	NAME-VARIANTS	DESIGNER	SPAN	LENGTH	YEAR
O-3		Wright			—
Biplane—Cancelled					
O-4		Martin			—
Biplane—Cancelled					
O-7	(O-2)	Douglas	40'	30'	'25
Biplane—Single Packard A-1500 450-hp engine—One built					
O-8	(O-2)	Douglas	40'	30'	'25
Biplane—Single Wright R-1454 400-hp engine—One built					
O-9	(O-2)	Douglas	40'	30'	'25
Biplane—Single Packard A-1500 450-hp engine—One built					
O-10	(OA-1)	Loening	45'	34'10"	'26
Biplane amphibian—Single Wright V-1460 480-hp engine—One built					
O-11	Falcon (O-1)	Curtiss	38'	28'4"	'27
Biplane—Single Liberty V-1650 420-hp engine—66 built					

NUMBER	NAME-VARIANTS	DESIGNER	SPAN	LENGTH	YEAR
O-12	Falcon (O-11)	Curtiss	38'	28'4"	'28

Biplane—Single P&W R-1340 410-hp engine—One converted

| O-13 | Falcon (O-1, O-11) | Curtiss | 38' | 27'2'" | '27 |

Biplane—Single Curtiss V-1570 600-hp engine—6 converted

| O-14 | | Douglas | 30'1" | 25' | '28 |

Biplane—Single Wright R-790 220-hp engine—One built

| O-15 | | Keystone | 37'3" | — | '28 |

Biplane—Wright R-790 235-hp engine—One built

| O-16 | Falcon (O-11) | Curtiss | 38' | 28'4" | '28 |

Biplane—Single Curtiss V-1570 600-hp engine—One built

| O-18 | Falcon (O-11) | Curtiss | 38' | 28'4" | '28 |

Biplane—Single Curtiss H-1640 600-hp engine—One converted

| O-20 | (O-19) | Thomas-Morse | 39'9" | 28'4" | '28 |

Biplane—Single P&W R-1690 525-hp engine—One built

| O-21 | (O-19) | Thomas-Morse | 39'9" | 28'4" | '28 |

Biplane—Single Curtiss H-1640 600-hp engine—One built

| O-22 | (O-2) | Douglas | 40' | 30' | '29 |

Biplane—Single P&W R-1340 410-hp engine—3 built

| O-23 | (O-19) | Thomas-Morse | 40' | 28'10" | '29 |

Biplane—Single Curtiss V-1570 600-hp engine—One built

| O-24 | | Curtiss | | | — |

Biplane—Cancelled

| O-25 | (O-2) | Douglas | 40' | 32' | '30 |

Biplane—Single Curtiss V-1570 600-hp engine—80 built

| O-26 | Falcon (O-1) | Curtiss | 38' | 27'2" | '29 |

Biplane—Single Curtiss V-1570 600-hp engine—One built

| O-28 | (O2U-3) | Vought | | | '29 |

Navy observation plane—One built

| O-29 | (O-2) | Douglas | 40' | 30' | '29 |

Biplane—Single Wright R-1750 525-hp engine—One converted

NUMBER	NAME-VARIANTS	DESIGNER	SPAN	LENGTH	YEAR
O-30		Curtiss			—
Monoplane—Twin engines—Cancelled					
O-33	(O-19)	Thomas-Morse	39'9"	28'4"	'29
Biplane—Single Curtiss V-1570 600-hp engine—One built					
O-34	(O-22)	Douglas	38'1"	30'	'29
Biplane—Single Curtiss V-1570 600-hp engine—One converted					
O-36	(O-35, B-7)	Douglas	65'3"	46'7"	'31
Monoplane—Twin Curtiss V-1570 600-hp engines—One built					
O-37	(O-10)	Keystone			—
Biplane amphibian—Cancelled					
O-38	(O-25)	Douglas	40'	32'	'31
Biplane—Single P&W R1690 525-hp engine—157 built					
O-39	Falcon (O-1)	Curtiss	38'	27'4"	'32
Biplane—Single Curtiss V-1570 600-hp engine—Ten built					
O-40		Curtiss			'32
Monoplane—Two built					
O-41		Thomas-Morse			'30
Biplane—Single Curtiss V-1570 600-hp engine—One built					
O-42		Thomas-Morse			—
Monoplane—Cancelled					
O-43	(O-31)	Douglas	45'8"	33'11"	'33
Monoplane—Single Curtiss V-1570 675-hp engine—25 built					
O-44	(B-11, OA-5)	Douglas	95'	69'10"	'32
Amphibian—Twin P&W R-1690 550-hp engines—One converted					
O-45	(B-10)	Martin	70'6"	44'9"	'33
Monoplane—Twin Wright R-1820 775-hp engines—One converted					
O-46	(O-43)	Douglas	45'8"	33'11"	'35
Monoplane—Single P&W R-1535 725-hp engine—88 built					
O-48	(O-46)	Douglas			—
Monoplane—Cancelled					

NUMBER	NAME-VARIANTS	DESIGNER	SPAN	LENGTH	YEAR
O-50		Bellanca	55'6"	—	'40
Monoplane—Ranger V-770 520-hp engine—3 built					
O-51	Dragonfly	Ryan	52'	—	'40
P&W R-985 450-hp engine—3 built					
O-53	(A-20)	Douglas			—
Monoplane—Twin engines—Cancelled					
O-54	Voyager	Stinson	34'	24'1"	'41
Monoplane—Continental O-170 65-hp engine—6 built					
O-55	Ercoupe	Erco			'41
Continental O-170 65-hp engine—One built					
O-56	(B-37)	Lockheed	65'6"	51'5"	'41
Monoplane—Twin Wright R-2600 1,700-hp engines—18 built					
O-61		AGA			—
Autogyro—Cancelled					

LIAISON

NUMBER	NAME-VARIANTS	DESIGNER	SPAN	LENGTH	YEAR
L-7	Monocoupe	Universal			'42
Single Franklin O-200 100-hp engine—19 built for export					
L-8	Cadet	Interstate	35'6"	23'6"	'42
Single Continental O-170 65-hp engine—9 built for export					
L-10		Ryan			'42
Single Warner R-55 180-hp engine—One drafted					
L-11		Bellanca			'42
Single P&W R-1340 450-hp engine—One drafted					
L-12	Reliant (C-81)	Stinson	41'9"	28'	'44
Single P&W R-985 450-hp engine—4 drafted					
L-14		Piper	35'6"	22'9"	'45
Single Lycoming O-290 130-hp engine—5 built					
L-15		Boeing	40'	25'10"	'47
Single Lycoming O-290 125-hp engine—12 built for Army					
L-16		Aeronca	35'	21'6"	'47
Single Continental O-190 85-hp engine—609 built for Army					

NUMBER	NAME-VARIANTS	DESIGNER	SPAN	LENGTH	YEAR

L-17   Navion   North Amnerican 33′5″   27′8″   '47
Single Continental O-470 225-hp engine—250 built for Army and ANG

OBSERVATION AMPHIBIAN

OA-5   (B-11, O-44)   Douglas   95′   69′10″   '31
Twin Wright R-1820 920-hp engines—One built

OA-6   Consolidated   —
Twin engines—Cancelled

OA-7   (OA-14)   Douglas   —
Monoplane—Twin Engines—Cancelled

OA-11   (OA-8)   Sikorsky   86′   52′   '42
Monoplane—Twin P&W R-1690 750-hp engines—One built

OA-12   (J2F-5)   Grumman   '42
Navy amphibian—7 built

OA-15   See Bee   Republic   —
Commercial amphibian—Cancelled

PHOTOGRAPHY

F-11   Hughes   '46
Twin boom—Twin P&W R-4360 3,500-hp engines—Two built

F-12   Republic   '46
Four P&W R-4360 3,500-hp engines—Two built

PURSUITS—WATERCOOLED

PW-2   Loening   39′9″   24′4″   '20
Monoplane—Single Wright-H 320-hp engine—8 built

PW-3   ORENCO   27′9″   23′10″   '21
Biplane—Single Wright-H 320-hp engine—3 built

PW-4   Gallaudet   29′10″   22′8″   '22
Biplane—Single Packard A-1237 350-hp engine—Two built

PW-5   Fokker   39′5″   27′2″   '22
Single Wright-H 300-hp engine—Ten built

NUMBER	NAME-VARIANTS	DESIGNER	SPAN	LENGTH	YEAR
PW-6		Fokker	29'6"	23'3"	'22

Single Wright-H 315-hp engine—One built

PURSUITS

P-4	(PW-9)	Boeing	32'1"	23'10"	'28

Biplane—Single Packard A-1500 510-hp engine—One converted

P-7	(PW-9)	Boeing	32'	24'2"	'28

Biplane—Single Curtiss V-1570 600-hp engine—One converted

P-8		Boeing	30'	23'4"	'28

Biplane—Single Packard A-1530 600-hp engine—One built

P-9		Boeing	36'7"	25'8"	'28

Monoplane—Single Curtiss V-1570 600-hp engine—One built

P-10		Curtiss	33'	24'6"	'28

Biplane—Single Curtiss V-1710 600-hp engine—One built

P-11	(P-6)	Curtiss	31'6"	23'2"	'29

Biplane—Single Curtiss H-1640 600-hp engine—3 built

P-13	Viper	Thomas-Morse	28'	23'6"	'29

Biplane—Single P&W R-1340 450-hp engine—One built

P-14		Curtiss			—

Biplane—Single engine—Cancelled

P-15		Boeing	30'6"	21'9"	'29

Monoplane—Single P&W R-1340 525-hp engine—One built

P-17	(P-1)	Curtiss	31'7"	22'10"	'29

Biplane—Single Wright V-1460 475-hp engine—One converted

P-18		Curtiss			—

Biplane—Cancelled

P-19		Curtiss			—

Monoplane—Cancelled

P-20	(P-11)	Curtiss	31'6"	23'6"	'29

Biplane—Single Wright R-1820 575-hp engine—One converted

P-21	(P-3)	Curtiss	31'6"	22'6"	'30

Biplane—Single P&W R-1340 410-hp engine—One built

NUMBER	NAME-VARIANTS	DESIGNER	SPAN	LENGTH	YEAR
P-22	(P-6)	Curtiss	31'6"	23''2"	'31

Biplane—Single Curtiss V-1570 600-hp engine—One built

P-23	(P-22)	Curtiss	31'6"	23'10"	'32

Last biplane Pursuit—Curtiss V-1570 600-hp engine—One built

P-24		Detroit-Lockheed	42'9"	28'9"	'31

2-seat monoplane—Single Curtiss V-1570 600-hp engine—Two built

P-25		Consolidated	43'11"	29'6"	'32

2-seat monoplane—Single Curtiss V-1570 600-hp engine—Two built

P-27	(P-25)	Consolidated			

2-seat monoplane—Cancelled

P-28	(P-25)	Consolidated			

2-seat monoplane—Cancelled

P-29		Boeing	29'1"	25'	'34

Monoplane—Single P&W R-1340 550-hp engine—3 built

P-31		Curtiss	36'	26'3"	'33

Monoplane—Single Curtiss V-1570 600-hp engine—One built

P-32	(P-29)	Boeing			

Monoplane—Single engine—Cancelled

P-33	(P-30)	Consolidated			—

2-seat monoplane—Cancelled

P-34		Wedell-Williams			—

Monoplane—Single engine—Cancelled

P-41	(P-35, P-43)	Seversky	36'	26'10"	'38

Single P&W R-1830 1,200-hp engine—One built

P-42	(P-36)	Curtiss	37'4"	30'6"	'39

Single P&W R-1830 1,050-hp engine—One built

P-44		Republic			—

Single engine—One seat—Cancelled

P-45	(P-39C)	Bell			—

Single engine—Became P-39C—Cancelled as P-45

NUMBER	NAME-VARIANTS	DESIGNER	SPAN	LENGTH	YEAR
P-46	(P-40)	Curtiss	34′4″	30′4″	'40

Single Allison V-1710 1,150-hp engine—Two built

P-48		Douglas			—

Single engine—Cancelled

P-49	(P-38)	Lockheed	52′	39′10″	'42

Twin boom—Twin Continental V-1430 1,350-hp engines—One built

P-50	(F5F-1)	Grumman	42′	31′11″	'41

Twin Wright R-1820 1,200-hp engines—One built

P-52		Bell			—

Pusher monoplane—Cancelled

P-53	(P-46)	Curtiss			—

Monoplane—Single engine—Cancelled

P-54		Vultee	53′10″	54′9″	'43

Pusher monoplane—Lycoming H-2470 2,300-hp engine—Two built

P-55	Ascender	Curtiss	41′	29′7″	'43

Pusher monoplane—Allison V-1710 1,275-hp engine—3 built

P-56		Northrop			'43

Pusher flying wing—Single P&W R-2800 2,000-hp engine—Two built

P-57		Tucker Aviation			—

Monoplane—Single engine—Cancelled

P-58	Chain Lightning	Lockheed	70′	49′7″	'44

Two-seat—Twin Allison V-3420 2,600-hp engines—One built

P-59	(original designation)	Bell			—

Pusher monoplane—Designation reused for P-59 Jet—Cancelled

P-60	(P-53)	Curtiss	41′4″	33′8″	'42

Single P&W R-2800 2,000-hp engine—5 built

P-62		Curtiss	53′6″	39′8″	'43

Single Wright R-3350, 2,300-hp engine—One built

P-65	(P-50)	Grumman			—

Twin engine—Tricycle gear—Cancelled

NUMBER	NAME-VARIANTS	DESIGNER	SPAN	LENGTH	YEAR
P-67		McDonnell			'44
Twin Continental V-1430 1,200-hp engines—One built					
P-68		Vultee			—
Single engine—Cancelled					
P-69		Republic			—
Single engine—One seat—Cancelled					
P-71		Curtiss	82'6"	—	—
Large pusher monoplane—Twin engines—Two seats—Cancelled					
P-72	(P-47)	Republic	41'	33'6"	'44
Single P&W R-4360 3,450-hp engine—Two built					
P-73	No record				
P-74	No record				
P-75		General Motors	49'	41'6"	'43
Single Allison V-3420 2,200-hp engine—13 built					
P-76	(P-39E)	Bell			—
Single engine—Tricycle gear—Cancelled					
P-77		Bell	27'6"	22'10"	'44
Single Ranger V-770 520-hp engine—Two built					
P-78	(P-51B)	North American	37'	32'3"	'41
Single Packard V-1650 1,350-hp engine—Two built					
P-79		Northrop	38'	14'	'45
Flying wing—Twin Westinghouse W-19 Jet engines—Twin fins—1 built					

RACER PLANES

R-4		Loening			'22
Monoplane—Single Packard A-2025 575-hp engine—Two built					
R-7	No record				

TRAINER-AIRCOOLED

TA-1		Elias			'21
Biplane—Single Curtiss-OX 90-hp engine—3 built					

NUMBER	NAME-VARIANTS	DESIGNER	SPAN	LENGTH	YEAR
TA-2		Huff-Daland			'21
Biplane—Single Curtiss-OX 90-hp engine—3 built					
TA-4		Engineering Division			—
Single engine—Cancelled					
TA-5	(TA-3)	Dayton-Wright	34'9"	25'8"	'22
Biplane—Single Lawrence J-1 200-hp engine—One built					
TA-6	(TA-2)	Huff-Daland	31'	24'8"	'23
Biplane—Single Lawrence J-1 200-hp engine—One built					

TRAINER-WATERCOOLED

NUMBER	NAME-VARIANTS	DESIGNER	SPAN	LENGTH	YEAR
TW-1		Engineering Division			'20
Biplane—Single Packard A-1237 350-hp engine—Two built					
TW-2		Cox-Klemin			'22
Biplane—Single Wright-E 180-hp engine—3 built					
TW-4		Fokker			'23
Monoplane—Single Curtiss-OX 90-hp engine—One built					

PRIMARY TRAINERS

NUMBER	NAME-VARIANTS	DESIGNER	SPAN	LENGTH	YEAR
PT-2	(PT-1)	Consolidated	34'6"	28'1"	'25
Biplane—Single Wright R-790 220-hp engine—One built					
PT-4	(PT-3)	Consolidated			—
Biplane—Single engine—Cancelled					
PT-5	(PT-3)	Consolidated	34'6"	28'1"	'29
Biplane—Curtiss R-600 190-hp engine—One built					
PT-7	Pinto	Mohawk			'30
Monoplane—Single engine—One built					
PT-8	Courier (O-17, PT-3)	Consolidated	34'6"	28'1"	'31
Biplane—Single Packard R-980 220-hp engine—Two built					
PT-10	Sportsman	Verville			'31
Commercial pilot trainer—4 built					

NUMBER	NAME-VARIANTS	DESIGNER	SPAN	LENGTH	YEAR
PT-12	(BT-7)	Consolidated	31'7"	26'11"	'32

Biplane—Single P&W R-985 300-hp engine—Ten built

| PT-24 | | DeHavilland | | | '42 |

200 built in Canada for WWII

BASIC TRAINERS

| BT-3 | (PT-9) | Stearman | 32' | 24'8" | '31 |

Biplane—Single Wright R-975 300-hp engine—One built

| BT-4 | (O-1) | Curtiss | 38' | 27'2" | '31 |

Biplane—Single Curtiss V-1150 435-hp engine—One built

| BT-5 | (PT-9) | Stearman | 32' | 24'8" | '31 |

Biplane—Single P&W R-985 300-hp engine—One built

| BT-6 | (PT-11) | Consolidated | 31'7" | 26'11" | '32 |

Biplane—Single Wright R-975 300-hp engine—One built

| BT-10 | (BT-9) | North American | 42' | 27'7" | '38 |

Monoplane—Single P&W R-1340 600-hp engine—One built

| BT-11 | | Airresearch | | | — |

Plastic plywood monoplane—Cancelled

| BT-16 | Valiant (BT-13) | Vidal-Vultee | 42' | 28'10" | '42 |

Monoplane—Single P&W R-985 450-hp engine—One built

| BT-17 | | Boeing | | | '42 |

Monoplane—Single P&W R-985 300-hp engine—One built

ADVANCED TRAINERS

| AT-2 | (AT-1) | Huff-Daland | 31'1" | 24'8" | '25 |

Biplane—Two seats—Wright 180-hp engine—One built

| AT-3 | (PW-9A) | Boeing | 32' | 24'2" | '26 |

Biplane—One seat—Wright V-720 180-hp engine—One built

| AT-13 | Gunner | Fairchild | 52'8" | 37' | '41 |

Monoplane—Twin P&W R-1340 600-hp engines—One built

| AT-14 | Gunner (AT-13) | Fairchild | 52'8" | 37' | '41 |

Monoplane—Twin Ranger V-770 520-hp engines—One built

NUMBER	NAME-VARIANTS	DESIGNER	SPAN	LENGTH	YEAR
AT-15		Boeing			'41

Twin P&W R-1340 600-hp engines—Two built

TRAINERS SINCE 1948

NUMBER	NAME-VARIANTS	DESIGNER	SPAN	LENGTH	YEAR
T-17	Kaydet (PT-17)	Boeing	32′2″	25′	'48

Biplane—Single Continental R-670 220-hp engine—Redesignation

T-30		Douglas			—

Monoplane—Single engine—Cancelled

T-31	(NQ-1)	Fairchild			'49

Navy trainer—Single Lycoming R-680 295-hp engine—One built

T-32		Consolidated			—

Monoplane—Twin engine—Cancelled

T-35		Temco			'50

Monoplane—Single engine—13 built for export

T-36		Beech			—

Monoplane—Twin engines—Cancelled

T-40	(C-140)	Lockheed			—

Twin turbojets—Cancelled

INDEX

A

A-1 (Attack) 14, 77, 210
A-1 (Ambulance) 124, 378
A-2 (Attack) 38, 63, 216, 373
A-2 (Ambulance) 21
A-3 12, 61, 62, 216
A-4 38, 216, 373
A-5 373
A-6 373
A-7 (Fokker) 63, 373
A-7 (LTV) 14, 78
A-8 12, 63
A-9 (Detroit) 64
A-9 (Northrop) 80, 374
A-9 (Amphibian) 43, 232
A-10 (Curtiss) 12, 13, 63, 374
A-10 (Fairchild) 14, 80
A-10 (Amphibian) 43, 232
A-11 12, 64, 250
A-12 13, 63
A-12 (Amphibian) 232
A-13 13, 65, 374
A-14 13, 65, 374
A-15 374
A-16 13, 65, 374
A-17 13, 65
A-18 13, 65
A-19 13, 66
A-20 13, 67
A-21 374
A-22 14, 73, 374
A-23 74, 374
A-24 13, 69
A-25 13, 70
A-26 13, 70, 102
A-27 13, 72
A-28 13, 72
A-29 14, 72
A-30 14, 73
A-31 14, 74
A-32 374
A-33 14, 75
A-34 374

A-35 14, 74
A-36 14, 75
A-37 (Hughes) 374
A-37 (Cessna) 14, 81
A-38 374
A-39 374
A-40 374
A-41 375
A-42 375
A-43 375
A-44 375
A-45 375
AC-1 165
AC-47 30
AC-119 31, 151
AC-130 31, 157
AD-1 77
AD-5 14, 77
AD-7 77
AG-1 201
AG-2 201
AO-1 38, 214
AR-234 346
AT-1 52, 285
AT-2 49, 285, 396
AT-3 45, 285, 396
AT-4 52. 285
AT-5 52, 285
AT-6 53, 290
AT-7 53, 292
AT-8 53, 291
AT-9 53
AT-10 53, 291
AT-11 53, 292
AT-12 53, 290
AT-13 396
AT-14 396
AT-15 397
AT-16 53, 290
AT-17 53, 291
AT-18 53, 292
AT-19 53, 223, 290
AT-20 54, 291

AT-21 54, 293
AT-22 54, 293
AT-22 54, 293
AT-23 54, 291
AT-24 54, 291, 298
AT-37 (Attack) 81
AU-23 56, 305
AU-24 56, 305
AVRO-504K 11, 282
A3D 117
A6M 328
AACP 171
Abel, Rudolph 303
ACLS 166
Acosta, Bert 271
Adaptable Annie 141
Administrator 41, 56, 224, 303
Aerodrome 276
Aggressor 195
Airacobra 47, 256
Airacomet 31, 47, 176
Airacuda 47, 253
Airbus 22, 128
Air Force One 160
Air Skimmer 311
Albatros 320
Albatross 43, 56, 233, 304
Albert, Marcel 336
Albrook, Frank 363
Aldrich Don 341
Allen, James 350 to 353
Allen, Lew 362
Alpha 379
Altair 379
Amin, Idi 157
AMST-STOL 30, 168
Anderson, James 363
Andre, Dan 196
Andrews, Frank 87, 90, 230, 363
Angel of Mercy 206
Appendices (a listing) 315
Apt, Mel 107, 307
Ard, John 270

Argus 139
Arigi, Julius 242
Armstrong, Frank 92
Arnold, Hap 87, 143, 173, 277, 278, 318, 319, 352 to 361
Ascani, Fred 182
Ascender 393
Atlas 20, 117
Autogyro 203
Avitruc 28, 153
AVRO Anson 54, 291
AWACS 161, 170

B

B-1 (Keystone) 376
B-1 (Rockwell) 19, 120
B-2 15, 86
B-3 16, 86
B-4 16, 86, 376
B-5 16, 86, 376
B-6 16, 86, 376
B-7 16, 86, 219
B-8 39, 218, 376
B-9 16, 86
B-10 16, 87
B-11 230, 377
B-12 16, 87
B-13 16, 377
B 14 16 377
B-15 16, 87, 148
B-16 377
B-17 16, 90, 102, 148
B-18 16, 88, 90, 138
B-19 17, 88
B-20 377
B-21 377
B-22 377
B-23 17, 88, 90, 139
B-24 17, 92, 144, 148
B-25 17, 94, 102
B-26 (Martin) 17, 96
B-26 (Douglas) 17, 102
B-26K (On Mark) 20, 103
B-27 377
B-28 377
B-29 17, 98, 103
B-30 377
B-31 377
B-32 17, 100
B-33 377
B-34 17, 101, 138
B-35 18, 104, 111
B-36 18, 102, 104, 147, 187
B-37 18, 101, 138
B-38 16, 377

B-39 377
B-40 16, 90, 377
B-41 378
B-42 18, 20 108
B-43 18, 108
B-44 106, 378
B-45 18, 109
B-46 20
B-47 18, 20, 110
B-48 20
B-49 18, 111
B-50 18, 106
B-51 20
B-52 18, 112, 120
B-53 20
B-54 378
B-55 20
B-56 18, 20
B-57 19, 114
B-58 19, 115, 120
B-59 378
B-60 19, 104, 116
B-61 19, 117
B-62 20, 117
B-63 20, 117
B-64 20, 117
B-65 20, 117
B-66 19, 117
B-67 20, 117
B-68 20, 117
B-69 20, 117
B-70 19, 118
B-71 378
B-72 20, 117
B-73 378
B-74 378
B-75 20, 117
B-76 20, 117
B-77 20, 117
B-78 20, 117
B-79 378
B-80 20, 117
B-81 378
B-82 378
B-83 20, 117
B-84 378
B-85 378
B-86 378
B-8720, 117
BC-1 (BC-2) 53, 285
BF-109 (ME-109) 323
BF-110 (ME-110) 326
BLR-1 16
BLR-2 16
Breguet-14 11, 59, 83, 282

BT-1 51, 69, 216, 286
BT-2 52, 216, 219, 286
BT-3 50, 286, 396
BT-4 38, 216, 396
BT-5 50, 286, 396
BT-6 50, 286, 396
BT-7 52, 286
BT-8 52, 286
BT-9 52, 287, 290
BT-10 396
BT-11 396
BT-12 52, 290
BT-13 52, 289
BT-14 52, 290
BT-15 52, 289
BT-16 396
BT-17 290, 396
BT2D-1 77
BVL-12 373
B5N 328, 329
Baird, Bob 342
Baker, Addison 94
Baker, Newton 357
Bakalar, John 363
Balchen, Bernt 126
Baldwin Airship 350
Balsley, Clyde 238
Baltimore 14, 73
Bantam 57, 308
Bar, Heinz 346
Baracca, Francesco 242
Baracchini, Flavio 242
Baranov, Mihail 335
Barkhorn, Gerhard 324
Barksdale, Gene 363
Barling Bomber 15, 85
Barnes, Bill 182
Barrett, Joe 350
Bataan 151
Bateson, Bob 333
Batten, Gene 273
Baumer, Paul 321
Beahan, Kermit 100
Bearcat 342
Beaver 41, 56, 223, 304
Bebe 11, 237
Beck, Paul 351, 352
Beckwith, Charlie 157
Behrens, Otto 327
Bell, Alexander Graham 276, 350
Bellevue, Charles de 194
Bennett, Floyd, 125
Bergstrom, John 363
Berry, Joe 333

Bettis, Cy 272
Betty (Bomber) 329
Beurling, George 331
Biggs, James 363
Bird Dog 41, 226
Bishop, William 241
Blackbird 33, 49, 196
Black Death 337
Blackhawk 375
Black Pony 242
Black Sheep 340
Black Widow 32, 48, 265
Blakeslee, Don 260, 263
Blue Bird 56, 303
Blue Streak 94
Boardman, Russell 128
Bobcat 26, 53, 142
Bock, Fred 100
Bock's Car 100
Boeing Model-247 141
Boeing Model-307 142
Boelcke, Oswald 320
Bolling, Ray 237, 355, 363
Bolo 16, 24, 88, 138
Bolt, John 341
Bong, Richard 178, 255
Bordelon, Guy 341
Borum, F.S. 136
Boston 67
Boyd, Al 178
Boyington, Greg 259, 340
Brandenburg, Ernest 323
Braun, Wernher 344
Breene, R.G. 273
Brereton, Lewis 91, 96, 258
Brett, George 95, 256
Bridgeman, Bill 307
Brindley, Oscar 353
Bronco 41, 227
Brookley, Wendell 363
Brooks, Sydney 363
Brow, Hal 272
Brown, George S. 362
Brown, Harold 119, 122
Brown, Roy 240, 321
Brown, Russ 178, 348
Brumoski, Godwin 242
Bubbletop 261
Buchner, Hermann 346
Budanova, Katia 335
Buffalo 30, 166
Bugsmasher 131, 234
Bullpup 20, 117
Burcham, Milo 177
Burge, Vernon 352, 353

Burges, George 330
Butcher Bird 327
Butler, Ken 94
Byrd, Richard 125, 126, 128

C

C-1 21, 124, 125, 216
C-2 21, 125
C-3 21, 125, 378
C-4 (Ford) 21, 125
C-4 (Lockheed) 30, 164
C-5 (Fokker) 21, 378
C-5 (Lockheed) 30, 164
C-6 (Sikorsky) 21, 126
C-6 (Beechcraft) 30, 165
C-7 (Fokker) 21, 378
C-7 (DeHavilland) 30, 165
C-8 (Fairchild) 21, 126
C-8 (Bell-DeHavilland) 30, 166
C-9 (Ford) 21, 126, 379
C-9 (McDonnell-Douglas) 30, 166
C-10 (Curtiss) 379
C-10 (McDonnell-Douglas) 30, 167
C-11 379
C-12 (Detroit-Lockheed) 27, 147, 379
C-12 (Beechcraft) 30, 168
C-13 379
C-14 (Fokker) 21, 127
C-14 (Boeing) 30, 168
C-15 (Fokker) 21, 127, 379
C-15 (McDonnell-Douglas) 30, 168
C-16 379
C-17, 27, 147, 379
C-18 379
C-19 379
C-20 379
C-21 22, 127
C-22 379
C-23 379
C-24 379
C-25 379
C-26 22, 128
C-27 22, 128
C-28 380
C-29 22, 128
C-30 22, 128
C-31 380
C-32 22, 128
C-33 22, 129
C-34 22, 129
C-35 22, 129
C-36 22, 129
C-37 23, 129
C-38 23, 130
C-39 23, 130

C-40 23, 131
C-41 23, 380
C-42 23, 380
C-43 23, 131
C-44 380
C-45 23, 131
C-46 23, 132
C-47 23, 128, 133
C-48 23, 135
C-49 23, 135
C-50 23, 135
C-51 24, 135
C-52 24, 135
C-53 24, 135
C-54 24, 136
C-55 23, 132, 380
C-56 24, 137
C-57 24, 138
C-58 24, 138
C-59 24, 138
C-60 24, 138
C-61 24, 139
C-62 380
C-63 24, 139
C-64 25, 139
C-65 380
C-66 25, 139
C-67 25, 89, 139
C-68 25, 140
C-69 25, 140
C-70 25, 140
C-71 25, 141
C-72 25, 141
C-73 25, 141
C-74 25, 141, 154
C-75 25, 142
C-76 380
C-77 380
C-78 26, 142
C-79 380
C-80 380
C-81 26, 143
C-82 26, 143
C-83 26, 144
C-84 26, 144
C-85 380
C-86 26, 144
C-87 26, 92, 144
C-88 380
C-89 380
C-90 381
C-91 381
C-92 381
C-93 381
C-94 381

C-95 26, 145
C-96 26, 145
C-96 26, 145
C-98 381
C-99 27, 104, 147
C-100 27, 147
C-101 27, 147
C-102 381
C-103 381
C-104 381
C-105 27, 87, 148
C-106 381
C-107 381
C-108 27, 90, 148
C-109 27, 92, 148
C-110 381
C-111 27, 148
C-112 28, 149, 381
C-113 381
C-114 137, 381
C-115 137, 382
C-116 27, 137, 149
C-117 27, 149
C-118 28, 149
C-119 28, 150
C-120 28, 151, 382
C-121 28, 151
C-122 28, 153, 310
C-123 (Chase) 28, 153
C-123 (Fairchild) 28, 153
C-123 (Stroukoff) 29, 153
C-124 28, 141, 154, 158
C-125 28, 155
C-126 28, 155
C-127 382
C-128 151, 382
C-129 382
C-130 29, 118, 156, 213
C-131 29, 157
C-132 382
C-133 29, 154, 158
C-134 29, 153
C-135 29, 159, 171
C-136 382
C-137 29
C-138 382
C-139 382
C-140 29, 161
C-141 29, 161
C-142 29, 163
CG-1 382
CG-2 382
CG-3 382
CG-4 34, 200
CG-5 382

CG-6 382
CG-7 382
CG-8 382
CG-9 383
CG-10 383
CG-11 383
CG-12 383
CG-13 34, 201
CG-14 383
CG-15 34, 201
CG-16 383
CG-17 383
CG-18 28, 153, 383
CG-19 383
CG-20 35
CH-3 209
CH-19 207
CH-21 207
CH-47 164
CO-1 385
CO-2 386
CO-3 386
CO-4 38, 214
CO-5 45, 214, 386
CO-6 386
CO-7 386
CO-8 386
COA-1 42, 229
CV-2 30, 165
CV-7 30, 166
C-X 169
Cadbury, Egbert 323
Cadet 40, 221, 223, 389
Caldwell, Clive 332
Camel 11, 240, 282
Campbell, Doug 239, 356
Campbell, Jesse 153
Canberra 19, 114
Canestoga 381
Cannon, John 363
Canso 232
Cape Cod 128
Caravan 380
Caribou 30, 165
Carr, Bruce 264
Carswell, Horace 364
Carter, Jimmy 119, 120, 172
Castle, Fred 364
Catalina 42, 43, 231
Chain Lightning 393
Chandler, Charles 277, 318,
  350 to 358

Chapman, John Jay 238
Chapman, Vic 238
Charity 329
Cheli, Ralph 96
Chennault, Claire 258, 364
Chicago 217
Chickasaw 37, 206, 210
Childress, Rollin 97
Chinook 164
Christie, Art 278
Chummy 49
Cinema 35
City of San Francisco 229
Clark, Harold 364
Clark, Mark 223
Cloudboy 50, 284
Closterman, Pierre 334
Cochrane-Odlum, Jackie 143, 147, 183, 189,
  251
Collier, Robert 351
Collins, Jim 96
Collishaw, Ray 241
Columbia 58, 313
Columbine 151
Commander 41, 56, 224, 303
Commando 23, 132
Compton, K.K. 94
Comstock, Hal 261
Condor 15, 22, 86, 128
Connally, James 364
Constellation 25, 140
Cornell 51, 54, 289, 296
Corps d'Armee 11
Corsair 340
Corsair-II 14, 78
Counter-Invader 20, 103
Coupe 26, 144
Courier 39, 218, 395
Craigie, Larry 173
Crawford, Frank 242
Crippen, Robert 314
Critchlow, Dave 112
Crockett, Davy 64
Crossbow 20, 117
Crossfield, Scott 307, 309
Crow, Dave 160
Cruise Missile 121, 308
Cub 35, 40, 220, 222
Culley, Stuart 241, 323
Curtiss, Glenn 280
Cyclops 15, 85

D
DB-1 12, 62
DB-45 109

DB-47 110
DC-1 128
DC-2 22, 24, 88, 128, 129, 130
DC-2½ 16, 23, 130
DC-3 23, 24, 130, 133, 135
DC-3A 25, 140
DC-3B 26, 144
DC-4 24, 27, 136
DC-5 381
DC-6 28
DC-9 30
DC-10 167
DC-130 156
DF-80 178
DH-4 12, 60, 83, 215, 355
DOS 217
DWC 38, 217
D.I (Albatros) 320
D.II (Albatros) 320
D.III (Pfalz Albatros) 320
D.III (Fokker) 321
D.V (Albatros) 320
D.VII (Fokker) 243, 321
D.VIII (Fokker) 321
DR.I (Fokker) 321
D-558-I 307
D-558-II 307
Dahne, Heinrich 347
Dakota 133
Dargue, Herb 229
Dart 116
Dauntless 13, 31, 69, 77
Dauntless Dotty 99
Davis, Doug 273
Davis, George 182
Davis, Sam 364
Dean, T.D. 343
Defender 35, 39, 40, 220, 222
DeGaulle, Charles 335
Delta Dagger 33, 116, 187
Delta Dart 33, 116, 191
Destroyer (A-38) 374
Destroyer (B-66) 19, 119
Destroyer (BF-110) 326
Dewdrop 151
Dirigible (Bomber) 322
Dittmar, Hieini 344
Dodd, Townsend 355
Dolina, Maria 338
Dolphin 22, 42, 127, 128, 230
Dollar Nineteen 151
Dominator 17, 100
Donaldson, E.M. 343
Doolittle, Jimmy 61, 91, 95,
    260, 264, 272, 356 to 360

Doomsday 31, 171
Dora-9 327
Dow, James 364
Dragon 17, 25, 88, 139
Dragonfly (AT-37) 14, 81
Dragonfly (O-51) 389
Dreidecker 321
Drew, Adrian 187
Droop Snoot 254
Dyar, Roger 261
Dyberg, Bob 210
Dyess, Bill 364
Dyna Soar 308

E

E-3 31, 161, 170, 171
E-4 31, 171
EC-121 152
EC-135 160
EC-137 31, 170, 171
ERB-47 111
E.28/39 343
Eagle 33, 196
Eagleston, Glenn 182, 264
Eaker, Ira 92, 125, 229, 360
Earhart, Amelia 129, 130
Early Trainers 276
Edwards, Glen 112, 364, 365
Eglin, Fred 364
Eielson, Ben 147, 364
Eisenhower, Dwight 135, 151, 152, 303
Electra 22, 23, 129, 131
Ellington, Eric 364
Elliott, L.G. 273
Ellsworth, Lincoln 147, 364
Ellyson, Ted 351
Elmendorf, Hugh 248, 364
Elton, Al 92
England, John 365
Engle, Joe 312
Engler, Howard 99
Enola Gay 100
Ent, Uzal 365
Enterprise 58, 113, 311, 312
Ercoupe 389
Everest, Frank 186, 307
Executive 25, 141
Expediter 23, 43, 55, 131, 234, 296
Extender 30, 167

F

F-1 43, 126, 234
F-2 43, 234
F-3 43, 234
F-4 (Fighter) 33, 193

F-4 (Photo) 43, 234
F-5 (Fighter) 33, 194
F-5 (Photo) 43, 234
F-6 43, 235
F-7 43, 235
F-8 44, 235, 332
F-9 44, 235
F-10 44, 235
F-11 236, 390
F-12 (Fighter) 33, 196, 274
F-12 (Photo) 236
F-13 44, 236
F-14 44, 236
F-15 (Fighter) 33, 196
F-15 (Photo) 44, 236
F-16 34, 198
F-17 198
F-24 31
F-38 31
F-40 31
F-47 31
F-51 31
F-59 31, 176
F-61 32
F-63 32
F-80 32, 177
F-82 32
F-83 31
F-84 32, 179
F-85 34, 105
F-86 32, 181
F-87 (P-87) 34
F-88 33, 34, 186
F-89 32, 183
F-90 34
F-91 34
F-92 34, 116
F-93 32, 34, 182
F-94 32, 184
F-95 32, 34, 182
F-96 32, 34, 180
F-97 32, 34, 185
F-98 34
F-99 34
F-100 32, 185
F-101 33, 186
F-102 33, 116, 187, 191
F-103 34
F-104 33, 188
F-105 33, 189
F-106 33, 116, 191
F-107 32, 34, 186
F-108 34
F-109 34
F-110 33, 34, 193

F-111 33, 121, 191
FA-26 71
FB-17 44, 235
FB-29 44, 236
FB-111 20, 121, 191
FG-1 340
FI-156 225
FM-1 (Airacuda) 47, 253
FM-1 (Wildcat) 339
FM-2 338
FOK F-1 321
FP-51 43, 235
FP-61 236
FP-80 44, 178, 236
FW-190 327
F3A-1 340
F4B 46, 248
F4F 338
F4U 259, 340
F5F-1 393
F6C 246
F6F 255, 341
F7F 342
F8F 342
F9F 348
Fagot 348
Fairchild, Muir 229, 365
Faith 329
Falcon 12, 38, 61, 62, 215, 373,
    386, 387, 388
Farrow, Bill 95
Faulkner, Ted 99
Fechet, James 358, 359
Feinstein, Jeff 194
Ferebee, Tom 100
Fernandez, Manuel 182
Fighting Falcon 34, 198
Finucane, Brendan 332
Fishbed 349
Fiske, Bill 260
Flak Bait 98
Flaming Coffin 12, 60
Fleetster 379
Florine Jo Jo 94
Floyd Bennett 126
Flying Boxcar 28, 31, 144, 150
Flying Classroom 55, 293, 298, 301
Flying Dump Truck 77
Flying Filling Station 148
Flying Fire Truck 157
Flying Fortress 16, 27, 44, 90,
    102, 235
Flying Machine 371
Flying Razor 321
Flying Tigers 134, 258

Flying Whitehouse 171
Flying Wing 18, 104, 111
Foch, Ferdinand 356
Fokker, Anthony 237
Fokker Trimotor 21, 125
Fonck, Rene 238
Forbes, Dan 112, 365
Ford, Edsel 125
Ford Trimotor 21, 125
Forsythe, Frank 191
Fort, Cornelia 143
Forwarder 24, 26, 139, 144
Foss, Joe 256, 339
Foulois, Ben 318, 350 to 359
Franco, Francisco 324
Frantisek, Josef 330
Freedman, Mort 153
Freedom Fighter 33, 194
Freedom One 161
Fritz, Hans 322
Fuchida, Mitsuo 328
Fullerton, Gordon 312

G

G-1 (Gyro) 36, 203, 220
G-2 (Gyro) 385
G-3 (Glider) 34
G-4 (Glider) 34
G-10 (Glider) 383
G-13 (Glider) 34
G-14 (Glider) 383
G-15 (Glider) 34
G-18 (Glider) 383
G-20 (Glider) 28, 35, 153
GA-1 12, 62
GA-2 12, 62
GB-1 131
GB-2 131
GC-130 156
GH-1 140
GH-2 140
GK-1 139
GMB 15, 21, 84
GMP 21
GRB-36 105, 180
G-IV (Gotha) 323
G4M 329
Gabreski, Frank 182, 262
Galaxy 30, 164
Gallagher, Jim 106
Gallagher, John 153
Galland, Adolf 324, 327, 346
Gamma 27, 147, 374
Gebaur, Art 180, 368
General's Jeep 223

Gentile, Don 264
Gentry, Jerauld 311
George, Harold 136, 365
Gerlach, Hans 225
Gibson, Guy 334
Gladiator 329
Glamorous Glennis 306
Glenn, John 314
Globemaster-I 25, 141
Globemaster-II 28, 141, 154
Godfrey, John 264
Goering, Hermann 174, 264, 332,
    323, 333
Goodfellow, John 365
Gooney Bird 23, 30, 133
Goose 42, 43, 231, 232
Graf Zeppelin 323
Grasshopper 26, 35, 39, 40, 145,
    220, 221
Gray Ghost 55, 302
Green, Herschel 262
Griffiss, Townsend 365
Grubbs, Manson 153
Gruzdev, Konstantin 336
Guardsman 46, 53, 251, 290
Gulayev, Nikolai 336
Gunn, Paul 68, 95
Gunner 54, 293, 396
Gunship (C-47) 30
Gunship (C-119) 31, 151
Gunship (C-130) 157
Gustav (BF-109) 324
Guynemer, Georges 238
Gypsy Rose Lee 259
Gyro-Glider 58, 311
Gyroplane 203

H

H-4 36, 205
H-5 36, 205
H-6 37, 205
H-12 37, 205
H-13 37, 206
H-19 37, 206
H-21 37, 207
H-43 37, 207
HAWK-75 (P-36) 46
HB-1 (HB-2, HB-3) 15, 85, 375
HB-29 104
HC-130 156, 209
HD-1 12, 242
HE-162 347
HE-176 344
HE-178 173, 347
HE-280 173, 345

HH-1 37, 208, 209
HH-3 37, 209
HH-19 37, 210
HH-21 37, 211
HH-43 38, 211
HH-53 38, 212
HU-16 56, 304
Hadrian 34, 201
Haise, Fred 312
Hall, Don 68
Hallmark, Dean 95
Halsey, Bull 95
Halverson, Harry 93, 125
Hamann, Major 160
Hamilton, Lloyd 365
Hanes, Harold 186
Hansell, Haywood 99
Hanson, Bob 340
Harmon, Ernest 216, 365
Harris, Cecil 342
Harris, Hal 85
Harrison, William Henry 177
Hartmann, Erich 264, 325
Hartney, Hal 271
Harvard 53, 290
Havoc 13, 43, 67, 234
Hawk 45, 46, 47, 52, 246, 251, 285
Hawk-75 251
Hawks, Howard 147
Hedman, Duke 258
Hegenberger, Al 125
Heglund, Svien 333
Heimatschutzer 345
Heinkel, Ernst 173
Hellcat 341
Helldiver 13, 70
Helt, Bob 274
Henry, Bill 342
Hercules 29, 31, 156
Herk 157
Herky Bird 157
Herndon, Hugh 128
Hickam, Horace 64, 365
Hill, Frank 260, 327
Hill, Pete 90, 365
Hindenburg (Zeppelin) 323
Hippe, Ken 339
Hirohito, Emperor 100
Hitler, Adolf 173
Hodgson, Walt 207
Holloman, George 365
Home Protector 345
Hoover, Bob 260
Hope 329
Hopkins, Jim 100

Horsa 35, 202
Hound Dog 20, 113, 117
Howard, Charles 365
Howard, Jim 264
Hudson 13, 14, 24, 53, 72, 139, 292
Hudson, Bill 270, 336
Huey 208
Hughes, Howard 147, 148
Hughes, Lloyd 94
Humphreys, Fred 277, 279, 351
Hunter, Frank 365
Hurribomber 330
Hurricane 326, 330
Huskie 37, 38, 207, 211
Husky Junior 50, 284
Hustler 19, 115
Hypersonic 57, 113, 309

I

IL-2 337
Immel, Perry 137
Independence 150
Intruder 19, 114
Invader 13, 17, 70, 102
Iroquois 37, 208, 209
Irwin, Walt 189

J

JB-45 109
JC-130 156
JKC-135 159
JM-1 96
JN-1 10
JN-2 372
JN-3 372
JN-4 10, 278, 280, 354
JN-5 373
JN-6 373
JRB-1 132
JRB-2 132
JRC-1 142
JRF-5 231
JRF-6 231
JRS-1 231
JU-52 380
JU-87 325
J2F-5 232, 390
J2M 328
J8N1 346
Jabara, James 182
Jackson, Joe 153
Jeep 53
Jenny 10, 278, 280, 373
Jerstad, John 94

Jetstar 29, 161
Joersz, Eldon 274
Johnson, Bob 262
Johnson, Clarence 175, 196
Johnson, Dick 182
Johnson, Gerald 365
Johnson, Howard 189
Johnson, James 260, 331
Johnson, Leon 94
Johnson, Lyndon 154, 161, 196, 274
Johnson, Tex 112
Jolly Green Giant 37, 209
Joltin' Josie 99
Jones, David C. 362
Jones, Q.B. 278, 279
Jordan, Joe 189
Josephine Ford 125
Jug 31, 47, 261
June, Harold 128
Junkin, Sam 260, 327
Jupiter 20, 117

K

KB-29 17, 106, 109, 180
KB-47 18, 110
KB-50 18
KC-10 30, 167
KC-97 27, 146
KC-124 154
KC-130 156
KC-135 29, 159, 167, 171
Kaishek, Chiang 141
Kamikaze 329
Kane, John 94
Kansan 53, 54, 292, 296
Kate 329
Kaydet 50, 51, 284, 289, 397
Kearby, Neel 262
Keesler, Sam 366
Kegelman, Charles 67
Kelly, George 278, 351, 352, 366
Kelly, Oakley 124
Kelly's Cobras 93
Kenly, Bill 318, 356 to 358
Kennedy, Jackie 161
Kennedy, John 103, 119, 161, 186, 187, 309
Kenney, George 95, 262
Khrushchev, Nikita 303
Kikka 346
Kincheloe, Iven 189, 307, 366
Kindley, Field 366
King Air 30, 165
Kingcobra 32, 48, 267

Kingsley, Dave 366
Kirtland, Roy 277, 278, 352, 366
Kites 263
Kittel, Otto 325
Kittyhawk 259
Kitty Hawk 119, 276, 371|
Klunkers 256
Knight, Pete 310
Komet 344
Kozhedub, Ivan 257, 336
Krafft, Heinrich 327

L

L-1 40, 221
L-2 40, 145, 221
L-3 (Liaison) 40, 222
L-3 (Zeppelin) 322
L-4 (Liaison) 40, 144, 222
L-4 (Zeppelin) 322
L-5 40, 223
L-6 40, 221, 223
L-7 389
L-8 389
L-9 41, 223
L-10 389
L-11 389
L-12 389
L-13 41, 223
L-14 389
L-15 389
L-16 389
L-17 390
L-19 41
L-20 41, 223
L-21 41, 224
L-26 41, 224
L-27 41, 224
L-28 41, 225
L-30 (Zeppelin) 322
L-53 (Zeppelin) 323
L-70 (Zeppelin) 323
LA-5 336
LA-7 336
LA-9 336
LAGG-3 336
LB-1 15, 85
LB-2 375
LB-3 15, 375
LB-4 376
LB-5 15, 85
LB-6 15, 85
LB-7 15, 85, 376
LB-8 15, 85, 376
LB-9 15, 85, 376
LB-10 15, 85, 86, 376

LB-11 15, 85, 376
LB-12 15, 85, 376
LB-13 15, 85, 376
LB-14 15, 85, 376
LC-126 28, 155
LNE-1 35
LUSAC-11 12, 241
LZ-48 (Zeppelin) 323
Lackland, Frank 366
Lafonte, Francois 238
Lahm, Frank 277, 350 to 360
Lambert, Larry 266
Laminar-Flow 57, 310
Lancaster 334
Lancer 47, 259
Land Racer 49, 272
Lang, Emil 325
Langley, Samuel 276
Lanphier, Tom 254
LARA 227
Larner, Ed 95
Larson, Don 366
Laughlin, Jack 366
Lawson, Ted, 95
LeMay, Curtis 99, 137, 362
Lent, Helmut 326
Le Pere 12
Liberator 17, 26, 27, 43, 54, 92, 144, 235, 293
Liberty Plane 60
Lifting-Body 58, 311
Liftmaster 28, 149
Liggett, Hunter 356
Lightning 31, 43, 47, 234, 254
Lindbergh, Anne Morrow 80
Lindbergh, Charles 80, 255
Lipscomb, Paul 264
Litvak, Lilya 335
Locklear, Ormer 281
Lodestar 24, 25, 137, 138, 139
Loewenhardt, Erich 241, 322
Loring, Charles 178, 366
Love, Moss 352
Love, Nancy 143
Lowry, Frank 366
Lowry, Norm 181
Lucky Lady-II 106
Lufbery, Raoul 237, 240, 356
Luke, Frank 240, 357, 366
Lundie, Jim 153

M

MB-1 (Bomber) 15, 84, 123
MB-2 (Bomber 15, 85, 358

MB-3 (Pursuit) 44, 244, 271
MB-6 (Pursuit) 48, 245, 271
MB-7 (Pursuit) 245
MB-9 (Pursuit) 49, 245
MB-10 (Pursuit) 245
ME-109 (BF-109) 173, 323
ME-110 (BF-110) 326
ME-163 173, 344
ME-210 327
ME-262 174, 345
ME-263 344
ME-310 327
ME-410 327
MIG-3 334
MIG-15 180, 182, 347
MIG-17 348
MIG-19 348
MIG-21 189, 194, 348
MIG-23 198
Model-A (Wright) 277, 350, 371
Model-AR (Dorand) 282
Model-B (Wright) 10, 277, 351
Model BE-2 (RAE) 282
Model-C (Wright) 277, 352, 371
Model-D (Curtiss) 10, 278, 351
Model-D (Engr Div) 373
Model-D (Gallaudet) 372
Model-D (Thomas) 372
Model-D.XI (Fokker) 45
Model-E (Curtiss) 278, 352, 371
Model-F (Burgess) 371
Model-F (Curtiss) 278, 352, 371
Model-F (Farman) 282
Model-F (LWF) 373
Model-F (Wright) 278, 352, 371
Model FE-2 (RAE) 282
Model-G (Caudron) 282
Model-G (Curtiss) 10, 278, 353
Model-G (LWF) 373
Model-H (Burgess) 10, 278, 352
Model-H (Sloan) 279, 354, 372
Model-HS (Wright) 278, 353, 371
Model-I (Burgess) 278, 353, 371
Model-J (Burgess) 371
Model-J (Curtiss) 278, 353, 372
Model J-1 (Standard) 281
Model-JN (Curtiss) 278
Model JN-1 (Curtiss) 10
Model JN-2 (Curtiss) 372
Model JN-3 (Curtiss) 372
Model JN-4 (Curtiss) 10, 278, 280, 354
Model JN-5 373
Model JN-6 373
Model-K (Burgess) 372

Model-L (Curtiss) 278, 354, 373
Model MS-12 (Morane-Saulnier) 282
Model-N (Curtiss) 278, 353, 372
Model-O (Curtiss) 372
Model-R (Curtiss) 278, 354, 372
Model-R (Wright-Martin) 279, 372
Model-S (Curtiss) 278, 354, 373
Model-S (Martin) 278, 372
Model-S (Sturtevant) 279, 354, 372
Model-SJ (Standard) 10, 281, 355
Model S-4 (Thomas-Morse) 10, 282,
Model-T (Martin) 278, 353, 372
Model-TT (Martin) 378, 354, 372
Model-V (LWF) 279, 354, 372
Model 1A2 (Sopwith) 11, 282
Model-8 (Wright) 351
Model-504K (AVRO) 282
MS-234 (Morane-Saulnier) 221
MacArthur, Douglas 90, 151
MacDill, Les 367
MacDonald, Charles 255
Mace 20, 117
MacFarlane, Willard 197
MacMillan Expedition 229
MacReady, John 124, 214, 241
Maitland, Les 125, 272
Malan, Adolph 332
Malmstrom, Einar 367
Mannock, Ed 241
Mao Tse Tung 178
Marauder 17, 54, 96, 102, 291
March, Peyton 367
Marquardt, George 100
Marseille, Joachim 324
Marshall, George 69
Martin Bomber 15, 16
Maryland 73, 374
Matador 19, 117
Mather, Carl 367
Maughan, Russ 245, 272
Maxwell, Bill 367
Mays, Willie 181
McCampbell, David 255, 342
McChord, Bill 65, 366
McClellan, Hez 128, 251, 366
McConnell, Fred 367
McConnell, John P. 362
McConnell, Joseph 182
McConnell, Tom 367
McCoy, Mike 367
McGuire, Tommy 255, 329, 367
Meatball 328
Melancon, Mark 181
Memphis Belle 92
Menoher, Charles 123, 318,
  357 to 359

Mentor 55, 298
Mescalero 55, 301
Messerschmitt, Willy 323
Meteor 173, 343
Meyer, John 264
Midget, 348
Miller, Glenn 139
Milling, Tom 277, 278, 352 to 357
Mills, Harry 271
Minesweeper 212
Minuteman 20, 117
Miss Veedol 128
Mitchel, John 367
Mitchell (B-25) 17, 44, 54, 94, 102,
  235, 291, 298
Mitchell, Billy 85, 94, 214, 238,
  272, 354 to 361
Mixmaster 20, 108
Mohawk 46, 251
Molders, Werner 237, 324
Mongol 349
Monocoupe 389
Monomail 379
Montgomery, Bernard 96
Monthan, Oscar 364
Moody, George 367
Moore, Joe 190
Moore, Woody 91
Morgan, Bob 92
Moseley, Corliss 271
Mosquito 44, 235, 332
Mould, P.W. 330
Muller, H.L. 278
Murray, Kit 307
Mussolini, Benito 225, 329
Mustang 14, 31, 43, 47, 75,
  235, 263
MX Missile 121
Myers, Joe 265, 346

N
NASA-905 58, 312
NBL-1 15, 85
NBL-2 85, 375
NBS-1 15, 85
NBS-2 375
NBS-3 375
NBS-4 86, 375
NF-104 189
NH-1 140
Nieuport-11 11, 237
Nieuport-12 11, 282
Nieuport-17 11, 238, 282
Nieuport-21 282

Nieuport-23 282
Nieuport-24 282
Nieuport-27 282
Nieuport-28 11, 238
NKC-135 159
NQ-1 397
Nagumo, Chuichi 69, 328
Nash, Slade 182
Navaho 20, 117, 308
Navigator 53, 54, 292, 296
Navion 390
Nell (Bomber) 329
Nellis, Bill 367
Nelson, Bill 192
Nelson, Erik 217
Nelson, Sam 80
Neptune 20, 117
New Orleans 217
Nicholson, James 330
Night Havoc 48, 268
Night Hurricane 330
Nightingale (C-70) 25, 140
Nightingale (C-9) 30, 166
Night Intruder 114
Night Lightning 254
Nightspit 331
Nishizawa, Hiroyoshi
  256, 329
Nomad (A-17, A-33)
  13, 14, 75
Nomad (OA-10) 65, 232
Noonan, Fred 130
Norseman 25, 139
Norton, Leland 367
Nowotny, Walt 345
Nungesser, Charles 238

O
O-1 (Curtiss) 38, 215
O-1 (Cessna) 41, 226
O-2 (Douglas) 38, 124, 216
O-2 (Cessna) 41, 226
O-3 386
O-4 386
O-5 38, 217
O-6 38, 217
O-7 38, 216, 217, 386
O-8 38, 216, 386
O-9 38, 216, 386
O-10 229, 386
O-11 38, 215, 386
O-12 38, 215, 387
O-13 38, 215, 387
O-14 38, 216, 387
O-15 387

O-16 38, 215, 387
O-17 39
O-18 38, 215, 387
O-19 39, 218
O-20 39, 218, 387
O-21 39, 218, 387
O-22 38, 216, 387
O-23 39, 218, 387
O-24 387
O-25 38, 216, 387
O-26 38, 215, 387
O-27 39, 218
O-28 387
O-29 38, 216, 387
O-30 388
O-31 39, 218
O-32 39, 216, 219
O-33 388
O-34 216, 388
O-35 39, 219
O-36 39, 219, 388
O-37 388
O--38 38, 216, 388
O-39 38, 215, 388
O-40 388
O-41 388
O-42 388
O-43 39, 218, 388
O-44 388
O-45 388
O-46 39, 218, 388
O-47 39, 219
O-48 388
O-49 39, 219
O-50 389
O-51 389
O-52 39, 220
O-53 389
O-54 41, 389
O-55 389
O-56 389
O-57 39, 145, 220
O-58 39, 220
O-59 40, 144, 220
O-60 40, 204, 220
O-61 389
O-62 40, 220
O-63 40, 221
OA-1 42, 229
OA-2 42, 230
OA-3 42, 127, 230
OA-4 42, 128, 230
OA-5 230, 390
OA-6 390
OA-7 42, 390

OA-8 42, 230
OA-9 42, 231
OA-10 42, 231
OA-11 42, 231, 390
OA-12 390
OA-13 42, 232
OA-14 42, 232
OA-15 390
ORENCO-D 44, 243
OV-10 41, 227
OV-101 58, 312
OV-102 58, 313
O2U-3 387
O'Donnell, Emmett 99
Offutt, Jarvis 367
Ohain, Hans 173, 347
O'Hare, Ed 339
Old, Archie 112
Old Boomerang 73
Olmstead, Bob 367
Orbital Vehicle 312, 313
Orion 380
Oscar 266
Oswald, Lee Harvey 161
Otis, Frank 368
Otter 302
Owl 39, 220
Ozier, Joe 188

P
P-1 45, 246
P-2 45, 246
P-3 45, 246
P-4 391
P-5 45, 246
P-6 46, 247, 273
P-7 391
P-8 391
P-9 391
P-10 391
P-11 247, 391
P-12 46, 248
P-13 391
P-14 391
P-15 391
P-16 46, 249
P-17 247, 391
P-18 391
P-19 391
P-20 247, 391
P-21 247, 391
P-22 247, 392
P-23 247, 392
P-24 250, 392
P-25 248, 250, 392

P-26 46, 249
P-27 392
P-28 392
P-29 392
P-30 46, 64, 250
P-31 392
P-32 392
P-33 392
P-34 392
P-35 46, 251
P-36 46, 251
P-37 47
P-38 47, 254
P-39 47, 256, 290
P-40 47, 257, 290
P-41 46, 47, 251, 259, 392
P-42 252, 392
P-43 47, 259
P-44 392
P-45 392
P-46 252, 393
P-47 47, 261
P-48 393
P-49 393
P-50 393

P-51 47, 75, 263
P-52 393
P-53 393
P-54 393
P-55 393
P-56 393
P-57 393
P-58 393
P-59 (piston engine) 393
P-59 (jet engine) 47, 173, 176
P-60 393
P-61 48, 265, 269
P-62 393
P-63 48, 267
P-64 48, 267
P-65 393
P-66 48, 268
P-67 394
P-68 394
P-69 394
P-70 48, 266, 268
P-71 394
P-72 394
P-73 394
P-74 394
P-75 394
P-76 394
P-77 394
P-78 394

P-79 394
P-80 48, 175, 177
P-81 34
P-82 48, 269
P-83 34
P-84 179
P-85 (F-85) 34, 105
P-86 (F-86) 181
P-87 (F-87) 34
P-88 (F-88) 33, 34, 186
P-89 (F-89) 183
PA-1 44, 245
PB-1 (Pursuit Biplace) 46, 249
PB-1 (Patrol Bomber) 90
PB-2 46, 64, 250
PBJ-1 95
PBN-1 232
PBO-1 72, 139
PBY-5 42, 231
PBY-6 232
PB2B 232
PB4Y 92
PC-121 151
PE-2 337
PG-1 (Pursuit) 45, 245
PG-3 (Glider) 34, 201
PN-1 44, 245
PR-XVI 44, 235
PT-1 50, 284
PT-2 50, 395
PT-3 50, 217, 284
PT-4 395
PT-5 395
PT-6 50, 284
PT-7 395
PT-8 39, 218, 395
PT-9 50, 284
PT-10 395
PT-11 50, 284
PT-12 50, 284, 396
PT-13 50, 284, 289
PT-14 50, 289
PT-15 50, 289
PT-16 50, 284
PT-17 50, 52, 289
PT-18 51, 289
PT-19 51, 289
PT-20 51, 289
PT-21 51, 289
PT-22 51, 289
PT-23 51, 289
PT-24 396
PT-25 51, 289
PT-26 51, 289

PT-27 51, 289
PV-1 101
PW-1 45, 245
PW-2 245, 390
PW-3 390
PW-4 390
PW-5 390
PW-6 245, 391
PW-7 45, 245
PW-8 45, 245
PW-9 45, 246
P2B-1 103
Packet 26, 143
Packplane 382
Paine, Top 368
Pangborn, Clyde 128, 141
Panther 348
Pantobase 29, 153
Parshin, G.M. 338
Pathfinder 128, 254
Patrick, Mason 61, 318, 357
   to 360, 368
Patterson, Frank 368
Pattle, Marmaduke 260, 330
Patton, General 174
Peacemaker (AU-23) 56, 305
Peacemaker (B-36) 18, 104
Pease, Harl 368
Peashooter 46, 249
Pelican 209
Perrin, Elmer 368
Pershing, John J. 279, 354, 355
Petain, Philippe 335
Petersen, Forrest 309
Peterson, Chesley 260
Peterson, Dave 197
Peterson, Pete 181
Petlyakov, Vladmir 338
Phantom 33, 193
Piccio, Ruggiero 242
Pickard, Percy 333
Pilgrim 379
Pinball 267
Pinto 395
Pioneer 343
Pirate 15
Platen, Magnus 322
Pikryshkin, Alex 257, 335, 337
Polando, John 128
Polar Star 147
Polbin, Ivan 338
Polifka, Karl 254
Pope, Harley 368
Popson, Ray 308
Porter, Phil 266

Post, Wiley 147
Powers, Gary 119, 196, 303
Powers, Tom 99
Preddy, George 264
Privateer 92
Provider 28, 153
Puff the Magic Dragon 135
Putnam, Paul 339

Q
QB-47 110
QF-63 32, 48, 267
QF-80 178
Quail 20, 113, 117
Quesada, Elwood 125

R
R-1 (Racer) 48, 271
R-1 (Rotary Wing) 385
R-2 (Racer) 48, 271
R-2 (Rotary Wing) 36, 203
R-3 (Racer) 48, 271
R-3 (Rotary Wing) 36, 203, 385
R-4 (Racer) 272, 394
R-4 (Rotary Wing) 36, 204
R-5 (Racer) 49, 272
R-5 (Rotary Wing) 36, 204
R-6 (Racer) 49, 272
R-6 (Rotary Wing) 36, 204
R-7 (Racer) 394
R-7 (Rotary Wing) 385
R-8 (Racer) 49, 272
R-8 (Rotary Wing) 385
R-9 (Rotary Wing) 385
R-10 (Rotary Wing) 385
R-11 (Rotary Wing) 385
R-12 (Rotary Wing) 36, 204
R-13 (Rotary Wing) 36, 204
R-14 (Rotary Wing) 385
RB-17 44, 235
RB-29 44, 104, 236
RB-36 104
RB-45 109
RB-47 110
RB-49 104, 112
RB-50 106
RB-57 19, 114, 115
RB-66 19, 117
RB-69 20, 117
RC-45 132
RC-121 152
RC-130 156
RC-131 157
RC-135 160
RF-4 193

RF-5 194
RF-51 43, 235
RF-61 32, 44, 48, 236, 266
RF-80 236
RF-84F 32, 105, 179
RF-86 182
RF-101 187
RF-105 190
RF-111 191
RH-53 153, 212
RP-63 267
RP-80 44
RY3 92, 145
R2C1 49, 272
R3C1 49, 272
R3C2 49, 272
R4D-8 382
R4Q 151
Racer 271
Raiden 328
Raider 28, 155
Rainmaker 268
Rall, Gunther 325
Ramjet 308
Randolph, Bill 285, 368
Ranger 206
Rascal 20, 110, 117, 308
Razorback 261
Reber, Sam 318 353 to 358
Recruit 51, 289
Red Baron 241, 320
Reese, Gus 368
Reidy, Tom 342
Reinhard, Wilhelm 321
Reitsch, Hanna 344-Reliant 26, 143,
   290, 389
Reliant 26, 143, 290, 389
Reporter 44, 236
Retchkalov, Grigori 336
Reynolds, Bob 333
Richards, John 61
Richthofen, Manfred 240, 320, 321
Richthofen, Wolfram 324
Rickenbacker, Eddie 240, 355 to 357
Ridgeway, Matt 223
Ritchie, Dick 194
Robin 379
Robins, Warner 368
Rockefeller, John 125
Rocketship 56, 57, 306
Rockne, Knute 125
Rockwell, Kiffen 238
Rodgers, John 352
Rogers, Joe 191
Rogers, Will 147

Rommel, Erwin 75, 225, 258
Roosevelt, Franklin 74, 137, 248, 259
Roosevelt, Teddy 276
Rose, Tricornot de 237
Rotating Wing 203
Royce, Ralph 237, 356
Rudel, Hans 326
Rudorffer, Erich 325, 346
Rufe (Floatplane) 328
Ruptured Duck 95
Rushworth, Bob 175, 309
Ryan, John 318, 356, 357
Ryan, John D. 362

S

SA-16 233
SALMSON 2A-2 11, 59, 282
SB-68 20, 117
SBD 13, 69
SB2A 374
SB2C-1 13, 70
SB3C 374
SC-54 137
SCA 312
SE-5 12, 241
SH-19 206
SH-21 207
SOPWITH Camel 11, 240, 282
SPAD-VII 11, 282
SPAD-XIII 11, 239, 243
SR-71 49, 196, 274, 302
Sabre Dawg 182
Sabre Jet 32, 181
Sabreliner 55, 300
Sacred Cow 137
Sadat, Anwar 195
Sakai, Saburo 256, 328
Salamander 347
Samaritan 29, 157
Santos-Dumont, Alberto 322
Saunders, LaVerne 99
Sayer, Gerry 343
Scaroni, Silvio 242
Schall, Franz 346
Schilling, Dave 179, 262, 368
Schnaufer, Heinz 326
Schneider, Walter 327
Schroeder, Rudolph 241
Schwalbe 174, 345
Scorpion 32, 183
Scott, Bob 259
Scott, Frank 353, 368
Scout 12, 241
Scriven, George 352 to 355

Sea Bee 390
Seafire 331
Sea Hurricane 330
Sea King 209
Sea Racer 49, 272
Sea Stallion 38, 212
Sebille, Louis 178
Selfridge, Tom 277, 350, 368
Sentinel 40, 220, 223
Sentry 31, 170
Sewart, Allan 368
Shady Lady 55, 302
Shafter, Bill 257
Shahan, Elza 254
Shangri-La 264
Shaw, Ervin 369
Shawnee 207
Sherman, Bill 278
Shiden (Fighter) 329
Shomo, Bill 264
Shooting Star 32, 44, 48, 55, 177, 236, 298
Shrike 12, 13, 63, 65, 70, 110, 308, 374
Shuttle Carrier Aircraft 312
Silver Star 55, 298
Sioux 36, 37, 204, 206, 210
Skalski, Stanislaw 331
Skeel, Burt 271
Skorzeny, Otto 225
Skybolt 20, 113, 117
Skycar 380, 381
Skychief 147
Skymaster (C-54) 24, 136
Skymaster (O-2) 41
Skymaster-II 27, 149, 226
Skyraider 14, 77
Skyrocket 307
Skystreak 307
Skytrain 23, 133
Skytrain-II 27, 149
Skytrooper 24, 135
Skyvan 151
Skywagon 56, 304
Slayton, Deke 313
Smith, Carroll 265
Smith, Charles 137
Smith, John 339
Smith, Lowell 61, 217
Smith, Paul 266
Smith, Roger 197
Smith, R.W. 189
Smith, Tuck 232
Snark 20, 117
Sophomore 52, 287, 290

Spaatz, Carl 125, 260, 318, 354 to 361
Space Shuttle 196
Spad 78
Spate, Wolfgang 344
Spatz, Hal 95
Speedster 381
Speed Vega 379
Speer, Bob 93
Spitbomber 331
Spitfire 47, 260, 326, 331
Sportsman 395
Springs, Elliot 241
Spyspit 331
Squier, George 354 to 356
Squirt 343
Staggerwing 131
Stalin, Premier 137, 335
Stallion 56, 305
Stanley, Bob 173, 176
Starfighter 33, 188
Starfire 32, 184
Starlifter 29, 161
Stead, Croston 369
Steakley, Ralph 99, 236
Stealth Aircraft 122
Stephens, Bob 196
Stiletto 57, 307
Stirling 334
Storch 225
Stork 225
Stormbird 174, 345
Strasser, Peter 323
Stratofortress 18, 112
Stratofreighter 26, 27, 146
Stratojet 18, 110
Stratolifter 29, 159
Stratoliner 25, 142
Stratotanker 29, 159
Stuka 325
Sturmvogel 174, 345
Sugita, Shoichi 256, 329
Super Constellation 28, 151
Super Courier 41, 56, 225, 304
Super Cub 41, 56, 224, 304
Super Dumbo 98
Super Electra 27, 139, 148
Superfortress 17, 18, 44, 98, 103, 106, 236
Superhog 109
Super King Air 30, 168
Super Sabre 32, 185
Suprun, Stepan 335
Swallow 174, 345
Sweeney, Charles 100
Sweet, George 351

T

T-1 (Transport) 21, 123
T-2 (Transport) 21, 123
T-3 (Transport) 124, 378
T-6 (Trainer) 54, 296
T-7 54, 296
T-11 54, 296
T-17 397
T-19 54, 296
T-28 54, 296, 298
T-29 55, 157, 296, 298
T-30 397
T-31 397
T-32 397
T-33 55, 178, 296, 298
T-34 55, 298
T-35 397
T-36 397
T-37 55, 283, 299
T-38 55, 299
T-39 55, 300
T-40 161, 397
T-41 55, 301
T-43 55, 301
TA-1 283, 394
TA-2 395
TA-3 49
TA-4 395
TA-5 49, 283, 395
TA-6 49, 283, 395
TA-20 67
TA-152 328
TB-24 54, 92
TB-25 54, 102, 296, 298
TB-26 54, 96
TB-32 17, 101
TB-45 109
TB-47 110
TB-50 106
TC-45 54, 55, 132, 296
TC-47 133
TC-54 137
TC-131 157
TF-15 196
TF-80 55, 178, 184, 296, 298
TF-86 182
TF-100 186
TF-102 188
TF-104 189
TG-1 35, 202
TG-2 35, 202
TG-3 35, 202
TG-4 35, 202
TG-5 35, 202, 222
TG-6 35, 202, 222

TG-7 383
TG-8 35, 202, 222
TG-9 383
TG-10 383
TG-11 383
TG-12 383
TG-13 383
TG-14 384
TG-15 384
TG-16 384
TG-17 384
TG-18 384
TG-19 384
TG-20 384
TG-21 384
TG-22 384
TG-23 384
TG-24 384
TG-25 384
TG-26 384
TG-27 384
TG-28 384
TG 29 385
TG-30 385
TG-31 385
TG-32 35, 202
TG-33 35, 202
TH-1 37, 208
TM-61 19
TP-1 (Pursuit) 45, 214, 249
TP-47 (Trainer) 261
TR-1 49, 275, 302
TRF-51 235
TW-1 283, 395
TW-2 395
TW-3 49, 283
TW-4 395
TW-5 49, 283
Tailfirst 119
Tailless 104, 112
Talon 55, 299
Tank, Kurt 327
Taylor, Ken 259
T-Bird 298
Tempest 333
Texan 53, 54, 290, 296
TFX 20, 191
Thacker, Bob 270
Thor 20, 117
Thud 190
Thunderbirds 181
Thunderbolt (P-47) 31, 47, 261, 343
Thunderbolt (British) 343
Thunderbolt-II (A-10) 14, 80
Thunderchief 33, 189

Thunderflash 32, 105, 179
Thunderjet 32, 179
Thunderstreak 32, 179
Thurlow, Hiram 148
Tibbets, Paul 100, 110
Tierney, Bob 266
Tigercat 342
Tiger Hawk 259
Tiger Mouth 259
Tiger Shark 259
Tiger-II 33, 195
Tilt-Duct 58, 310
Tilt-Prop 57, 310
Tilt-Wing 57, 310
Tin Goose 125
Tinker, Clarence 369
Titan 20, 117
Tito, Marshal 100
Tokyo Rose 99, 236
Tomahawk 259
Tornado 18, 109
Tower, Les 90
Towers, John 352
Trainers (WWII statistics) 294, 295
Travel Air 273
Traveler 23, 131
Travis, Bob 369
Trejo, Ed 153
Trent, Leonard 101
Trimotor 125, 378, 379, 381
Triplane (Tripe) 321
Trojan 54, 298
Truax, Tom 369
Truly, Richard 312
Truman, Harry 150, 270
Trusty 49, 50, 284
Tunner, Bill 137
Turner, Roscoe 141
Turner, Sullins 369
Tweety Bird 55, 299
Twin Courier 56, 303
Twining, Nathan 361
Twin Jenny 373
Twin Mustang 32, 487, 269
Tyndall, Frank 246, 247, 369

U

U-1 302
U-2 55, 274, 275, 302
U-3 56, 303
U-4 56, 303
U-5 56, 303
U-6 56, 304
U-7 56, 304
U-9 224

U-10 56, 304
U-16 56, 304
U-17 56, 304
U-21 30, 165
UC-36 22, 130
UC-37 23, 130
UC-40 23, 131
UC-43 23, 131
UC-45 23, 132
UC-61 24, 139
UC-64 139
UC-70 25, 140
UC-71 25, 141
UC-72 25, 141
UC-78 26, 142, 291
UC-81 26, 143
UC-83 26, 144
UC-86 26, 144
UC-95 26, 145
UC-96 26, 145
UC-100 27
UC-101 27
UH-1 208
UH-13 37, 210
UH-19 207
Udet, Ernst 241, 322, 325

V
V-1 (Buzz Bomb) 324, 332,
    333, 343
V-2 (Buzz Bomb) 324, 333
VC-6 30, 165
VC-9 166
VC-47 149
VC-97 146
VC-117 149
VC-118 150
VC-121 151
VC-131 157
VC-135 160
VC-137 29, 160, 171
VCP-R 48, 271
VCP-2 45
Valencia, Gene 342
Valiant 52, 289, 396
Valkyrie 19, 118
Vance, Leon 369
Vandenberg, Hoyt 151, 361, 369
Vanguard 48, 268
Vaughn, George 241
Vega 27, 147, 379
Vengeance 14, 74
Ventura 17, 18, 101, 138
Vertijet 57, 308
Vigilant 39, 40, 219, 221

Villa, Pancho 279, 354
Vinogradov, P.S. 337
Violet Lightning 329
Viper 391
Voisin Fighter 237
Volksjager 347
Voll, John 262
Voodoo 33, 186
Voss, Werner 321
Voyager 41, 223, 389
Vraciu, Alex 342
Vultee 147

W
WB-29 104
WB-47 110
WB-50 106
WB-66 118, 310
WC-135 160
Wagner, Boyd 256
Walker, Joe 113, 175, 309
Walker, John 351, 352
Walker, Ken 369
Walmsley, John 103
Walsh, Ken 340
Ward, Ed 350
Warhawk 31, 47, 257
Warmer, Ben 91
Warsitz, Erich 344, 347
Waters, Andy 110, 154
Wayne, Bob 178
Webb, Jim 369
Weiss, Robert 327
Welch, George 259
Wendel, Fritz 324
Westover, Oscar 358 to 360, 369
Wheeler, George 369
Whistling Death 340
White, Bob 175, 309
White, Tom 361
Whitehead, Ennis 229
Whiteley, John 230
Whiteman, George 259, 369
White Wing 276, 350
Whittle, Frank 173
Wichita 53, 291
Widgeon 42, 232
Widow Maker 96
Wildcat 338
Wilhelm, Kaiser 237
Wilkins, Hubert 147
Williams, Al 272
Williams, Charles 369
Williams, Roger 128
Wilson, H.J. 343

Wilson, Woodrow 356
Wingate's Raiders 134, 201
Winnie Mae 147
Winslow, Alan 356
Wolfe, Ken 99
Wood, Jack 94
Wooden Wonder 332
Woods, Timber 329
Workhorse 37, 207, 211
Wright Flyer 10, 371
Wright, Orville 277, 350 to
    361, 370
Wright, Wilbur 350 to 352, 370
Wurger 327
Wurmheller, Josef 327
Wurtsmith, Paul 370

X
X-1 57, 306
X-2 57, 107, 307
X-3 57, 307
X-4 57, 308
X-5 57, 308
X-6 105
X-7 308
X-9 110, 308
X-10 308
X-11 308
X-13 57, 308
X-14 57, 308
X-15 57, 113, 175, 196, 309
X-15A-2 57, 310
X-17 308
X-18 57, 310
X-19 57, 310
X-20 308
X-21 57, 310
X-22 58, 310
X-23 308
X-24 58, 113, 311
X-25 58, 311
X-28 311
XS-1 56, 104, 306

Y
YAK-1 335
YAK-3 335
YAK-7 335
YAK-9 335
YAK-15 347
Yale 52, 290
Yamada, Sadayoshi 329
Yamamoto, Isoroku 90, 255,
    328, 329
Yamschikova, Olga 348

Yancey, Lew 128
Yankee Doodle 35
Yeager, Chuck 104, 175, 306
Yevstignev, K.A. 336
YIC 379
Young, John 314

Z

Zeke 328
Zempke, Hub 262
Zeppelin (Dirigible) 322
Zeppelin, Ferdinand 322
Zero 328

Zero-sen 328
Zerstorer (Destroyer) 326
Ziegler, Skip 307, 308

3